Request For Feedback

In a book such as this, there is always room for improvements, and I am eager to make them, but I need your help. Therefore, I ask for and sincerely desire your feedback. What works? What doesn't work? What do your children like, dislike? What needs further clarification? What should be added? What can be omitted? What neat ideas or techniques do you have that you are willing to share? I will gratefully acknowledge your contributions as they are incorporated in the next edition of this volume.

Ongoing Support (no charge)

I wish to support your teaching and your children's learning in any way that I can. Toward this end, I host a Yahoo group site: http://groups.yahoo.com/ Sign in and go to: K5science/ Through this site, you can post whatever questions, problems, and ideas that you or your children may raise and discuss them freely with other teachers. I will be on hand and it will be my pleasure to help in any way that I can. You can remain anonymous to the degree you desire, but entry into the group is subject to approval to avoid spammers. There is no charge and there will be no solicitations. If you wish, you may also contact me personally at

bnebel@erols.com

or 410-744-34912 during business hours EST

Sincerely yours,
The author,
Bernard J. Nebel, Ph.D.

Another Book By The Author To Aid Your Teaching

A general coverage of both teaching strategies and information for K-5 teachers and homeschoolers is contained in "Nebel's Elementary Education: Creating a Tapestry of Learning." Especially strong in science, this text also provides suggestions and information for the teaching of economics, government, reading and writing, and character education. It may be purchased at amazon.com/

Safety

None of the exercises described in this text entail exposing you or your children to risks beyond those confronted in everyday life. Nevertheless, certain safety precautions are included within the text of lessons where appropriate. This said, it is taken for granted that parents, caregivers, and teachers remain responsible for any for any and all safety issues concerning yourselves and your charges. Neither the author nor the publisher of this volume assume any responsibility for untoward events or injuries that may occur.

Acknowledgements

This volume is an outgrowth of a previous work, "Nebel's Elementary Education: Creating a Tapestry of Learning" (NEE) (see www.pressforlearning.com/). This volume is basically an expansion into specific lesson plans of information presented in the science chapters of the NEE text for the K-2 level. Therefore, all those who aided and supported me in the production of that volume deserve acknowledgement here as well. Deserving particular mention in addition or again are: Carol Richey for researching and providing me with the lists of children's books for correlated readings; Mary Soto and Deanna Schmidt for field testing; Deborah Keene and Kristin Roloff for proofreading; numerous colleagues, friends, and anonymous reviewers who offered comments and criticisms along the way; Faye Oriowski and others at Outskirts Press for seeing this manuscript into print; and most of all, my loving wife Maggie who has endured long hours of my being in my "cave."

About the Author

Bernard J. Nebel is a professor emeritus of environmental science education at the Community College of Baltimore County, Catonsville, Maryland. He was one of the pioneers of environmental science education, and authored a still widely used text, Environmental Science, Prentice Hall, 1981 and subsequent editions. In addition to teaching and research, Dr. Nebel maintained an active avocation in teaching younger children. Addressing the problems of science education at the K-5 level is now the focus of a second career.

Some years ago, a committee investigating elementary science education concluded that the arena of science is too diverse to be covered in any systematic way. The best that can be done, they deemed, is to give children a sampling that (presumably) demonstrates what science is about. Thus, science at the elementary level has become a matter of teachers (or committees) making more or less arbitrary selections from the vast smorgasbord of topics available. Try to imagine the result if math were taught this manner. Does more need to be said regarding why science education is in such disarray and achievement is so low?

Nebel breaks through the limiting view of the old conventional wisdom. In this volume, he constructs the smorgasbord into a balanced diet. The result is pathways of easy-to-conduct lessons that build logically and systematically toward a broad, solid, integrated foundation of scientific knowledge, conceptual understanding, and thinking skills. Child-centered, hands-on activities, and techniques that engage children in their own learning are used throughout. Most of all, Dr. Nebel, offers ongoing, on-line support at no cost (above). A second volume that will continue the pathways and foundation building through the 5th grade is in progress.

Table of Contents

Preface, Introduction, and Directions for Using This Text

This book is about much more than the 41 lesson plans presented. The pathways and techniques described will enable you to guide children toward developing a solid foundation of scientific understanding that will include:

- coverage of all the major fields of science
- a broad, integrated structure of factual knowledge supported by child-centered explorations
- an understanding of key concepts and principles supported by personal observations, experience, and logical reasoning
- skills of inquiry becoming habits of mind for life-time learning
- meeting standards
- integrating reading, writing, geography, and other subjects

The distinction of this text that makes these outcomes achievable is two-fold. First, major ideas and concepts are digested into easy-to-follow, incremental steps (individual lessons) that build logically and systematically. By following this logical, steppingstone-like sequence, each new lesson provides a natural review and reinforcement of what has gone before and leads to a solid, integrated knowledge and understanding of the whole.

DO NOT treat this volume as a smorgasbord for random selections. As in math, lessons are arranged and designed to impart knowledge, understanding, and skills in a logical, systematic order.

Second, to develop skills of inquiry, each lesson involves children in first-hand observing, organizing, and thinking their way toward rational conclusions. Interest and motivation are maintained by using children's own findings and experiences as the center of study. This is to say, lessons are child-centered and maximize experiential learning.

Science standards will be met as natural outcomes of engaging children in acquiring a broad, integrated foundation of understanding, not by teaching to the standards. Specifically, the lessons cover all the science standards (K-2) for the state of California and more. If your state should include a topic that is not covered in one or more of the lessons, the overall plan will reveal where it may be inserted and integrated most effectively. The need for special equipment and materials is minimized, and interest and motivation are maximized by centering lessons on things/events that children interact with in a personal way.

Every lesson includes a list of age-appropriate children's books that may be used to synergistically develop reading skills while reinforcing the science lesson at hand. Writing about the lesson is always promoted, and many lessons have a strong geographic component in addition to the science.

In summary, this text is designed to: provide a steppingstone-like sequence of lessons that facilitates systematic building of concepts; maintain interest by centering lessons on what children regularly experience; engage children's own thinking in reaching rational conclusions; and integrate other subjects. Together, these aspects put into action the latest findings concerning how children learn most effectively and efficiently.[1] A further discussion of these findings and how this text corresponds is given in Chapter 1. Techniques for bringing children to exercise and develop their skills of inquiry and logical thinking are discussed in Chapter 2.

Most significantly, using this text does not require any particular background on the part of teachers. Users will find that instructions go far beyond just monitoring students through a given activity. Additionally, they provide guidance for keeping children's attention and thoughts focused on the subject, and they give directions for leading them to use their own rational thinking in reaching conclusions. Users report that they learn and gain insight along with their students, and I may add that they are excellent role models in doing so. (Also, please note my offer of ongoing support; see the front matter.)

In the remainder of this Introduction and Preface, we will discuss the organization and format of the lessons in more detail.

Organization of Lessons

The steppingstone sequence of lessons noted above is actually four pathways to be pursued more or less in tandem. This is not really different from noting that math is divided into numerical operations on one hand and shapes and angles on the other, and development of both are carried along together. Thus, I have broken science into four categories that together cover all the major fields. The categories are: a) Nature of Matter; b) Life Science; c) Physical Science; and d) Earth and Space Science. Each of the categories or threads begins with a lesson suitable for kindergarten or grade one and builds systematically from there as shown in the flowchart on facing pages 12-13. A summary of the concepts developed in each thread is also given (pages 7-10). Therefore, DO NOT treat this volume as a smorgasbord for random selections. In each thread, as in

---

[1] Donovan, Suzanne M. and Bransford John D., Editors. "How Students Learn: Science in the Classroom." National Research Council, Washington, D.C.: The National Academies Press, 2005

math, lessons are arranged to impart knowledge, understanding, and skills in a logical, systematic order.

You do have the option and flexibility of which thread to start with and which to turn to next, but it will be important to rotate among the threads so that each is carried forward at each grade level. The flexibility in moving from one thread to another allows you to tailor lessons to the interests of students, times of year, weather conditions, or special opportunities as appropriate while still adhering to the logical development within each thread. Then there will numerous points at which lessons in different threads will be mutually supporting and should be integrated.

Use the flowchart as a guide. Pursue the lessons within each thread in the order indicated by the arrows. This is because the lessons in each thread build in logical increments, each lesson acting as a steppingstone to the next. At various points on the flowchart, you will note double headed arrows connecting boxes from side to side. These indicate lessons that are mutually supporting and which should be presented in close proximity and integrated. There are also arrows connecting threads. These are points where the same concept applies to both, and integration between threads should be emphasized. There is no demarcation between grade levels because each of the threads is a seamless progression and students will move as fast as they do. This calls for teachers at successive grade levels to communicate as to how far students have gotten so that the next teacher can move them on without breaking the sequence. (The same threads will be continued in the volume for Grades 3 to 5.)

In following each of the threads there will be a combined building of factual knowledge, conceptual understanding, and development of observational and thinking skills. Again, an overview of principles and concepts being developed in each thread is presented with the flowchart (pages 7-1318-19). It will be important to refresh these concept overviews on repeated occasions so as not to "lose sight of the forest for the trees." By the same token, this volume may be useful for older students whose science background is spotty or deficient.

DESIGN AND PRESENTATION OF LESSONS

There is no such thing as a lesson plan that will teach itself; student mastery will invariably depend on the teacher's implementation as well as the plan. The following gives an outline of lessons and stresses points that will aid your implementation.

All the lessons in this volume have the same format:

> **Overview**
> **Time Required**
> **Objectives**
> **Required Background**
> **Teachable Moments**
> **Methods and Procedures**
> **Questions/Discussion/Activities to Review, Expand, and Assess Learning**

Bernard J. Nebel, Ph.D.

> To Parents and Others Providing Support
> Connections to Other Topics and Follow-Up to Higher Levels
> Re: National Science Education (NSE) Standards
> Books for Correlated Reading

Overview gives a very brief description of what students will be learning in the lesson and its importance in advancing their knowledge and understanding. In introducing a lesson, you may use it to help orient students' thinking.

Time Required is a rough estimate of the time it will probably take you to present the core of the lesson in order to help you plan your weekly schedule. However, remain flexible. Children may become fascinated and ask a multitude of questions extending the topic and relating it to other examples. Go with this so far as is practical. On the other hand, students may be utterly disinterested and just can't/won't get with it. Force-feeding lessons is counterproductive. Hence, we recommend letting it go, switching to another thread, and coming back to it on another occasion, perhaps from a different angle. At the very least, you have planted the seed; it may take time to germinate.

In every case, presenting the core of the lesson in the allotted time should not be considered as the whole of the lesson. The steppingstone sequence requires some review of previous material to be sure that students connect new material with what they already know (see "Required Background" below). Then reviews, particularly finding additional examples that connect the lesson to children's real-world experience, are always invaluable (see "Questions/Discussion/Activities to Review, Reinforce, Expand, and Assess Learning" below). A few lessons are ongoing in nature. For example, in Lesson B-4A students embark on a project of learning to identify common flora and fauna of their region, and in Lesson D-6 they begin tracking seasonal changes. These will require a few minutes attention on a weekly, if not daily, basis. Planning a "preclass warm-up session" for such purposes is described in detail in Chapter 1. The amount of time you choose to allot to such reviews and enhancements will be entirely up to you. Only recognize, as is described in Chapter 1, that they are critical in leading students to maintain interest and develop thinking and interpersonal skills, as well as mastering certain concepts. They also serve to reveal points of confusion or misunderstanding and provide opportunities for clarification.

Objectives will give you a more detailed overview of the learning outcomes that should be expected through the lesson. But don't teach to the objectives. The mastery of objectives should be a natural by-product of the lesson and engaging students in discussion and reviews.

Required Background lists preceding lessons, the steppingstones if you will, leading up to the lesson at hand. These should be reviewed to the extent necessary to be sure your students are up to speed before proceeding. Such reviewing and making connections with what has gone before is a crucial aspect of learning.

Materials. Budgetary restraints are fully appreciated. Therefore, lessons have been designed to use a minimum of special materials and equipment. Materials that are called for are mostly common items that will be found in most kitchens or are available at regular grocery, home supply, or office supply stores. (There is no intention to use this text as a means of selling you a lot of additional stuff.) How much/many of each item you need will depend on your own decisions regarding having your students work individually, in small groups, or conducting the activity as a class demonstration.

Teachable Moments. The inclusion of this section is probably unique but it is important. We can all appreciate how our learning is most efficient and meaningful when it pertains directly to what we are involved in at the time. This section suggests games, activities, or occasions that capture children's attention and make launching into the given lesson particularly appropriate. Prepare yourself to teach the given lesson, but then anticipate or create a teachable moment in which to present it.

Methods and Procedures takes you and your students through the steps of the exercise, but there is more than just a cookbook recipe. A most important element underlying learning is keeping students' minds engaged and thinking about what they are doing or observing and the rational conclusions that may be derived (see Chapter 1). For this purpose you will find numerous points at which you are instructed to pose questions and intersperse question and answer (Q and A) discussion. Instructions are sufficiently detailed to enable you to guide students toward rational interpretations in the course of doing the exercise. That is, interpretation is fully integrated with procedures. Moreover, you do not need to have previous background regarding interpretations yourself. Instructions will guide you, as well as your students toward the accepted conclusions.

Questions/Discussion/Activities to Review, Reinforce, Expand, and Assess Learning. Learning becomes long-term and facilitates thinking only as it is applied to different examples and used in reasoning and problem solving. Similarly, solid understanding of a concept develops gradually, with repeated exposures reflecting different angles and facets. This section provides a list of some things you can do to bring students to review, reflect, and apply their learning to keep it alive and growing. The list should not be viewed as all-inclusive, however. You may add ideas and students will doubtlessly raise questions, examples, and thoughts of their own.

You will note that "Make a book illustrating ____ " is a recurring item. Making simple, paper-fold "books" is described in Appendix A (page 380). As well as bringing students to reflect on the lesson at hand, the "books" provide a convenient handle for assessing understanding, promoting writing skills, and showing where additional assistance is needed. In grades three and above (the next volume in this series) this activity will evolve into keeping more formal notebooks.

Conducting small-group Q and A discussion sessions is another recurring item in this section. The importance, utility, and techniques of conducting such sessions are discussed in detail in Chapter 2.

Bernard J. Nebel, Ph.D.

To Parents and Others Providing Support. This section provides a listing of specific things parents and/or caregivers can do to promote their children's mastery of the learning, understanding, and skills. Note that suggestions listed are not simply monitoring kids filling out work sheets. They are designed to engage caregivers in helping their children relate their school learning to their real-world experience and, thus, reinforce and extend the lesson.

At the end of Lesson B-2, you will find a suggested letter to parents/caregivers asking for their support in getting their children prepared for that lesson. It is hoped that this will serve to initiate ongoing communications keeping them informed of the lessons you are doing and gaining the supportive actions as suggested in this section. Note that a major goal of the National Science Education Standards is to create a learning community that goes beyond the boundaries of the school. A most important part of this community will be parents/caregivers and children will benefit immensely from their direct involvement. If you are a homeschooler, it will be natural to combine the suggestions given here with the previous section.

Connections to Other Topics and Follow-Up to Higher Levels comes back to the idea of a steppingstone sequence. The current lesson should always be viewed, not as an end in itself to be checked off as done, but as a steppingstone toward continually advancing understanding. Every lesson should end, not with all the questions answered, but with additional questions hanging out begging for further investigation.

Re: National Science Education (NSE) Standards. Here you will find a listing of the NSE content standards to which the given lesson relates. If you are not already familiar with these standards, which are strongly promoted by the National Science Teachers Association (NSTA), you will note that they are very general, visionary, and nonspecific in nature. Therefore, there is no way that any given lesson can be seen as giving students a total mastery of any given standard. Thus, you will find many lessons will relate to the same standard(s), but this is as it should be, because all the NSE standards speak to heading toward a goal as opposed to a specific bit of knowledge.

Furthermore, content standards are only one part of the NSE standards. In total, they pertain to all aspects of science education. Specifically, they include: "Science Teaching Standards," "Standards for Professional Development of Teachers of Science," "Standards for Assessment in Science Education," "Science Education Program Standards," and "Science Education System Standards," as well as "Science Content Standards."[2] In the course of conducting the lessons in this volume and continuing on to the volume for grades 3-5, you will be drawn to implement most of these standards, as well as having your students meet specific content standards.

Of course, you are undoubtedly under pressure to implement the more specific science content standards of your state in response to the "No Child Left Behind Act

---

[2] "NSTA [National Science Teachers Association] Pathways To the Science Standards: Elementary School Edition," 2nd edition. Arlington, VA: NSTA press, 1996.

of 2002." We have taken heed that the lessons in this volume address all the content standards for the State of California (Grades K-2). As these are generally considered to be among the most rigorous, it is probable that standards for other states are addressed as well. If not, the framework of lessons presented will allow the insertion of additional lessons at a logical and timely point.

Again however, the point of this volume is, not to teach to given standards, but to build a broad integrated foundation of scientific knowledge and understanding. Meeting specific content standards should be a natural byproduct of this process. With this perspective, you will find many lessons in this volume that go beyond the particular standards of your state. They are, nevertheless, critical in building the foundation desired.

Books For Correlated Reading. Here you will find a listing of grade-appropriate children's books that pertain to the given lesson. You may use any one or more of these books (or others) to integrate your teaching of reading with review, reinforcement, or enhancement of the given science lesson.[3]

---

[3] I am greatly indebted to Carol Ritchey for researching and compiling this list.

FLOWCHART OF LESSONS AND OVERVIEW OF CONCEPTS PRESENTED

(Threads A, B, C, and D are to be conducted in tandem.)

A-THREAD: Nature of Matter

Organizing things into categories aids both memory and understanding. It comes into play in many areas of science and other subjects as well. Students will begin to exercise this skill and discover its benefits in Lesson A/B-1 as they group familiar household and classroom items into categories. The skill is put to use and the concept of matter is introduced as they sort materials into solids, liquids and gases and observe that these are interchangeable states of matter (Lesson A-2). Basic attributes of matter—having mass and occupying space—are conveyed as students experimentally determine that air is "real stuff" (Lesson A-3).

Activities in Lesson A-4 bring students to reason and understand that all matter has a particulate nature, and this is another of its basic attributes. They are further brought to reason and understand that the distinction between solids, liquids, and gases is in the degree to which particles are attracted to one another. Recognizing and distinguishing different materials (metals, plastics, wood, etc.) on the basis of their properties, including magnetic properties is the focus of Lessons A-5 and A-5A

Students' concept and understanding of the particulate nature of matter and its different states depending on relative attraction among particles are brought into play and reinforced as they use them in interpreting their observations and experiences regarding air pressure, vacuums, and the Earth's Atmosphere (Lesson A-6), evaporation and condensation (Lesson A-8), dissolving, solutions, and crystallization (Lesson A-9) and other phenomena. How metals and other resources are obtained from various earth materials is not neglected.

All of this creates a foundation that leads seamlessly toward understanding more sophisticated concepts of chemistry and chemical reactions that are introduced at higher levels

B-THREAD: Life Science

The skill of categorization (Lesson A/B-1) is further exercised as students sort various items (both living and nonliving) picked up or sighted in and around their homes into

8

three basic categories: living/biological; natural nonliving; and human-made things (Lesson B-2). The confusion between "actively living" and "no-longer living" things is solved by introducing the term, biological. At the same time, this lesson guides students to be cognizant of the attributes we use to make distinctions among these categories. As students recognize that the living/biological category includes both plants and animals, they are guided to derive attributes that are common to all living things. Separating fact from fiction is not neglected.

Drawing on their recognition that everything requires a source of energy to "make it go" (Lesson C-1), students are brought to recognize that the key distinction between plants and animals hinges on where the "living thing" obtains its source of energy—animals from food; plants from sunlight (Lesson B-3). In addition to forming the basis for dividing living things into two major kingdoms, this provides the crux for moving into observations and considerations of food chains and adaptations (Lesson B-5). (Introduction to the three additional kingdoms is saved for later.)

The concept of life cycles is introduced in Lesson B-4 by having students cite and discuss familiar examples. This leads seamlessly into further identification and study of flora and fauna common their area (Lesson B-4A). Along the way, the concept of a species as a kind of plant/animal capable of mating and reproducing its own kind is developed and the distinction between species and breeds/varieties becomes clear (Lesson B-4B).

Watching various "critters" move, an adjunct of Lesson B-4A, is always a subject of fascination to children. Lesson B-6 brings students to observe and understand how each movement of their body, bending the arm at the elbow for example, is caused by certain muscles pulling between two points of the skeleton to cause bending at the joint. Students will learn that they share this basic body-movement design with many other sorts of animals (fish, amphibians, reptiles, birds, and mammals) and this is the basis for categorizing them as vertebrates. But students will observe that there are other designs as well—skeletons outside (insects, crabs, and other arthropods) and no skeletons at all (worms, etc.). The movements of these are observed and analyzed as well with the conclusion that they also depend on contraction of given muscles (Lesson B-7). Students also learn here that the different body designs are the basis for categorization of animals into different phyla.

The understanding that body movements are performed by particular muscles pulling between specific points on the skeleton is used as the cornerstone for further studies of anatomy and physiology. Students are first led to observe that movements need to be exactly coordinated and controlled to accomplish a given act, raising a spoon to the mouth for example. In turn they are guided to witness that the action of muscles is preceded by information from sensory organs and decisions made by a processing center before the muscles act. Thus, and major parts and role of the nervous system are revealed (Lesson B-8).

Other major internal organs and their roles are brought into the picture as students consider the need for energy to make the muscles "work." Thus, the digestive system, lungs, heart, and kidneys are considered, not in terms of anatomy, but in terms of providing energy, i.e., fuel to burn, oxygen to burn it, disposal of wastes, and a means of transporting these things from one place to another (Lesson B-9). This treatment provides students with a conceptual foundation to which any degree of additional details can be easily added.

In parallel lessons students will observe basic plant anatomy and consider how each part of that structure functions to support the plant's needs (Lesson B-10). Lesson B-11 engages students in observing, not just germination, but also how seeds and seedling responses are adapted to the environment. Lesson B-12 moves on to a consideration of how agricultural practices (soil and water management) must be adjusted to support plant's needs.

C-THREAD: Physical Science

The concept of energy is generally omitted from the curricula of lower grades because common wisdom says that it is too abstract for young students to master. This omission, however, allows false, naïve notions to grow in children's minds and these become increasingly difficult to erase. In contrast, Lesson C-1 makes the idea of energy concrete by simply drawing children's attention to their experience that something is always required to make things go, work, or happen. That "something" may be electricity, heat, light, or a push/motion. Children readily accept that these are categorized as forms of energy and further observe that one form of energy may be changed into another form through various means, such as a light bulb changing electricity into light and heat. Sound is portrayed as a special form of movement energy—vibrations (Lesson C-2).

Through familiarity with such things as batteries, putting fuel in cars, and rolling things down ramps, students are brought to understand the distinction between kinetic and potential energy (Lesson C-3) and the distinction between matter and energy (Lesson C-4). Additional parameters of energy, namely inertia and friction, are conveyed through experiential learning in Lessons C-5 and C-6. The basic physical "law" that every action has an equal and opposite reaction is conveyed to children by simply putting it in terms of the easily observable and demonstrable fact that "push pushes back," i.e., one cannot push on something without its pushing back with equal force (Lesson C-7). Magnets and magnetic force, so important in physical science, were addressed in Lesson A-5A, but may be reintroduced here.

In the second volume of this series, this thread will extend into investigating how motors, engines, and other electrical and mechanical devices work. Through this work, students will become experientially familiar with the concept that energy cannot be created out of nothing. There must always be a source delivering at least the amount of energy used. In turn, this provides the basis for examining energy resources, present and future, and their pros and cons. Along the way, math skills will be called into play on many occasions.

D-THREAD: Earth and Space Science

Drawing on common experience that unsupported objects fall down and using a globe as a centerpiece, children are brought to appreciate that there is a force we call gravity that attracts everything toward the center of the Earth. The terms, horizontal and vertical, are introduced and seen in relation to the direction of gravitational force (Lesson C-1). (Gravity being the force that holds planets and other bodies in orbit is saved for a later lesson.) In Lesson C-2, children are brought to understand that the regular occurrence of day and night results from the Earth's rotation. Further aspects relating time to the rotation of the Earth are explored in Lesson C-5.

In the meantime, students learn to draw maps as simple bird's-eye depictions of what is on the ground and go on to relate various maps to their surroundings and to respective areas or points on the globe (Lesson D-3). They learn how we designate N, E, S, and W on the globe and how this, in turn, determines how we draw and read maps (Lesson D-3A). Additionally, these lessons, plus understanding the rotation of the Earth, enable students to determine directions by the position of the rising and/or setting sun.

Lesson D-4 embarks students on an ongoing project through which they will attain appreciation for various major ecosystems, such as forests, grasslands, tundra, and deserts, and where they are located in the world. Likewise, they will identify the locations of mountains, major rivers, and so on. Lesson D-6 also starts children on an ongoing project. Here, their activities will convey appreciation for the seasonal changes (temperature, day-length, and rainfall as well as changes in the world of living things) that occur in their region, and they will relate these changes to the progression of the Earth in its orbit around the sun. These studies provide a foundation for all future environmental and ecological studies.

Lesson D-7 returns to the topic of weight and gravity. Students witness how falling objects do not show weight and that is extended to an understanding of how weightlessness in space results from being in orbit, not in any change in the object or elimination of gravity. Students extend this to understanding the orbiting of heavenly bodies in general, and come to recognize the distinction between weight and mass.

All four threads are integrated as students examine fossils (Lesson D-8) and relate them to rocks, minerals, and soil (Lesson A-10) and the physical processes of erosion and sedimentation. An integrating overview is also provided as students consider how all human-made materials and items are derived from resources that come from natural non-living or biological worlds (Lesson D-9)

In the second volume of this series (in progress) each of these threads will be continued in the same systematic manner toward higher levels of understanding. Many of the lessons will draw on measurements and calculations thus integrating math and science.

Bernard J. Nebel, Ph.D.

FLOWCHART FOR PRESENTATION OF LESSONS.
A. NATURE OF MATTER

PURSUE
B. LIFE SCIENCE

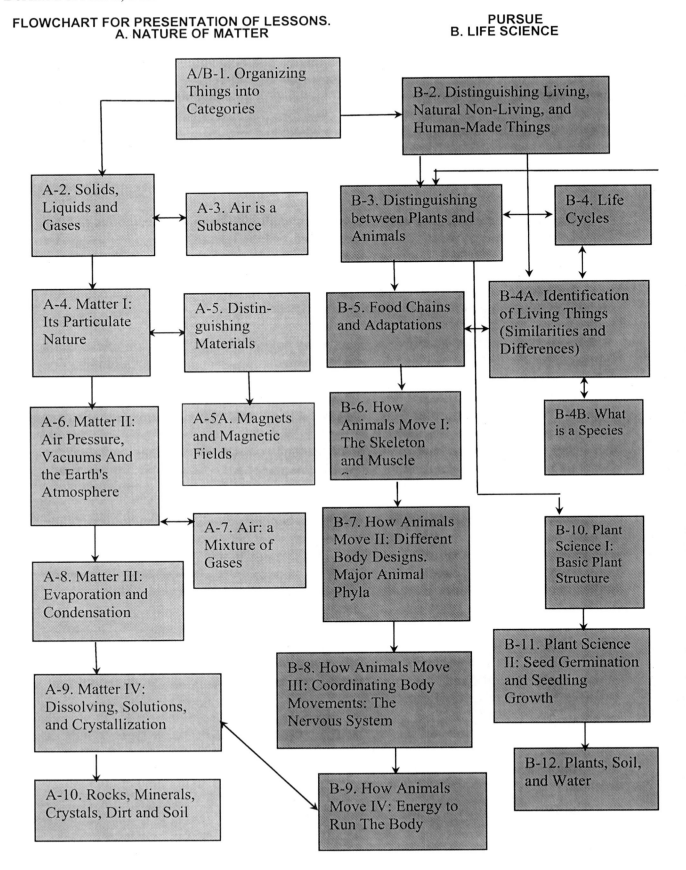

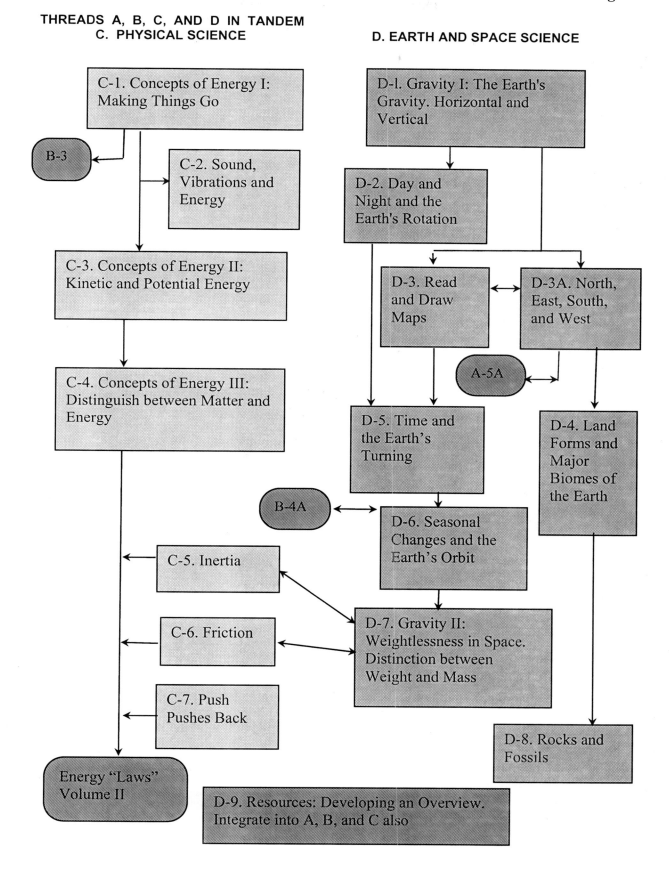

THREADS A, B, C, AND D IN TANDEM

C. PHYSICAL SCIENCE

D. EARTH AND SPACE SCIENCE

C-1. Concepts of Energy I: Making Things Go

B-3

C-2. Sound, Vibrations and Energy

C-3. Concepts of Energy II: Kinetic and Potential Energy

C-4. Concepts of Energy III: Distinguish between Matter and Energy

C-5. Inertia

C-6. Friction

C-7. Push Pushes Back

Energy "Laws" Volume II

D-l. Gravity I: The Earth's Gravity. Horizontal and Vertical

D-2. Day and Night and the Earth's Rotation

D-3. Read and Draw Maps

D-3A. North, East, South, and West

A-5A

D-5. Time and the Earth's Turning

D-4. Land Forms and Major Biomes of the Earth

B-4A

D-6. Seasonal Changes and the Earth's Orbit

D-7. Gravity II: Weightlessness in Space. Distinction between Weight and Mass

D-8. Rocks and Fossils

D-9. Resources: Developing an Overview. Integrate into A, B, and C also

Chapter 1

Teaching According to How Students Learn

This book is designed to aid users (inservice teachers, parents, and homeschoolers) in guiding their students and children toward a high standard of scientific literacy. This high standard will include: a) having a broad knowledge covering the basics of all the major natural sciences, b) having that knowledge organized into frameworks of conceptual understanding that serve for critical thinking and evaluating information, and c) having the mental skills of acquiring new information in a scientific manner. These mental skills are often designated by the single word, *inquiry*; they include observing, questioning, examining, organizing, recording, critical thinking, hypothesizing, testing, theorizing, and communicating. Importantly, this book is for people who may not, themselves, have much background in science. Lessons provide sufficient instructions and background information for teachers to learn along with their students and be excellent role models in doing so. Teachers with more background in science are likely to gain new insights. It follows logically, however, that these objectives will be achieved most effectively by adjusting teaching techniques to correspond to how children learn.

TEACHING ACCORDING TO HOW CHILDREN LEARN

Much research in recent years has been directed at how students learn. A number of basic principles have come out of this research.[4] For contrast, however, let us first take a look at methods that have been shown to be quite ineffective, but which are still commonly used.

What Does Not Work

There is a strong natural inclination to believe that if we drill students sufficiently in learning certain basic facts, these facts will go into and remain in their permanent memory banks and be available to use in later thinking and problem solving. An aspect of this belief is the idea that once we have covered a topic and students have tested as knowing it, we do not have to deal with it again. Unfortunately this idea proves to be erroneous. Facts learned but not reviewed and put into use in further learning are quickly forgotten. Worse, what is remembered is apt to become jumbled and confused. Even

---

[4] Donovan, Suzanne M. and Bransford John D., Editors. "How Students Learn: Science in the Classroom." National Research Council, Washington, D.C.: The National Academies Press, 2005.

more troublesome is that many students demonstrate that they know a given fact but still are not able to apply it in reasoning or deriving the answer to a question. The clear conclusion is that memorizing scientific facts does not lead to conceptual understanding. Even more pronounced, it does not, by itself, lend to critical thinking, problem solving ability, or any other skills of inquiry that are the backbone of scientific thinking.

A second even more popular method of teaching science is to have children perform certain exercises, activities, or experiments. There is an assumption that, if children perform the manipulations called for in the activity, they will come away with an understanding of the concept demonstrated. Again, we are grossly disappointed when sometime later students show that they have no memory of even doing the exercise, much less any idea of the principle it was intended to convey. This failure makes sense in terms of our own experience. We have probably all followed a cookbook recipe without considering how or why the ingredients and steps led to the outcome. Indeed, we may have followed the steps while our minds were elsewhere.

If rote memorization and even doing activities do not lead to scientific understanding, what does? Here we turn to the principles of "How Students Learn."[5] .

Principles Of How Students Learn

> **Principle 1. There are two parts to developing real understanding. There is learning of factual information, but understanding comes only as facts are integrated together into a broader, conceptual context.**

Take understanding a sport such as baseball, for example. There are many "facts" involved: bases, pitchers mound, batter's box, strikes, balls, throwing, hitting, catching, running, outs, runs, innings, etc. But it would be absurd to say that we had any understanding of baseball if we only knew the facts. It would be equally absurd to think we could grasp the concept of playing baseball without knowing the facts. It all makes sense, and we gain an understanding of baseball as we fit the facts together into the context of how the game is played. Then, the facts and the concept become mutually supporting. There is a synergy between the two. Imagine trying to memorize all the facts of baseball apart from the context of the game or trying to understand the concept of the game without the facts. It is in the context of the game that the facts are organized, remembered, and provide an understanding. Reflect on any hobby, avocation, or profession in which you may be engaged, and you will find the same synergism between knowing facts and having those facts fit into a broader framework of understanding.

Most importantly, the ability to use information in constructive thinking, communicating, and problem solving only comes as factual knowledge is integrated into conceptual understanding. Just imagine a person trying to describe baseball if she/he only knew facts. Answering any question about baseball immediately calls on knowing certain facts and knowing how those facts fit into the context of the game.

---

[5] ibid.

Turning to the teaching of science, this principle says that we must guide children in observing and learning certain facts, and additionally, we must guide them in integrating those facts into broader concepts. Teaching facts without integrating them into broader concepts is wasted effort; thinking that we can have students understand concepts without supporting facts is naïve.

Principle 2. New understanding is constructed on a foundation of existing understanding.

Principle 2 is really an extension of what we have just said regarding integrating facts into a framework of conceptual understanding. Taking our baseball analogy one step further, suppose we wish to convey the distinction between a "walk" and a "run." If the person already understands the basic concept of the game, conveying the distinction will be relatively straightforward as it fits into and expands on what they already know. On the other hand, if they are unfamiliar the concept of baseball, attempting to make the distinction would come across as meaningless and nonsensical.

Putting this idea in terms of teaching, it says that it is extremely important to make connections between what students are already familiar with and the new information. Material that students are unable to connect into what they already know is unlikely to be learned. Even worse, it is likely to turn students off to school and learning. We can probably appreciate this phenomenon from our own experience. New information is meaningful and delights us when it connects and adds to what we already know. On the other hand, if new information does not fit into our foundation of existing understanding, we are most likely to dismiss or quickly forget it. The cliché is, "There must be hooks upon which to hang new information if it is to be retained."

Said another way, we must pay constant attention to helping students construct and expand that framework of integrated facts, ideas, and concepts noted above. On this basis, it is understandable why a series of science activities on different topics that have little if any connection to one another does not serve to yield scientific literacy.

Principle 3. Effective learning depends on students self-monitoring what they know, and don't know, and striving to fill in gaps.

We can see and appreciate this mind activity through our own experience in preparing for a test or to give a presentation. Our minds indulge in self-talk, "This part is clear. I have that down. But this part is vague. I need to work on this more. Here is something that I don't understand at all. It doesn't make sense. I need to ask someone for help or do further reading/research to get this." It is this kind of self-monitoring that directs our efforts in the proper places and increases the efficiency and effectiveness of our learning. The technical word for it is *metacognition*. Of course, this kind of self-monitoring is manifested in experienced students who have really learned to study. We can hardly expect young children to do this automatically. Therefore, as teachers, we must guide children in developing this behavior.

The cornerstone for developing this metacognitive behavior is to create a classroom atmosphere in which students feel safe in asking questions and expressing their ideas. Moreover, they should feel that their questions and ideas are appreciated and given due consideration. Having students express their existing views regarding a topic at the outset of its study is particularly important. It is through such expressions that we can assess their current understanding and misunderstanding and focus our teaching accordingly. Testing after the topic has been covered only shows if students have mastered it or not. It does nothing to aid and steer them along the way. Without correction and steering along the way, students are very apt to maintain their erroneous concepts even in the face of correct information.

While we may see the virtue of determining the mindset of students before plunging into a topic, the opposite is too often the case. Questions do delay and interrupt the teacher's presentation and, besides, they may be off topic. Therefore, questioning is commonly discouraged implicitly or explicitly. The model student is seen as the one who just sits quietly, absorbs information, and does not ask questions. This is very unfortunate because inhibiting a child's questioning undercuts this major principle of their learning. Questions are an outward manifestation of inner thinking. It we wish to nurture children's aptitude for thinking, we must honor and respect their questioning. We expand on this and other techniques of guiding students to think in Chapter 2.

Principle 4. Learning needs to connect to real-life experience.

This is a facet of Principle 2, but we make it a principle by itself for emphasis and clarity. Much literature describes the inability of children below the teen years of age to grasp abstract concepts. Rather, they are concrete thinkers. They are able to think most effectively about what is in their own direct experience. The lesson for teaching science is straightforward. It means that lessons must always be centered around and supported by what children observe, experience, and can relate to directly. Attempts to have students learn facts or concepts apart from their direct experience will be disappointing.

Teaching Skills Of Inquiry

The above four principles apply to learning in all areas. When it comes to science, there is a host of additional mental attributes that we wish students to master. These attributes of scientific thinking, as noted above, are often grouped under the single word, *inquiry*. They include: observing, questioning, examining, organizing, recording, critical thinking, hypothesizing, testing, theorizing, and communicating.

The first and most important thing to recognize is that these are *skills*, mental skills to be sure, but skills nevertheless. Like physical skills, they are only developed by practice, practice, practice. One or even a few exercises that call for observation, for example, will not develop the habit and skill of observation any more than a few piano lessons will enable one to play a piano concerto. The same can be said regarding the other aspects of inquiry. Skills are not something to be memorized in the head; they are "habits" that can only be developed and maintained through constant exercise and practice.

HOW THIS BOOK IS ALIGNED WITH PRINCIPLES OF LEARNING AND IMPARTING INQUIRY SKILLS

As explained in the Preface and Introduction (pages 1-78-9) the lessons in this text are organized such that gaining factual information is integrated with developing broad, conceptual understanding. This is in harmony with Principle 1. Further, by having lessons build in a logical, steppingstone fashion and conducting reviews of past lessons as a part of each presentation, teachers can assure that new information is constructed on a foundation of existing understanding (Principle 2). Incorporating Q and A discussion into lessons, as will be explained in greater length in Chapter 2, will promote a monitoring of what students know and don't know and invite filling in gaps (Principle 3). In keeping with Principle 4, all lessons center around things and/or events that are a part of children's everyday experience. A particular technique of promoting students making connections between their classroom learning and real-world experience is the following.

Making Connections Between Classroom Learning And Real-World Experience

As just noted, an extremely important aspect of aiding children's learning is in helping them make connections between what they are learning in school and what they are seeing, finding, and experiencing in the real world. Thus, you will find that each lesson calls for children, with the aid of their parents/caregivers, to find examples of additional connections between the topic under study and their real world. To implement this happening, we strongly advocate devoting the first 30 minutes or so of each class day (or at least two or three times a week) to a sharing, show-and-tell, directed-talk time—referred to in the following as a preclass period.

It often takes about 30 minutes to get children settled down and oriented toward their schoolwork. Note that this "settling down time" generally amounts to having children suppress their thoughts and excitement of the real world and focus their attention on schoolwork, which they may see as having little if anything to do with their real world. Making such a separation, even if it is only in children's minds, is counterproductive to how they learn most effectively. Instead, we want to make every effort to aid them in relating their schoolwork to what they experience in their real world. The preclass period can accomplish this goal and provide for other learning objectives as well.

As was just said, the follow-up of lessons in this text consistently calls for students, with the help of parents and other support-givers, to be on the lookout for examples that connect or apply their classroom learning to what they see, find, or experience in the real world. The preclass time provides an opportunity for them to share their discoveries and insights. These preclass sessions will foster children's learning in numerous ways.

- First and foremost, they serve to build bridges between the classroom and real world.
- They provide an ongoing review, reinforcement, and amplification of material.
- They provide opportunities to bring in additional examples to extend and enhance learning.
- They provide opportunities for integration of diverse lessons.

- They expose misunderstandings and provide opportunities for clearing them up.
- They provide ongoing opportunities to fit factual material into the broader picture of conceptual understanding.
- They provide a platform for stressing careful observation, noting similarities and differences, rational thinking, and other aspects of inquiry.
- They provide a platform for teaching aspects of character education and development, both in regard to giving presentations and in regard to attentive, respectful listening and questioning.
- They provide a tool for assessing numerous aspects of each child's progress.
- They are an important aspect of creating an atmosphere where children take responsibility for their own learning (see Chapter 2).

The actual content and conducting of preclass sessions will become clear as you see them in connection with specific lessons.

FINDING JOY IN LEARNING AND BECOMING SELF-MOTIVATED LEARNERS

It is widely recognized that in today's rapidly changing world, it will be necessary for students to become life-long learners. Therefore, in addition to all the above, we want students to become self-motivated learners and enjoy learning. Of course, the two are almost one in the same; the student who enjoys learning will be self-motivated to do so. But, how can we bring students to enjoy learning? This is elusive.

On the one hand, it is abundantly clear that some people truly enjoy learning. A professional in any academic area will describe the joy and satisfaction he/she finds in exploring, finding, learning new aspects, and gaining new insights pertaining to their work. On the other hand, too many students become turned off from school, come to see it as drudgery, and can hardly wait to be finished with it. For them, it is clear that learning has not translated into joy. Can we find a way that will bring students, at least a higher proportion, to enjoy learning?

One widely used tactic has been to incorporate the learning into games, projects, and contests. Unfortunately, this approach has several failings. First, the activities often consume an inordinate amount of time while the informational content is slight. Second, children tend to get caught up in the activity itself and miss the point of learning entirely. Third, competitions give the elation of winning to only one person/team; others are disappointed. Last and most serious, the whole idea that games will make children enjoy learning commits a logical fallacy, namely a reversal of cause and effect. The fact that learning may lead to joy does not mean that fun and games will lead to learning.

A second common tactic is to offer rewards or prizes of one sort or another for achievement. Alfie Kohn, in his book "Punished by Rewards"[6] describes how this practice is actually counterproductive. In essence, it leads students to "jump through the hoops" simply to get the reward. It actually short circuits finding the intrinsic joy that

---

[6] Kohn, Alfie. "Punished by Rewards." Boston, MA: Houghton Mifflin Co., 1993.

should come from the learning itself. It may even subvert basic learning as students readily discover how to do the minimum to get the reward.

We need to come back and reexamine the problem. Exactly what aspect of learning is it that generates elation and joy? I think we can gain insight from our own experience. A feeling of exhilaration rarely comes in the process of simply learning factual information. Just recall those hours of drudgery to get facts for an exam into your head. Joy comes specifically when pieces or ideas that our minds have been working with suddenly come together into a picture of greater understanding. We witness this even in mundane examples of doing jigsaw, crossword, or other sorts of puzzles. Our minds struggle with the pieces; then there is a sudden rush of delight as we discover how they fit together or a certain piece fits into the whole. The same is true in mystery stories. There is the mystery that intrigues our mind. Then there is a series of clues. Finally, joy and satisfaction come as we see how the pieces fit together and reveal "who-done-it." This phenomenon applies to children as well. There is nothing more rewarding to teachers than seeing a child light up with joy as they suddenly realize how facts she/he has been struggling with fit together into a picture of understanding, or see how to solve a problem.

In conclusion, our minds seem to be wired such that we like confronting problems, wrestling with them, and then an effervescence of delight is produced as the pieces come together to resolve it. Indeed, the elation is such that we tend to keep coming back for more puzzles to solve. In short, the phenomenon produces self-motivation. Of course, there is an implicit assumption that the problem or puzzle is appropriate to our level of skill. If it is too easy, it is a bore. If it is too difficult, we will give up and leave it. There is no joy in either. The joy comes when there is a challenge, but one we can master.

This has great implications for our teaching. It says that, so far as possible, we should conduct our teaching in a manner that presents students with a problem, a "mystery" to be solved, invites them to grapple with it, and provides only the hints necessary for them to reason out the solution for themselves. Simply telling them the answer spoils the fun in the same way that giving away the ending of a mystery story spoils the fun of reading it.

Happily, science lends itself especially well to this format of teaching. The essence of science is observing or experiencing something and turning that into a mystery to be solved. What is it? How does it work? How does it relate to other things? What are the similarities and the differences? Pondering and perhaps testing different possibilities comes next. Finally, there is the "ah-ha" experience as there is a breakthrough of insight into how the pieces fit together or how a particular piece fits to give a better picture of the emerging whole.

The organization of lessons in this text is designed to facilitate this progression. As already described, the sequence of lessons in each thread is planed to gradually give students pieces of the "mystery" and fit them together into a picture of expanding understanding. Lessons themselves (see "Methods and Procedures" of individual lessons) frequently ask you to invite students to question, ponder, and discuss, that is, grapple with the "puzzle" and, with minimal guidance, figure it out for themselves.

But, this latter aspect cannot be scripted. Students will ask the questions, ponder, and have ideas as they do. Therefore, in terms of promoting and guiding this process, much will fall to the individual teacher. A number of pointers can be given, however. Chapter 2 to is devoted to these pointers. Let us only say here that it is a skill that you will develop with practice. Don't expect to do it perfectly all at once.

Chapter 2

Guiding Students to Think

A perennial frustration among teachers is getting students to THINK. Testing often shows that many students, although they have learned certain facts and definitions, still fail to apply this learning in deriving the answer to a thought question or solving a problem. The cliché is, "They fail to put two and two together." This is no minor problem. Having students develop thinking ability is at the core of educational objectives. It is the center of problem solving, critical analysis, acquiring further information, distinguishing logic from illogic, rational from irrational, and other thought processes basic to life. How can we teach students to think? It would be great if we had a magic potion, but we don't. Still, there are basic techniques of teaching that will go a long way toward this objective. We devote this chapter to their consideration.

DISTINGUISHING LEARNING AND THINKING

First, we need to be clear that learning and thinking involve rather different mental functions. By analogy, learning is putting material into the hopper of a mill. Thinking is turning the mill on to grind, sift, and churn that material, organize it in various ways, see similarities and differences, see gaps that need to be filled in, and above all, use it in problem solving. Learning facts, putting material in the hopper so to speak, does not require any particular thinking or reasoning. It can be achieved with sufficient repetition and drill, while the thinking part of the brain remains in neutral. Witness how we are capable of learning "sentences" of nonsense words that involve no thinking regarding meaning whatsoever.

In short, learning by itself does not automatically lead to thinking. But thinking does promote learning. As the facts and ideas are mulled over, organized, and structured in various ways, learning is greatly enhanced. As with every attribute, some children have the natural aptitude to put their thinking in gear for all its benefits. These children become our outstanding students. As educators, however, we need to encourage and stimulate more, if not all, of our students to engage their minds in productive thinking.

How does one get students to turn on, use, and develop their thinking ability? There are three features to this process. First is creating and maintaining an atmosphere in which students feel free and safe in asking questions. Second is posing questions that will bring students to reflect and contemplate. Third is to promote question and answer (Q and A) discussion that will guide their thinking. There is more to say about each of these steps.

Bernard J. Nebel, Ph.D.

STIMULATING THINKING

Maintaining an Atmosphere Open to Questioning

The most efficient and effective learning is motivated by an inner quest for information and understanding. The central pillar of this quest is a questioning attitude. Happily, most children are born with this attitude. Our role as teachers should be to encourage it, not extinguish it. Questions should be acknowledged, appreciated, reflected upon, and answered as appropriate. Avoid actions and statements that may discourage questions and admonish students to behave similarly toward the questions of their classmates.

Of course, questions from students can become problematic. There may be so many that you don't get one answered before there is another. Some questions will lead far from the topic at hand, and some may be off-topic entirely. Others may lead into a prolonged discussion for which there is not enough time. Therefore, while we wish to create and maintain an atmosphere that encourages questioning, there will be times that it must be ordered and controlled.

One can ask that questions be held during certain portions of a presentation, but then make time for questions at the end. If there are a number of questions on top of one another, they should be written on the board and then addressed in turn, or students may select which they wish to explore first.

Invariably, there will be many different kinds of questions from children, and different kinds of questions call for different kinds of responses. The following is a listing of different sorts of questions and suggested responses.

Students' questions and suggested responses

Type 1 Questions: "What is that? What's it for?" Such questions are straightforward and deserve a straightforward answer. The child is learning the names of and functions of different items.

Type 2 Questions: Much the same as Type 1 questions, perhaps with "What are you doing?" and "Why?" added in, but here you are quite sure that the child knows the answer. Recognize that children may frequently ask questions to get your attention and engage you in conversation without any particular desire for information. Such questions may be simply returned: "You tell me!" Indeed, this may become a kind of game, but note that it is getting the child to practice the skills of thinking and explaining. The child also gains the pride of showing that he or she does know or can derive the answer. Then again, the answer may show a gap or mistake in understanding and one can help the child clarify that point.

Type 3 Questions: Some children will pretend they have a question, but then launch into telling their own story. Call them on this before they go too far. Do you have a question about _____, or are you telling a story? Stories are for story time. Now is for _____.

Type 4 Questions: These are questions that seem totally unrelated to the topic at hand. Rather than dismissing or discouraging such questions outright, an action that discourages questioning in general, it is better to ask, "What connection are you making between this (the topic at hand) and your question?" We just might be surprised to find that the student is making some sort of logical connection; he is striving to structure his knowledge, and this should be praised. Or, perhaps the student will confess that his mind was just wandering, and he is not making a real connection. Again, such questions should not be reprimanded. They serve to tell us that we lost that student and probably others. The off-topic question provides an opportunity to refocus. "Thank you for asking that; now please ask another question about _____ (the topic being discussed)."

Type 5 Questions: These are questions that are sound, sincere, and on-topic, but in all honesty, we are clueless regarding an answer. These are the most beautiful questions of all. Far from dreading them, dodging them, or giving flip answers, we should receive such questions as nuggets of gold. The question tells us that the child is probing into an area of knowledge that we perhaps have never considered. It provides a real opportunity to be a role model in being excited about learning on into adult life.

The way one responds to questions of this fifth category is of critical importance. Every parent and teacher should take a workshop in learning how to say, "I don't know." The way these three words are said can convey so many undertones of additional meaning:

1. Annoyance: Don't bother me with such questions.
2. Embarrassment: I know I should know, but I don't and please don't embarrass me with such questions.
3. Dismissal: I don't know and I don't care. Such questions are unimportant.

You can see that each of these is a negative way of saying, "I don't know." Each will tend to discourage questioning, which in due course discourages learning. But there is a way of saying, "I don't know," that is inspirational. Pause, consider, reflect, and show interest conveying a sincere attitude, verbally expressed or not, that says: "Wow! That is really a neat question. That's important, but I never really considered that before. Do you suppose we can find the answer, or do you think you could find the answer for me?" Such questions may be recorded in a special place. Some of these questions may soon be answered. Others may provide intrigue for a lifetime.

Posing Questions that Bring Students to Reflect

A teacher may ask two sorts of questions in the course of teaching: one is to determine if students have mastered certain information. Such questions call for a simple recitation of facts, definitions, or other information that students have hopefully committed to memory. Such questions and questioning are not our concern here. The second form of questions, which are our concern, is to convey an invitation toward thinking and reasoning. Questions and the way they are posed need to portray a sincere desire for information and understanding.

> I wonder how _____?
> How might we separate _____ into categories?
> What are your reasons for that choice?
> How do you think _____ may be related to _____?
> How does _____ differ from _____?
> How might we go about solving this problem?
> What do you think causes _____?
> What do you think will be the effect of _____?

Note that the questions are framed such that they cannot be answered by a simple "yes" or "no" or another simple memorized answer. They should be framed to draw students to look more closely, ponder, and reflect; in other words, to think.

The tone of voice will be as important as the question itself. It should convey a sincere desire for information and understanding, an invitation to think and try different "solutions." Note that this is quite different from a tone of voice that is used in testing if students know certain information or have drawn the same inference that you have in mind. Questions and a tone of voice that express a quest for understanding invite students to think, explore, and try different solutions. Questions and a tone of voice asking for recitation simply put them on the spot as they the demand simple recall or coming up with a precast answer, a test they may pass or fail. In the former, you are being a guide in the quest for learning; in the latter you are being an authority figure demanding that they learn certain facts and think in a given channel, an attitude that does little to foster an intrinsic desire for seeking knowledge. Worse, many students will freeze up under the glare of "test questioning" and not be able to answer even when they know the information. In turn, their discomfort may lead them to dislike the whole schooling experience.

Again, questioning should express musing and wondering, a sincere desire to learn, and understand. It is perfectly acceptable if you don't have answers for all your musings. Indeed, a student may look at you and ask, "What's the answer?" Your response should be the same as for the probing questions coming from the student—interest, intrigue, enthusiasm. "I don't know. It just occurred to me. Do you think it is worth investigating?" In short, be a role model demonstrating the freedom to ask questions, whether or not the answers are known. Consider how far a scientist would get if he or she only asked questions to which answers were already known. The questioning, questing attitude that you display will draw the same out of students. It is especially appropriate to science, but it may apply to other subjects as well. Recognize that thinking and pondering do require time. Learn to be patient in allowing "think-time." More will be said about this later.

Finally, none of us will ever know all the answers. We should become comfortable in accepting that fact. This is not to say we should be happy idiots content to know nothing. But with all our knowledge and experience, we should make no attempt to create an atmosphere in which we try to put ourselves across as having all the answers. Rather, we should be role models who exemplify that learning is a lifelong process, and the driving force of that process is questions—questions we ask ourselves and questions posed by others; questions that bring us face to face with what we don't know; questions that reach beyond the limits of our existing knowledge and lead us to expand our consideration and understanding.

Extending Questions to Thought Provoking Discussion

With an atmosphere that is open to students' questions and sprinkled with the teacher's queries as described above, it will be natural for questions to extend into further discussion. The goal of this discussion should be to generate thinking that engenders learning, which generates more questions, which generate more thinking, in an ongoing cycle. Conducting and guiding Q and A discussion in a manner that draws students to ponder the situation and use their thinking to reach rational answers and further questions is a skill that will develop with practice. Don't expect to be a master of it at the outset. General pointers for conducting such discussion are the following.

We have already spoken of the importance of initial questions requiring more than a simple "yes" or "no" answer or regurgitation of a learned fact or definition. They should require students to reflect on observations, consider similarities and differences, causes and effects, and/or other parameters of the topic. Don't expect a student to immediately give a well-formulated answer to such a thought question. Above all, don't just drop them and call on another student. That can only leave them feeling dumb, which is the opposite of helping them learn to think. Rather, allow a full 10-15 seconds of think time; then ask, "Do you need some help?" If the answer is "yes," give pointers that direct their attention at the particular features or facts that are significant and aid them so far as necessary in step(s) of reasoning that go from these to a logical conclusion.

When a right answer is presented, our natural inclination is to give praise—"very good," and move on to the next point. Note that this does nothing for other students in the class. They may not have even paid attention to the answer, much less learned anything from it. Instead ask, "Do you agree with Johnie's answer?" "Why do you agree?" "What additional facts/experience indicate that Johnie's answer is acceptable?" The same holds for off-track answers. Rather than just saying "no" and calling on another student, ask in the same manner, "Do you agree with Johnie's answer?" "Why do you disagree?" "What are additional facts/experience that make you believe differently?" As contrary evidence is presented, interrupt and ask Johnie if he would like to revise his answer and point out that this is a common and okay thing in science. We commonly are confronted with information that requires a change in our thinking.

Another approach to dealing with a student's off-track answer is to ask, "If that, then what?" In other words, try to guide Johnie to see that his answer would imply, by logical cause-effect reasoning, an impossible conclusion. As Johnie begins to see the error of his answer, ask him if he wishes to change it. Again, keep in mind that the objective is to help students straighten out their thinking, not trap them into appearing dumb to themselves or others.

As discussion proceeds, additional questions are bound to arise. This should indicate that you are being successful. Discussion is generating thinking that is leading to more questioning. Depending on the type of question (see above), it may be addressed immediately, or, if it would seem to take the discussion too far astray, it may be written down for later consideration. Really, every discussion session should end not with all the questions answered but with additional questions hanging out begging for further investigation. You may actually diagram this on the board as a circle with the basic topic or finding in the center and arrows extending out like spokes on a wheel to additional questions that have been raised. So far as possible, facilitate students pursuing such questions as independent investigations and reporting their findings to the class. The preclass period described in Chapter 1 (page 19) may serve for this purpose.

In the course of every Q and A discussion and in the context of other reports as well, be watchful for breaches of logic. Call attention to them immediately, point out the fallacy involved, and encourage students to be on the lookout for such violations as well. Common logical fallacies are described in the "Baloney Detection Kit" included at the end of this chapter.

THIS BOOK AND Q AND A DISCUSSION

Maintaining a class atmosphere that is open to questions means that discussion may ensue at any time and in relation to any subject. Further, it may occur with the class at large, with small groups as students are engaged in a particular activity, as a one-on-one with a student as she/he is working independently, or in special small groups set up specifically for Q and A discussion (described in the next section). Whatever the occasion, the pointers outlined above should be followed so far as possible.

Pertaining to this text, you will find the instruction to conduct Q and A discussion at many points within each lesson. It will be especially pertinent in reviewing past lessons that bring students up to speed for the current lesson. (Note that each lesson has a section, "Required Background," that lists preceding lessons that the current lesson builds on.) Within the "Methods and Procedures" section of each lesson, you will find instructions to conduct Q and A discussion regarding particular points. This is directed at stimulating students to think about the meanings and implications of what they are doing and/or observing. Indeed, at any time students are engaged in working independently or in small groups circulate about the room and conduct one-on-one Q and A to keep students' minds focused on what they are doing, what they are observing, and conclusions being derived.

Finally, we strongly recommend that teachers arrange their class and schedules to accommodate small-group Q and A discussion sessions. Each lesson contains a section of questions suggested for small-group discussion. (See the "Questions/Discussion/ Activities To Review, Expand, and Assess Learning" section within each lesson.)

Specific suggestions for organizing and conducting small-group discussion sessions are given below.

FORMATION AND MANAGEMENT OF SMALL-GROUP DISCUSSIONS

It is really impossible to conduct a meaningful discussion with more than eight or nine students at a time; four to six is ideal. With more, a number of students will not get a chance to participate, a number will tune out, and others may indulge in disruptive behaviors. One can get around this by dividing your class into groups and rotating, one group having discussion while other groups are working independently on something else. Making "books" (see Appendix A) is always a constructive, independent activity, but reading, working at activity centers, math, or anything else may serve to keep students occupied.

You will want to call on volunteers (parents, seniors, older students doing service, etc.) or aids to monitor those doing independent work so that you may give full attention to the discussion group. Some volunteers may be skilled enough to lead discussion groups. (Enlisting volunteers to help out is actually a step toward meeting National Science Education Teaching Standard A, "... nurture a community of science learners.")

Turning to the management of the discussion group itself, begin by setting down rules of conduct. No disruptive behavior; no butting in while someone else is speaking; no comments that degrade another such as, "That's stupid." Let children participate in making rules. Actually write them on paperboard and post them. Don't start with more than three or four basic rules. Rules may be added or modified as situations require.

Most of all, emphasize that the discussion session is NOT a competition. No one will be getting points or gold stars for trying to prove that they know more than someone else. The purpose is to share ideas, questions, and think together as a team. If anything is to be rewarded, it should be respectful, thoughtful, supportive behaviors.

With young children, a discussion period will probably last no more than fifteen minutes, but be flexible and play it by ear. As children gain experience, they may well go longer.

Beyond adhering to the basic rules, you will almost always have two extremes of students who present particular challenges: those who are anxious to always be in there with a response and those who are shy and reticent to speak at all. A general technique that helps both, as well as everyone else, is to pose the question but then insist on a silent "think time" of at least 30 seconds (use a watch) before anyone raises their hand. You may softly repeat the question during the think time. If the overly eager student butts in too much, he/she may be required to sit outside the discussion circle and only observe for the remainder of the discussion. You may also use a system of tokens. Give each student a number of tokens (perhaps 5) at the start of discussion. Each student must put one of their tokens in the pot for each time they speak. When they are out of tokens they are out of talk. Emphasize that attentive, respectful listening is as important as speaking.

Bernard J. Nebel, Ph.D.

For the extremely reticent student, she/he may be taken aside before the discussion session and told in a very gentle way: I really miss hearing your voice in discussion. What can we do to make it easier for you to speak? You may go so far as to give them an upcoming question and help them prepare an answer.

As discussion proceeds, avoid responding to a student's answer with a simple "good" for a correct answer or a "no" for a wrong answer, and moving to another student. In both cases, pause, and ask the group or another student questions such as: Do you agree with that? How so? What are your reasons (supporting evidence/evidence to the contrary? Or, for wrong answers call "think time" and ask another question that invites the student to think about the implication(s) of her/his answer. "If that, then what?" For example, if a student answers that evaporation is a process of water simply disappearing, ask, "Do things such as marbles just disappear or do they go somewhere?" And, "If water particles can just disappear, why doesn't a wet rag in a container dry just as rapidly as one hung out?" Give the student a moment to reflect. Then ask, "Do you wish to change your answer?" Of course, other students need to obey the rule of maintaining a respectful silence during this process.

The point is to not to leave the student with a wrong answer looking and feeling dumb, but to stimulate their insight and reasoning toward seeing their own error and making a correction. As students gain experience, they may learn to conduct discussions with each other.

It is more than likely that students will ask a number of their own questions in the course of discussion, and such questioning should be allowed and encouraged. Questions represent thinking. However, review again the various types of questions and responses discussed above. In every case it is appropriate to hold a think time in which you consider the question and an appropriate response. In many cases, you may ask the student him/herself, and then other students, to give a stab at the answer.

Again, the goal is not competition; it is having students support one another in gaining understanding. (National Science Education, Teaching Standard B, "Challenge students to accept and share responsibility for their own learning.")

With this emphasis on scheduled discussion sessions, don't loose sight of the fact that bits of discussion may and should occur spontaneously on many other occasions as well. Show-and-tell sessions at the beginning of the day are particularly good for creating teachable moments and inviting discussion. The same rules of respect for each other and the speaker should apply. Likewise, discussion may be encouraged out of school as well. It will be well to communicate the value and techniques of discussion with parents/caregivers and encourage them to pursue discussion in the family/home environment.

If this all seems like too much trouble, review the National Science Education eaching Standards.7 You will note that skillfully conducted discussion is an avenue toward most, if not all, of those standards. Nothing else will serve as well. Always bear in mind that thinking is a skill. It is only developed through practice. The practice for thinking is thinking. Again:

**THINKING DOES NOT COME FROM LEARNING;
LEARNING COMES FROM THINKING**

Baloney Detection Kit[8]

Beginning level elementary students, having had relatively little real-world experience, are quite prone to blend and sometimes confuse imagination with reality. They are also prone to accept prejudices and certain false beliefs as facts. Need it be said that such prejudices and false beliefs can very well stay embedded throughout adulthood? Therefore, as parents and teachers, we have the especially important task of imparting the skill of using and recognizing logical, as opposed to illogical, thinking. This skill is the one thing that enables us to separate and distinguish reality from "baloney," both in our own thinking and in that of others. It is the most important critical thinking skill we can carry with us in real life.

The following is a listing and description of common logical fallacies that people are prone to fall into, wittingly to win a point right or wrong, or unwittingly as the mind does not always follow a logical course. It will be unproductive to have students attempt to memorize this list. As we have discussed, memorization does not, necessary, lead to using the information in thinking. Rather, we encourage being watchful for these fallacies coming up in the course of discussion, listening to speeches, and anywhere else. Stop and draw students' attention to what was said, explain the fallacious reasoning, and how/why it makes the conclusion invalid. This is one of those skills of mind that will need much practice, but gradually students will become able to catch the fallacies themselves.

1. Forceful declarations do not substantiate facts or truth.

We often hear people say, "It's a fact that…," "Everyone knows that…," or "People say that…." Indeed, such a declaration may be a fact, but it may not be. Students should learn that some people promote their own particular prejudice, opinion, or mistaken belief by forcefully declaring it to be a *fact* or *the truth* or labeling it as something "everyone

---

[7] "NSTA [National Science Teachers Association] Pathways To the Science Standards: Elementary School Edition," 2nd edition. Arlington, VA: NSTA press, 1996. (pages 131-137)

[8] (Adapted from: Sagan, Carl. "The Demon Haunted World." New York, NY: Random House, 1995.)

knows." Conversely, people may try to counter factual information by declaring it to be stupid, idiotic, or worse. Very simply, facts and the truth are what they are. Try as some people might, the real facts are not subject to alteration by human pronouncement. The best we can do is look beyond the declaration and try to examine the basic reality.

"Show me the evidence!" should be our request to any assertion. This need not be complex. For example, a young student may declare that frogs live in that pond. "Show me the evidence" may lead to the child's presenting you with a frog she or he caught there. On the other hand, a child may declare that a dragon lives in that pond. "Show me the evidence" may yield, "So and so told me."

"That is not real evidence," we would reply. Real evidence might be dragon scales, scorched bushes at the side of the pond from his fiery breath, or perhaps dragon footprints. "Did you see any of these?" If the child answers yes, then ask, "Is a dragon the only possible explanation for the evidence? Could the scorched bush be the result of somebody's campfire? Could the scales really be clam shells?"

Simplistic as this example may seem, it is still bringing the youngster to understand the concept that a simple declaration should not and does not suffice. Evidence must exist that can stand up to scrutiny. It is the evidence (or lack thereof) that should lead us to believing whether something is true, not the force of someone's pronouncements. As youngsters mature, the problems addressed will become more sophisticated; yet "Show me the evidence!" and reaching a conclusion based on evidence remains the basic concept.

It is important to teach this lesson repeatedly, as instances come up in formal and informal discussion because a large portion of the public (and young people in particular) are prone to reaching decisions on the basis of the force of pronouncements, rather than by looking at evidence. Consider smoking. The evidence for deleterious health effects from smoking is overwhelming, yet many young people are more persuaded by the declarations of cigarette advertising. Global warming is another, even more serious issue. "Show me the evidence!" would reveal overwhelming data that substantiates the existence of global warming trends. Yet effective remedial actions are being blocked by dogmatic pronouncements from parties with vested interests swaying public opinion and votes. Indeed, any number of examples might be cited where winning a point is a matter of emotional persuasion rather than evidence. How might the world be changed if people were trained to seek and demand evidence?

2. Every effect has a rational cause.

This concept is so embedded in adult thinking that it needs little supporting argument. We automatically assume that there must be a rational cause for a crash, a fire, a disease. But emphasis should be placed on the word *rational*. Rational cause means that it is amenable to our understanding in terms of real things and/or processes that, in turn, will lend themselves to investigation and understanding. For example, up until the early 1800s, many people believed that evil spirits were responsible for disease. Disease was treated by means that were presumed to rid the body of evil spirits. Such treatments were uniformly ineffective and some (e.g., bleeding the patient) undoubtedly did more harm than good. It was only as disease began to be understood in terms of rational causes, namely bacteria and viruses, that real progress began in combating disease.

We certainly do not fully or even partially understand the causes of all events or phenomena. Still, the belief in rational causes is more responsible for the advancement in understanding than any other single idea. It invites us to carry on systematic investigations until we find that cause. The very fact that investigations are continually unveiling increased understanding gives validity to the concept itself. On the other hand, beliefs in supernatural causes have not yielded any such increase in understanding. Therefore, beliefs in the supernatural are uniformly without merit.

3. Two events occurring in sequence do not imply a causal relationship.

Recognizing that every effect has a cause, there is a strong tendency to jump to a conclusion and designate the wrong factor as the cause. This is especially true in the case of two events occurring one after the other. People are all too prone to conclude that the first is the cause of the second. Beware of baloney! Point out to students that given everything that is going on in the world, it would be impossible not to have all kinds of unrelated events occurring in a sequential time frame. Therefore, two events occurring one after the other does not by itself imply any causal relationship. Additional evidence must support a causal relationship, namely the existence of a logical mechanism or process that can explain how the first event caused the second.

For example, suppose someone rang the doorbell and suddenly all the lights went out. One might jump to the conclusion that ringing the doorbell was the cause of the lights going out. Further checking may reveal that it was only coincidence. The real cause was a utility pole down the street being struck by a car. On the other hand, further checking may indeed reveal a connection. Perhaps a short had occurred in the doorbell. In either case, however, only further checking can reveal the true cause.

4. The universe is governed by natural laws and principles.

Interminable experimentation has revealed and confirmed that things behave according to certain principles. We have come to refer to these principles as NATURAL LAWS. Consider the movement of objects and the Laws of Motion, for example. We should emphasize to students that these natural laws have nothing in common with the laws that humans pass. Gravity, for example, does not exert its force because humans decided to pass a Law of Gravity. Gravity, so far as all our observations and testing reveal, is a basic force, an attraction between all bodies, that is constant through the universe and remains constant in time. It is all these observations that lead us to the conclusion that gravity is a fundamental principle that operates and governs the functioning of the universe. Thus, we refer to it as a natural law. The same can be said for other natural laws.

To be sure, some natural laws have been amended over time to include special circumstances. For example, Einstein showed that in special conditions, matter could be converted to energy and this is now utilized in nuclear weaponry and nuclear power plants. Thus, the "Law of Conservation of Matter" was amended to except these special situations. However, all our experimentation and tests show the principle of conservation of matter still holds for all traditional chemical and biological reactions.

Since natural laws, so far as all our observations and tests reveal, are basic principles underlying the functioning of the universe, any speculation or theory that involves making an exception to a natural law should be greeted with a high degree of skepticism at the very least. Experience shows that it is more than likely to turn out to be baloney.

5. **Inability to think of an alternative does not make the reason at hand right, nor does proving one idea wrong make another idea right.**

Some students (and adults) are prone to declare, "This must be the cause because I can't think of any other possibility." Again, beware of baloney. Our inability to think of an alternative cause does not make the cause we think of true. By the same token, if we can think of two possible causes, proving one false does not automatically make the other true. In general, a true cause cannot be determined by elimination or failure to think of alternative causes. A true cause needs to be verified by setting up test situations, or making further observations that demonstrate its validity, or by showing the mechanism or process by which it produces the effect.

6. **All the data and/or observations must be considered.**

Some persons are prone to base a conclusion on certain supporting evidence while ignoring other facts or observations that contradict that conclusion. An overall concept, theory, or conclusion must take into account all observations. If there is only one *confirmed* observation that is inconsistent with a concept or theory, that is enough to throw the conclusion into question. (Note, however, the emphasis on *confirmed*, i.e., additional studies by different parties that demonstrate the validity of that observation. Sometimes a report that seems to contradict an established theory is itself fallacious.) Similarly, if a conclusion depends on an argument with multiple steps, each step must stand the test of rigorous logic. If one link in a chain is fallacious, that is enough to invalidate the conclusion. Even more, if the starting premise of an argument is in error, all the rest of the argument is worthless.

7. **Beware of generalizing from the particular to the universal.**

Statements such as "The only good bug is a dead bug!" are derived from erroneous reasoning: Some bugs are bad, therefore all bugs are bad. It is generalizing from the particular, the fact that some bugs are bad, to the universal, all bugs are bad. The fallacy is evident in the fact that there are many beneficial insects as well as harmful ones. If people are asked to explain their prejudices, they will often reveal the same sort of logical fallacy. Certain people of a given nationality behaved that way, therefore all people of that nationality will behave that way. In fact, any nationality will include people with all sorts of traits and behaviors, and exceptions to any generalization will abound. Therefore, beware of baloney whenever you detect a generalization being based on only one or two examples.

This is not to say that generalizations cannot be made. The natural laws that were noted under number 4 are basically generalizations. But note that they are not generalizations based on just a few particular examples. They are based on overwhelming numbers of experiments and observations, with no exceptions being found. This is the essence of inductive reasoning, the reasoning that takes one from particular observations or tests to an overall theory that explains all such observations and tests. Even then, the theory is at first tentative. It only gains status as being true and useful when the outcomes of further experiments can be reliably predicted on the basis of the theory. At any time, outcomes that are contrary to a theory can force its modification, if not its abandonment.

The question becomes: How many tests or observations must be made before a generalization can be considered valid? For example, how many apples out of a barrel must we taste and find sour before concluding that all the apples are sour? This is not a simple question to answer. It gets into the whole science of statistical analysis, which is beyond our scope here. Suffice it to say here that generalizations based on one or a few particulars should at the very least be subject to scrutiny. To what degree, if any, do exceptions occur?

8. Quantities must add up.

When an idea is basically logical but the quantities don't add up, beware of baloney. For example, consider the speculation of running a full-sized car on a flashlight battery. The logic is that the car requires energy and a flashlight battery will provide energy. However, in this case, the energy need is totally out of proportion to the source making the idea nonsensical. Students need more than just the link between one thing and another; they need to consider the relative quantities involved.

9. Attacking the speaker, not the argument.

An *ad hominem* attack, meaning *to the man,* is attacking the speaker rather than identifying fault with what she or he is saying. The classic example is declaring the speaker to be an idiot or worse. We should look at any *ad hominem* attack not just as baloney, but as very rancid baloney.

10. Declaring that one theory is as valid as another because they are both theories.

Some persons will declare one theory is just as valid as another because they are both theories; not proven fact. First order baloney! Theories are not equal. We need to look at supporting evidence. If one theory is supported by a great deal of evidence and a second theory has little or no hard supporting evidence, the first is more likely than the second to represent a true picture of reality.

11. "Believe or suffer!" or "Believe and be rewarded!"

Some persons are fond of trying to win their point by declaring that dire consequences will befall you if you don't believe a given conclusion, or great rewards will come to you if you do believe, all in the absence of any substantiating evidence. For example, one can validly declare that dire consequences will befall a person that doesn't believe in gravity and is about to walk off a cliff. Enough evidence for consequences exists. However, chain letters that simply declare good things will happen if you send them on and bad things will happen if you break the chain are utter baloney.

12. Beware of predicting outcomes on the basis of unsubstantiated causes.

Young people are especially prone to engage in reckless behavior, believing that nothing bad will happen because they are basically good, invulnerable, or whatever. We all know the potential consequences. Similarly, any number of people buy lottery tickets because they feel lucky. Again, they are predicting an outcome without any cause. In short, predicting outcomes on the basis of anything short of valid cause-effect reasoning is more than likely to be baloney.

Again, be watchful for any of these logical fallacies coming up in the course of discussion or anywhere else. Each occurrence provides a teachable moment for emphasizing the point of logic.

Thread A

Nature of Matter

Pursue Threads B, C, and D in Tandem

For a flowchart of the lessons and an overview of concepts presented see pages 8-13.

Lesson A/B-1

Organizing Things Into Categories

Overview:

The ability to organize is a life skill of such major importance that it hardly needs mentioning. As well as physically sorting items into categories, it includes a way of thinking that is tantamount to clear, logical reasoning. This lesson will bring children to recognize how organizing and placing things into categories is used in many aspects of everyday life. It will further lay a critical foundation as it sets children on a path of organized thinking.

Time Required:

Introductory discussion (10-15 minutes)
Recalling and noting examples at various times when and where organization is conspicuous (3-5 minutes per occasion)
Games/activities (20-30 minutes as desired)

Objectives: Through this exercise, students will be able to:

1. Recognize and use the following words in their proper context: ORGANIZE, ORGANIZATION, CATEGORY(IES).

2. Recognize and describe how items are arranged in categories in homes, stores, libraries, and other situations in everyday life.

3. Use organization in thinking/memory exercises.

4. Recognize how different purposes may require different systems of organization.

5. Organize an assortment of miscellany into logical categories according to different criteria.

6. Older children should demonstrate organizational skill in their everyday lives, as well as in thinking and writing reports.

Required Background: No special background is required.

Bernard J. Nebel, Ph.D.

Materials:

No special items or equipment are required for this lesson. It will involve simply noting the organization that is conspicuous in homes, stores, libraries, and other situations.

An assortment of miscellaneous items from "junk" drawers or closets may be used for various activities and assessment.

Teachable Moments:

Teachable moments will occur as you show children how your classroom is arranged and where various things are kept and as you introduce and children to the school library and media center. Cleaning up, sorting things out, and putting things away offer additional teachable moments. Parents/caregivers will find teachable moments in any visit to a supermarket or department store.

Methods and Procedures:

In pointing out and explaining to students how your classroom is arranged, introduce them to and use the words ORGANIZE(D) and CATEGORY(IES). For example, you might say: Let's look at the way I have our room ORGANIZED; that is, arranged. The first CATEGORY, or kind of things, is books. They are all kept on the shelves here. The second CATEGORY is art and drawing supplies; they are kept in the cabinets here, and so on, as suits your particular room.

Similarly, on visiting the school library and media center, state that you want them to see how it is ORGANIZED. K through grade 1 books are one CATEGORY; they are here. The CATEGORY of video discs is there, etc.

In Q and A discussion, have children reflect on the CATEGORIES of things in their home bedrooms and how they are ORGANIZED. Coach them as necessary in citing categories; clothing, toys, books, shoes, and how they are kept (supposed to be kept) each in its particular chest of drawers, closet, open shelves, and so on.

On subsequent occasions, have children visualize their home kitchens, name the CATEGORIES of things kept there (dishes, eating utensils, pots and pans, food stuffs, etc.) and how they are ORGANIZED, each kept in particular drawers and cabinets. Likewise, you may have children visualize a familiar supermarket, recall, and name categories (fresh produce, fresh meats, breakfast cereals, canned goods) and how the store is ORGANIZED so that each category is in a particular section.

Note that what students may cite as a category may be quite variable and still be correct. The key point for all to understand is that a category includes two or more different items that share a certain similarity or purpose. Most categories can be subdivided into more specific categories, and/or different categories may often be combined into a single larger category. For example, the category of fresh produce might be divided into subcategories: fruits, vegetables. Or, fresh produce may be combined with other foodstuffs into the single category we call groceries.

As children master the meaning and concept of categories and organization, use Q and A discussion to have them ponder its purpose and usefulness. Why do we put things in categories and organized them accordingly? If necessary, they may be prompted, "How would customers in a supermarket find things if they were not arranged in categories? How would the store manager be able to keep track of what items needed to be restocked if they were not organized?" Encourage children to look for and cite additional examples of how things are organized according to categories.

The following activity is an excellent way to have children exercise and demonstrate their organizational skills. For each individual or small group, prepare an assortment of miscellaneous items from around the room/home: various sorts of pencils/markers, scraps of different colored paper and cloth, play things, books, buttons, beads, string/cord, paper clips, staples, thumbtacks, etc., whatever is handy. The challenge for each student/group is to sort their miscellany into categories.

Again, it will be evident that the same miscellany may be sorted according to different criteria: color, size, use, material from which it is made, or other. It is advantageous to let students choose among themselves how they will sort their miscellany, but they may need input in resolving disputes that arise. Hopefully different groups will choose alternative criteria. Then at the end, each group can give a show-and-tell describing the criterion used for sorting and the resulting categories. Again, students should be impressed that there is no single right way to do it.

Making Categories and Organization into a Way of Thinking

Here is the fundamental purpose behind categorization and organization that should be made clear to children. Human brains are simply incapable of handling a lot of different things at the same time. The discordant information may be tucked away in our minds, but it is often impossible to get it out unless it is organized into certain patterns of thought. Without organization, many bits of information may be simply lost and forgotten. Learning, recall, and thought are greatly enhanced by organization.

A game that both children and adults enjoy and which demonstrates this principle is the following. Twenty miscellaneous items from around the house are placed on a tray. People are allowed to look at the tray for a total of 60 seconds; then the tray is removed or covered and each person lists as many of the items as they can in two minutes. The problem most people confront is that the mind gets stuck on one or two items and other items are lost from immediate recall.

There is a secret to developing greater success at this game, and you can help children learn the secret. It is to mentally sort things into categories. For example, there may be utensils from the kitchen, items from a desktop, an assortment of small toys, and some coins. By mentally putting the things in categories, one reduces the number of things one must recall directly. Assign the categories to the fingers of your hand. Then, as you recall the categories from your fingers, most if not all the associated items in each category will be remembered as well.

We are often awed by "memory experts" who demonstrate their skill on shows. Indeed, they may have remarkable minds, but they have also developed the skill of making associations among all the bits of information. That is, they organize the bits into patterns that make recall easier.

Such games/activities will emphasize, again, that there is no single right way to organize things. The choice will be according to purpose at hand. Kids generally enjoy challenges of sorting the same miscellany according to different criteria.

The final objective of this lesson, which will be ongoing, is to have kids look for and find the overall pattern of organization in any subject they address. Coach them as necessary in achieving this objective. Once they recognize the overall pattern of organization, further learning, to say nothing of recall, will be greatly enhanced.

As students progress, organized thinking should gradually become a habit of mind. For example, writing a report is a task that overwhelms many students. A large part of the feeling of overwhelm is because all the ideas tend to bounce around like the balls in a random number drawing machine. You will need to coach students that the first step is to jot down thoughts and ideas as they occur. Then it is a matter of sorting them into categories, arranging the categories, and fleshing them out.

But let children know that any given organization is not cast in stone. Changing organizational structures to better meet the challenges at hand is an ongoing activity of humans. Making changes does not diminish the importance of organizational skills.

Questions/Discussion/Activities To Review, Reinforce, Expand, and Assess Learning:

Continue to use the words CATEGORY(IES) and ORGANIZATION as they apply in every day situations. For example, in cleaning up after an activity or at the end of the day, say, "Let's get the room ORGANIZED; put everything back in its proper CATEGORY." Incidentally, assigning individual students to pick up and put away a given category of items will avoid their crowding and pushing at storage locations.

Encourage children to notice how items are organized and/or displayed in categories in various stores, zoos, museums, and other places they may visit. Have children share these observations/experiences in ensuing classes or small group discussions. In each case, have students describe how items are grouped into categories and displayed in given locations.

In small groups, have children take turns organizing the same set of miscellanea according to different criteria: color, use, material from which it is made, etc.

To Parents and Others Providing Support:

In addition to reviewing and repeating any of the games/activities described above it will be helpful to:

While in any store, call your children's attention to how similar items are grouped into CATEGORIES and displayed in certain locations. As children gain the concept, ask them to describe the organization that they observe in an unfamiliar store, library, museum, or other location.

Enlist children's support in sorting laundry into certain categories and putting it away accordingly.

Coach children to organize things in their rooms into categories and put them away accordingly. (Let children have a say in how the organization is done.)

Cleaning out a messy closet or drawer offers a practical exercise in sorting things into categories and organizing each to a given location. Discuss this with your children as you help them do it.

At bedtime, in addition to other routines you may have, reminisce about the day's activities. To aid recall, coach your child in terms of thinking of things in categories: fun and games, meal times, learning activities, etc.

For older children facing the writing of a report, coach them in first putting down ideas and organizing them into categories.

Connections To Other Topics And Follow Up To Higher Levels:

Organizational skills introduced here will come into play and be reinforced in many lessons that follow, particularly: A-2, "Solids, Liquids, and Gases;" A-5, "Distinguishing Materials;" B-2, "Living, Natural Nonliving, and Human-made Things;" B-4A, "Identification of Living Things."

Food groups and nutrition provide another practical and important application of organization. Stress how foods are divided into categories, the basic food groups. Stress the nutritional importance of each, and the nutritional risks of too much of that category known as "junk foods."

Again, organizational ability is a life-skill that can and should be applied to anything collected and to any topic addressed including business, social, and governmental organizations.

Re: National Science Education (NSE) Standards

This lesson is a steppingstone toward developing students' understanding and abilities aligned with NSE, K-4:

Bernard J. Nebel, Ph.D.

Unifying Concepts and Processes
 • Systems, order, and organization

Books for Correlated Reading:

Marks, Jennifer L. *Sorting by Color*. Capstone Press, 2007.
_____. *Sorting by Size*. Capstone Press, 2007.
_____. *Sorting Money*. Capstone Press, 2007.
_____. *Sorting Toys*. Capstone Press, 2007.

Pluckrose, Henry Arthur. *Sorting* (Math Counts). Children's Press, 1995.

Priddy, Roger. *Counting Colors*. Priddy Books, 2004.

Wong, Nicole. *"L" Is for Library*. Upstart Books, 2006.

Lesson A-2

Solids, Liquids, and Gases

Overview:

This exercise will acquaint children with the fact that there are three basic forms or states of matter: solids, liquids, and gases. They will learn fundamental characteristics of these states and they will recognize that some things, most notably water, will change among these states depending on temperature. This will lay the foundation for understanding numerous additional aspects of both natural and human-made materials and how they interact to give observed effects.

Time required:

Part 1.   Identification Of Solids, Liquids, And Gases (activity, 20-30 minutes, followed by periodic 5 minute reviews)

Part 2.   Changes Between Solid, Liquid, And Gas (demonstrations with interpretive discussion, 20-30 minutes, plus periodic reviews)

Objectives: Through this exercise, students will be able to:

1. Understand and use the following words in their proper context: GAS, LIQUID, SOLID, MATTER, STATES OF MATTER.

2. Identify gases, liquids, and solids as they are encountered in everyday life.

3. Describe the key attributes of gases, liquids, and solids.

4. Recognize and refer to gases, liquids, and solids as different STATES OF MATTER.

5. Tell how water and other substances may change between solid, liquid, and gas depending on temperature.

6. Tell how many things, especially living things, are complex combinations of solids, liquids, and gases.

Required Background: No special background is required.

Bernard J. Nebel, Ph.D.

Materials:

Part 1. Identification of Solids, Liquids, and Gases
Water and various other liquids in clear containers from around the kitchen and bathroom (cooking oil, liquid detergent, syrup, shampoo, etc.)
Various solid items from around the house (toys, pencils, dishes, coins, etc.)
Three medium-sized cardboard boxes respectively labeled: SOLIDS, LIQUIDS, and GASES. Alternatively, areas on the floor can be marked and labeled.

Part 2. Changes Between Solid, Liquid, and Gas
Water in a container
Ice cubes in a bowl

Teachable Moments:

Introduce this exercise as a new game that kids may be invited to play at any convenient time.

Methods and Procedures:

Part 1. Identification of Solids, Liquids, and Gases

Set out three boxes (or mark areas on the floor) labeled respectively SOLIDS, LIQUIDS, GASES, and a variety of items as listed above. Tell students that you want to teach them a new game of "Solids, Liquids, and Gases."

Start children out by placing some items in the appropriate box and making the statement yourself: Water is liquid. (Be sure you draw your children's attention to the liquid in the container and not the container. You may do so by pouring a small amount of the respective liquid into a bowl.) A pencil is a solid. Air is gas. (Blow or fan your hand in the "gases" box.) Let children continue, taking turns. Items can be taken out of the boxes and "recycled" at any point. Coach children as necessary with conversation such as the following:

Water is a liquid. Anything that flows or runs like water is a liquid. Some liquids such as syrup may flow slowly, but if it flows at all we call it a liquid. Hence, liquids need to be kept in containers.

Air is a gas (actually a mixture of different gases as we shall see in a later lesson). Anything that behaves like air is a gas. Some children will have trouble appreciating that air is a real substance, because it seems to be just "emptiness." Draw their attention to how we breathe it in and out, how we can blow bubbles with it, and how we blow up balloons and tires with it. We also feel it blowing on our faces. (Lesson A-3 will go further in demonstrating the nature of air.)

There are many sorts of gases and most will mix with the air such that they cannot be

46

seen, but they can be smelled. All smells are something in a gas form. The smell of perfume, for example, is particles of the liquid perfume becoming a gas, moving through the air, and being detected by your nose. Add a safety lesson here by pointing out that some gases are poisonous to breath. If they detect a strange unpleasant odor, they should get away from it into fresh air and tell adults about it. In this regard, make special mention of the gas burned in gas stoves and furnaces.

Solids are any and all things that have a distinct size and shape. They will sit by themselves without the need of a container. But there are many sorts of solids. Some are hard and brittle, some soft and flexible, some are relatively light, others are heavy and so on. But, if it keeps its shape when it is left alone, it is a solid.

Some students may note that things like sugar and breakfast cereal also flow as you pour them and you need to keep them in containers. Should they be called liquid or solid? Point out that solids in small particles (powder or granules), will behave somewhat like a liquid. But here is the difference. Spilled sugar and such things can be swept into a pile that keeps its shape. If we push a liquid into a "pile," it immediately reverses and spreads out again. Thus, if it can be pushed into a pile that maintains its shape, it is a solid.

You can continue this exercise beyond the items you have collected by just pointing to or naming things children are familiar with. As children gain familiarity, small groups may play a game of "20 Questions" by themselves, the first questions determining whether it is a gas, liquid, or solid.

Children will demonstrate their understanding of gases, liquids, and solids by their ability to identify various things accordingly. But be careful in your questioning. If you point to a bottle of fluid, be specific that you are referring to the fluid and not the bottle, which is, after all, a solid. The same goes for balloons and other containers filled with gas.

Indeed, there are many things that may be combinations of gas, liquids, and solids. Living things are an especially important example. Living organisms, ourselves included, are complex combinations. Children may enjoy the challenge of identifying gas, liquid, and solid parts of the body.

Part 2. Changes Between Solid, Liquid, and Gas

Have kids classify an ice cube. They will most likely identify it as a solid. But then watch it melt. What does it become? Water, a liquid! Conversely, liquid water becomes a solid, ice, when placed in the freezer. While kids have undoubtedly experienced this, they may never have given it thought. Be prepared to demonstrate melting of ice and freezing of water with them.

So, is water really a solid or a liquid? Emphasize that solid, liquid, and gas refer to the present state of the material. Thus, ice is a solid; water is a liquid. The conclusion is that some things will change between solid and liquid states. What does the change depend on? Use some Q and A discussion to bring kids to recognize that temperature is almost always involved.

Likewise, have kids note that other things may change between liquid and solid as well. Consider butter, wax, or bacon grease, for example. Again, bring their attention to the fact that temperature is involved. You may demonstrate these as you see fit. (DO NOT HEAT WAX IN A PAN DIRECTLY ON THE STOVE. IF IT GETS TOO HOT IT IS SUBJECT TO EXPLODING.)

As kids are comfortable with the idea that many things may change between liquid and solid, introduce them to the idea that things may change from liquid to gas as well. Again, water is the prime example.

As something wet becomes dry, what happens to the water? Kids will usually draw a blank. Simply inform them that it is going off in the air as a gas (water vapor). This is pursued further in Lesson A-8, after we have introduced the particulate nature of matter, Lesson A-4. For now, it is most prudent to just introduce the concept. This will sow the seed for those later lessons.

Many other things (both solids and liquids) may change to gas as well. Again, consider smells of various things. How does your nose detect a smell? For your nose to detect perfume, for example, there must be particles of the perfume along with the air that you breathe in. This is to say that the perfume turns to a gas and mixes with the air. The same may be said for other solids or liquids that smell: coffee, spices, alcohol, bacon, etc.

States of Matter

Instruct students that all solids, liquids, and gases are referred to as MATTER. We refer to solids, liquids, and gases as different STATES OF MATTER. Said another way, there are three fundamental STATES OF MATTER: solid, liquid, and gas. Further, the term MATTER applies to all living or biological material, as well as nonliving material, all human-made materials, and those that occur naturally. Don't miss the opportunity to relate this lesson to the previous lesson, A/B-1. In this lesson we are learning that all the "stuff" that the world and everything else is made of is called MATTER. Then we are separating matter into three major categories: solids, liquids, and gases.

Don't expect kids to comprehend the breadth of the term "matter" right away. The idea of matter as the basic construction material of the universe and the distinction between matter and energy are concepts that will come gradually as they are addressed again in future lessons.

Questions/Discussion/Activities To Review, Reinforce, Expand, and Assess Learning:

Make a book (Appendix A, page 381) illustrating the three states of matter.

Create an activity center where children can repeat the activity of placing various things in their proper category: solid, liquid, or gas.

Have children identify various things as gas, liquid, or solid and give the reason for their choice.

In small groups, pose and discuss questions such as:

.

What is the one word that embraces all solids, liquids, and gases?

The world is made of different states of matter. Describe how this is so.

Describe how you and/or other living things are made of different states of matter. (Cite parts or areas of your body that are solid, liquid, and gas.)

Could an actively living organism be entirely one state of matter? Describe how this would be impossible.

Give examples illustrating how a substance may change between solid, liquid, and gas states. What is generally involved in causing the change?

As students progress, it may be noted that most materials will go into different states if temperatures are extreme enough. Metals and even rock become molten (liquid) at very high temperatures. Gases will become liquid and even solid at very low temperatures. Students may be familiar with liquid oxygen and dry ice, for example.

To Parents and Others Providing Support:

It is evident that this whole exercise can by done/reviewed one-on-one with your child.

In the kitchen and bathroom, have children identify the things you and they use as solid, liquid, or gas.

At dinnertime, have children identify various parts of their meal as solid, liquid, or gas. (Any smell is a gas mixed with air.)

Have children identify components of their bodies that are solid, liquid, or gas. (Five-year-olds invariably get a kick out of noting that their urine is a liquid.)

Play "20 Questions," starting with: Is it gas, liquid, or solid, or a combination?

Connections to Other Topics and Follow-Up to Higher Levels:

This lesson provides a foundation for and leads naturally into further investigations regarding the nature of materials, changes, and cycles.

Re: National Science Education (NSE) Standards

This lesson is a steppingstone toward developing students' understanding and abilities aligned with NSE, K-4:

Unifying Concepts and Processes
 • Systems, order, and organization

Bernard J. Nebel, Ph.D.

Content Standard B, Physical Science
  • Properties of objects and materials

Content Standard D, Earth and Space Science
  • Properties of Earth materials

Books for Correlated Reading:

Curry, Don L. *What Is Matter?* (Rookie Read-About Science). Children's Press, 2005.

Garrett, Ginger. *Solids, Liquids, and Gases* (Rookie Read-About Science). Children's Press, 2005.

Mason, Adrienne. *Change It!: Solids, Liquids, Gases and You* (Primary Physical Science). Kids Can Press, 2006.

Zoehfeld, Kathleen Weidner. *What is the World Made Of? All About Solids, Liquids, and Gases*. HarperTrophy, 1998.

Walker, Sally M. *Matter* (Early Bird). Lerner, 2005.

Lesson A-3

Air Is a Substance

Overview:

In their preschool years, children experience solids and liquids as real things, but they often consider that the air surrounding them is just empty space. Even after Lesson A-2, "Solids, Liquids, and Gases," they still may consider gas as a state of nonexistance. Therefore, the fact that air and other gases are real substances, and share attributes of solids and liquids, needs experiential emphasis. This exercise will provide that emphasis. In turn, it is a basic building block in the foundation for understanding innumerable aspects of science.

Time Required:

The three parts of this lesson can be integrated into a single lesson.

Part 1. Introductory Discussion (15-20 minutes)
Part 2. Does Air Take Up Space? (activities plus interpretive discussion, 15-20 minutes)
Part 3. Does Air Have Weight? (activity, 30-40 minutes, plus interpretive discussion, 10-15 minutes)

Objectives: Through this exercise, students will be able to:

1. Show/tell how we can demonstrate that air occupies space.

2. Show/tell how we can demonstrate that air has weight.

3. Cite facts that show that air is a real substance.

4. Despite their many differences, solids and liquids share two attributes in common. Explain what these two attributes are.

5. Explain why gases, including air, are one of the categories of matter. Tell what attributes gases share with solids and liquids.

Bernard J. Nebel, Ph.D.

Required Background:

Lesson A-2, Solids, Liquids, and Gases

Materials:

Glass of water and a straw
Empty plastic soda bottle
Dishpan/sink of water
Rubber party balloons (at least two of the same size)
12-inch straight edge (ruler)
Paper clips
Tape
String or thread

Teachable Moments:

In the course of reviewing Lesson A-2 or as students are blowing bubbles, blowing up a balloon, feeling wind, or breathing hard, pose the question, "Is air (or gas) "real stuff" like liquid or solid, or is it really nothing at all?"

Methods and Procedures:

Part 1. Introductory Discussion

As suggested above, use an opportunity to pose the question "Is air (or gas) 'real stuff' like liquid or solid, or is it really nothing at all?" But, be prepared to be met with blank or confused looks, and to shift the question to, "How are these solids (any two or more items at hand) different?" Children may cite differences in color, size, shape, and so on. Do the same with different liquids and contrasting liquids and solids.

Okay, they are different in so many ways. Are there any ways in which all solids and liquids are alike? Are there certain attributes that apply to all of them? Our objective is to have students recognize that all solids and liquids take up or occupy space and they all have weight.[9] But they are unlikely to see this by themselves.

Guide them to this conclusion by using body language and other hints that demonstrations that two solids or liquids cannot be in the same place at the same time. One pushes the other out of the way. They may mix but each still takes up its own space. Likewise, demonstrate that they all have weight. If the two are mixed, the weight of the total is the sum of the weights separately. (The opportunity to have students prove this

---

[9] Of course, having weight assumes that gravity is present. The more technically correct term would be mass, which is independent of gravity. However, at this level it is impractical to get into this distinction. Therefore, it is most practical to keep the discussion in terms of weight, which will be experientially familiar. The distinction between weight and mass will be addressed later in Lesson D-7.

experimentally and practice weighing and math skills in the process is evident.) Thus, guide students to the conclusion that there are two attributes common to all solids and liquids regardless of other differences. They all occupy space and they all have weight.[9]

Does gas/air also occupy space and have weight? Some students may say yes; others, no. Take the opportunity to point out that in science answers are not determined by any sort of vote or opinion. They are determined by evidence derived from investigation. Hence, invite students to investigate to find out.

Part 2. Does Air Occupy Space?

As you blow up a balloon, ask, "Why is air pushing out the sides of the balloon? Would this occur if air were empty space?" Bring students to reflect that air must be taking up space in the balloon and hence is pushing out the sides. Likewise, have kids blow air through a straw into a glass of water and have them ponder what is occurring. Air is pushing the water aside creating bubbles that rise to the surface.

Another fun demonstration is to take an open, empty, clear plastic soda bottle, and stick it in a pan of water, open-end down. Children will observe that water does not enter the bottle. Why not? Is air taking up space in the bottle holding the water out? Let's test this idea. Poke a small hole in the bottom of the bottle so that the air can escape. Indeed, water now enters the bottle as the air exits. With these or similar demonstrations and discussion, guide students to conclude that air, like liquids and solids, does occupy space.

Part 3. Does Air Have Weight?

Air certainly appears to have no weight; hence, children will generally answer "no" to the question. Again stress that investigation is the key. Take a 12-inch ruler, three paper clips and proceed as follows: Tape a paper clip to the ruler at the 6-inch point such that one end of the paper clip protrudes just beyond the edge of the ruler. Similarly, tape a paper clip at the zero and another at the 12-inch mark so that their ends just protrude to the opposite side of the ruler.

Attach a string to the center loop and use it to dangle the straight edge off the edge of a table where it will hang freely. This is your balance. With two equal pieces of string, attach a new, deflated balloon lightly to each of the end loops. Make whatever adjustment may be necessary to have the balance hang level with the two empty balloons attached.

Now blow up one of the balloons to its full capacity and tie off the opening without using additional string. (Additional string would add to the weight.) Reattach the inflated balloon to its loop on the balance. Children will observe that the inflated balloon now pulls its end of the balance down somewhat indicating that it is heavier. Draw children to reason how the balloon and everything else is the same. Therefore, the heavier weight of the inflated balloon must be the weight of the air. (It is heavier because it is somewhat compressed; hence the inflated balloon contains more air than the same volume of normal air. But it usually suffices to simply note that the inflated balloon is heavier.)

Children will likely ask questions to the effect: If air has weight, why don't we feel its weight baring down on us? Explain that its weight pushes equally on all sides of us; for example, it is pushing up under our hand with the same force that it is pushing down from the top. Therefore, we usually don't feel its weight. But here is a simple activity.

Have kids suck in their cheeks. Explain that what they are really doing by sucking is reducing the amount and hence the weight of the air pushing from inside their mouths outward. It is the weight of the air outside that is pushing their cheeks in.

Review and reemphasize the conclusions of their experimentation: Air/gas does occupy space and it does have weight. Thus, air has the same basic attributes as liquids and solids. Air is real stuff; it is one of the three basic forms of matter addressed in Lesson A-2—a gas (actually a mixture of different gases). Further, have children consider points such as:

When we feel wind or fan ourselves, what is it that we are feeling?
When we breathe, what is it that we are pulling into and out of our lungs?
When we blow out a candle, what are we sending against the flame?

Invite children to think of further examples by which they experience air as "real stuff."

Integrate this into the preceding lesson, A-2. Review and reemphasize that matter is the basic construction material for every thing in the world (and beyond as well). There are three basic categories of matter: solids, liquids, and gases. These categories are grouped together as forms of matter because they all have the attributes that they take up space and they have weight (when gravity is present).[10] The fact that forms of matter may change one into another—freezing and melting of ice, for example—provides further evidence solids, liquids, and gases are related.

Some children may bring up the question of helium-filled balloons. It would appear that helium provides a "negative weight." Explain that this is similar to a block of wood or other floatable object under water. It is not weightless; it only weighs less than the water; hence, it floats to the top of the water when released. Likewise, helium is not weightless; it only weighs less than the air. Thus, the helium filled balloon is floating toward the top of the air/atmosphere. The same principle holds for hot-air balloons. Hot air weighing less than cooler air results in the balloon floating upwards.

Questions/Discussion/Activities to Review, Reinforce, Expand, and Assess Learning:

Make a book illustrating the attributes of air.

---

[10] These two attributes, taking up space and having weight, stem from the fact that all forms of matter have a particulate nature (atoms and molecules) and this becomes a third common attribute. This is addressed in Lesson A-4.

Set up an activity center where students can repeat any of the activities described.

In small groups, pose and discuss questions such as:

How can you show that air takes up space?
How can you show that air has weight?
Why can't you feel the weight of air pushing down on you?
How is air like liquids and solids?
What attributes do all solids, liquids, and gases have in common?
Why do we group solids, liquids, and gases as three categories of matter?

To Parents and Others Providing Support:

Kids love to play in water with containers. Facilitate and allow their playing in the bath or sink with clear plastic jars or bottles. Their "play" will provide experiential confirmation that water does not enter a vessel unless air is able to escape. Periodically, you may ask, "Why not?" And expect the answer, "Because air is taking up the space."

Help them make a balance as described and demonstrate again that air has weight. They will probably wish to weigh additional things on the balance as well. Of course, a balance does not weigh things as such; it only compares the weight of the two things on the opposite ends. Still, they will be gaining experiential understanding in the process. More sturdy balances, including teeter-totters, may be use or constructed as desired.

Use experiences of wind as teachable moments to reinforce the idea that air is real stuff that takes up space and has weight.

Periodically ask/discuss what two traits are common to all solids, liquids, and air/gases (all MATTER).

Connections to Other Topics And Follow-Up To Higher Levels:

The Earth's Atmosphere
The Particulate Nature of Matter
Different Gases In Air
Numerous aspects of chemistry and other sciences

Re: National Science Education (NSE) Standards

This lesson is a steppingstone toward developing student's understanding and abilities aligned with NSE, K-4:

Unifying Concepts and Processes
• Evidence, models, and explanation

Content Standard A, Science as Inquiry
• Abilities necessary to do scientific inquiry

Bernard J. Nebel, Ph.D.

    • Understanding about scientific inquiry

Content Standard B, Physical Science
    • Properties of objects and materials

Content Standard D, Earth and Space Science
    • Properties of Earth Materials

Books for Correlated Reading:

Branley, Franklyn M. *Air Is All Around You* (Let's-Read-and-Find-Out Science, Stage 1). HarperCollins, 2006.

Cobb, Vicki. *I Face the Wind* (Science Play). HarperCollins, 2003.

Dorros, Arthur. *Feel the Wind.* (Let's-Read-and-Find-Out Science, Stage 1). Scott Foresman, 1990.

Fowler, Allan. *Can You See the Wind?* (Rookie Read-About Science). Children's Press, 1999.

Sherman, Josepha. *Gusts and Gales: A Book About Wind* (Amazing Science). Picture Window Books, 2004.

Simon, Seymour and Nicole Fauteux. *Let's Try It Out in the Air: Hands-On Early-Learning Science Activities.* Simon & Schuster, 2001.

Stille, Darlene. *Air: Outside, Inside, and All Around* (Amazing Science). Picture Window Books, 2004.

Lesson A-4

Matter I: Its Particulate Nature

Overview:

In Lesson A-2, students learned that solids, liquids, and gases are all referred to as MATTER; they are different STATES or forms of matter. Then, in Lesson A-3, they learned that all matter has two common attributes. It occupies space and it has weight. Here, students will discover another attribute of all matter: it is comprised of particles. Further, they will learn that the difference between gases, liquids, and solids is not in their particulate nature, but in the degree to which particles attract together. This particulate nature of matter is a cornerstone for understanding virtually all science, technology, and much more.

Of course, the fundamental particles are atoms and molecules, but introducing these words and what they imply requires a degree of abstract thinking that is generally beyond the capacity of young minds. Therefore, we keep the discussion in terms of just "particles," a concept children can experience and grasp. Later on, this conceptual understanding can be easily refined into the more specific understanding of atoms and molecules.

Time required:

Part 1. All Matter Is Made Of Particles (activity, 30-45 minutes, plus interpretive discussion, 15-20 minutes)

Part 2. The Difference Between Solids, Liquids, and Gases (discussion with interpretive game/activity, 30 minutes)

Objectives: Through this exercise, students will be able to:

1. Demonstrate and explain how any gas, liquid, or solid may be divided again and again into smaller and smaller bits or PARTICLES.

2. Tell how a third attribute of all matter is its particulate nature, the individual particles being much too small to see with the unaided eye.

3. Describe and/or demonstrate with models how the distinction between solids, liquids, and gases comes from the degree to which their particles attract and hold together.

4. Tell how the individual particles of matter are in constant motion and how the degree of motion changes with temperature, while the attractive force between particles remains constant.

5. Explain freezing and thawing in terms of the two factors noted in 4 above.

Required Background:

Lesson A-2, Solids, Liquids and Gases
Lesson A-3, Air Is a Substance

Materials:

Marble-sized balls of clay or dough
A straw and soapy (sudsing) water in a glass
Water in a spray bottle
Pieces of metal and wood and a file
A variety of powder or granular substances (sugar, flour, salt)
A bit of soil
Optional—A hammer, chunks of clay or sandstone, and safety goggles

Teachable Moments:

Invite students to take another look at solids, liquids, and gases.

Methods and Procedures:

Part 1. All Matter is Made of Particles

Conduct a review of Lessons A-2 and A-3. Students should be familiar with these two facts: All matter (all solids, liquids, and gases) occupies space, and all matter has weight. Go on to say that in this lesson we are going to discover another attribute common to all matter. All matter is made up of teeny tiny bits that we PARTICLES. Let's show how this is true.

For solids of various sorts (paper, wood, metal), demonstrate and let children demonstrate for themselves how they can be cut or filed into tiny particles. They may also crush a lump of hardened clay by hitting it with a hammer. (WEAR SAFTY GOGGLES TO DO THIS.) Smear a bit of soil on white paper and note the fine particles. Also, "dissolve" a lump of clay in water. The cloudiness imparted to the water is given by clay particles. In each case, have children visualize that if they could put the filings or chips (particles) back together, they would end up with what they started with.

Many things we use, sugar, flour, and salt for example, are already ground into tiny particles. Then we put them together to make cookies. Are cookies made of particles? Consider all the crumbs.

For liquids, squirt some water from a spray bottle. What is happening? We are breaking the water into tiny particles. Spray repeatedly at the same place on your hand. What is happening now? The particles of water are joining back together of reform liquid water. Let kids do this for themselves.

For air (gas) let children use a straw to blow bubbles in a glass of soapy water. There will be large bubbles, but especially note the tiny bubbles. Instruct children to think about each tiny bubble as a "package" of air particles. Conspicuously, the bubbles break, allowing the particles to mix again.

As children see that everything they have worked with can be broken into particles and potentially put back together, guide them to name and reflect on other sorts of materials. Could they similarly be cut, filed, crushed, or otherwise divided into particles and put back together? Yes! With Q and A discussion, guide them to make the generalization that all matter, whether gas, liquid, or solid, is made of particles. That is, all matter has a PARTICULATE NATURE.

Emphasize that the particulate nature applies to all materials: parts of living things, human-made things, and natural nonliving things (rocks, water, air, soil). Yes, even we ourselves are made of particles—many different kinds of particles making up bone, skin, blood, etc. put together in very complex ways, but particles nevertheless. Go on to make the point that if we had fine instruments and magnifying lenses, we could divide even the tiniest particles that we can see into still tinier particles.

What is the conclusion? There are fundamental particles of water, sugar, salt, iron, aluminum, glass, wood, air, and everything else. The fundamental particles of a substance are unique and cannot be subdivided further without totally destroying the nature of the matter we started with. But, the fundamental particles of a substance are incredibly tiny, much too tiny to be seen individually. Even the smallest grain of sugar, for example, is comprised of many individual sugar particles.

Stress that the fundamental particles of a given substance are the same. A larger lump of clay, for example, does not contain larger clay particles; it only contains more of them, just as the amount of sugar in a container is a matter of the number of granules present.

Guide students in reviewing/summarizing the three attributes of matter that we have now uncovered: All matter has weight; all matter occupies space; and all matter is made up of particles. One may point out that each particle occupies a certain space and has a certain weight (mass). Thus, it is the particulate nature of matter that leads to the attributes of occupying space and having weight.

Part 2. The Difference Between Solids, Liquids, and Gases

Review the concept that all matter is made of particles. Then pose the question, "What, then, makes the difference between a substance being a solid, liquid, or a gas?" Let kids ponder this for a moment, but don't expect them to reason out the answer by themselves.

You will have to explain, probably more than once, that the difference is NOT in their particulate nature. It is in the degree to which the individual particles of the substance are attracted to one another. In SOLIDS, the particles attract very strongly, so strongly that they form a single tight mass. You may demonstrate this by pressing together a group of clay or dough balls. It should be evident that such a mass will have the attributes of a solid.

In LIQUIDS, there is an attraction, but a relatively weak attraction between particles. It is such that each particle may easily break its connections with its present "partners" and make connections with nearby particles. Thus, the particles stay together but are free to move over, under, and around one another. This is what allows the liquid to flow. You can demonstrate this by pouring sugar from a container. Here, it is gravity that is holding the granules together, but have students note that the freedom of the granules to move about one another allows the mass to flow much like a liquid. In a real liquid, the individual particles are too small to be seen but their freedom to move and slide over and around one another allows the mass to flow. It also allows objects to move through, as a fish or you may swim through water.

In a GAS like air, explain that there is no attraction between the individual particles. Indeed, there is empty space between the particles. Each particle may bump into and bounce off other particles and objects, but each is still free to go its own way. You can demonstrate this idea by blowing on a bit of flour or other powder and watching the "cloud" of particles disperse into the air. Kids can wave their hands through the cloud and note that they cannot feel it at all. This is because all the particles in the cloud, being separate with space between, are very easily pushed aside by any passing object.

Of course, you do feel the air as wind blowing in your face, and you see it as leaves and scraps of paper being whisked along by a breeze. Yes, the individual particles are separate, but a whole mass may move in a single direction. The wind in your face is really the pressure of countless particles on the move and pelting your face as they go by. Similarly, by fanning or blowing, you are sending a mass of air particles in a given direction. Use occasions of seeing things being blown about to discuss the result in terms of air particles moving as a mass. Students may note that the flow of water and other liquids can be interpreted in the same way.

As a final point of this lesson, explain that there is a natural tendency of all the tiny, fundamental particles that make up a substance to jiggle and move about. This is true even in solids; it is just the attraction between particles that holds them in place and makes their mass a solid. Now, this tendency of the particles to jiggle and move about increases with temperature, but, the strength of the attraction between particles, remains unchanged.

Challenge students to reflect and relate these two factors, attraction on the one hand and tendency to move about on the other. Can they explain how/why melting and freezing are related to temperature? Take ice for example. The reasoning should go as follows: As temperature decreases the jiggling and moving of the particles slows. At some point their motion becomes so slow that the attraction between particles takes hold and keeps them in place, i.e., freezing occurs. Melting is the reverse. As temperature rises, the tendency to jiggle and move increases. At some point the jiggling of the particles is great enough to break the rigid connections and the particles become free to move about one another, i.e., melting occurs.

Young children may have trouble with this reasoning, as it is somewhat abstract. Don't belabor it. The concept will become more clear as they model it in the following game and the same concept comes into play regarding evaporation and condensation (Lesson A-8). At the very least, it provides a first exposure to a concept that underlies much of chemistry.

A game/activity that illustrates the three states of matter, and which kids love to play is this. Each child is assigned to play-act being a particle. When the teacher shouts "SOLID," all should join together in a group hug. When the teacher shouts "LIQUID," children should separate to handshake distance and then move about and through the group briefly taking and dropping "handshakes" with each other. It should look something like a square dance in motion. When the teacher shouts "GAS," each person should clasp their own hands behind their backs and move randomly about the entire room. As a refinement, everyone should actually move in a straight direction, only changing direction as they run (gently) into another person/particle, the wall, or other objects, much like balls on a pool table.

With these instructions and some practice, the leader calls out SOLID, LIQUID, or GAS in random order and children respond accordingly. Then, guide children in visualizing and explaining how their group behavior in each of these three phases represents the behavior of particles in solids, liquids, and gases. By asking each child to play-act being specifically a water particle, you may further use the words freeze, melt, evaporate, and condense.

Questions/Discussion/Activities to Review, Reinforce, Expand, and Assess Learning:

Make books illustrating that all matter (solids, liquids, and gases) is made of particles. Further illustrate how relative attraction between particles makes the distinction between them.

Play the solid-liquid-gas game described above. Children may learn to play this on their own, one of them being the leader.

Take the opportunity when pouring drinks, handling solids, seeing wind blow things about, and so on, to talk about the particulate nature and how the particles of the substance are behaving with respect to one another, i.e., their relative attraction and movements.

In small groups, pose and discuss questions such as:

What are the three attributes of all matter?
How can you demonstrate each of the three attributes?

If all solids, liquids, and gases are comprised of particles, explain what makes the distinction between them?

To what degree are particles attracted to one another in this ___? (point alternately to solids, liquids, and air)

What happens to the behavior of particles as temperature increases? —decreases?

What is going on as a substance melts? —freezes? What two factors are involved how do they interact?

To Parents and Others Providing Support:

Around the kitchen, bathroom, and out of doors, when attention is focused on something, ask questions such as: What state of matter is it? Is it made of particles? How are those particles attracted together, or are they not attracted at all? Particularly, do this in reference to water freezing and ice melting.

Play the "solid-liquid-gas" game described above with friends or family.

Connections To Other Topics And Follow Up To Higher Levels:

Lessons A-6, A-8, and A-9 build directly on this exercise.
Lesson C-3, Matter and Energy
Any and All Aspects of Chemistry
The Earth's Hydrological (water) Cycle
Most aspects of Biology (both plant and animal) and Nutrition
Soil Science
Pollution and many other Environmental Issues

Re: National Science Education (NSE) Standards

This lesson is a steppingstone toward developing students' understanding and abilities aligned with NSE, K-4:

Unifying Concepts and Processes
  • Evidence, models, and explanation

Content Standard A, Science as Inquiry
  • Abilities necessary to do scientific inquiry
  • Understanding about scientific inquiry

Content Standard B, Physical Science
  • Properties of objects and materials

Content Standard D, Earth and Space Science
  • Properties of Earth materials

Books for Correlated Reading:

Curry, Don L. *What Is Matter?* (Rookie Read-About Science). Children's Press, 2005.

Mason, Adrienne. *Touch It!: Materials, Matter, and You* (Primary Physical Science). Kids Can Press, 2005.

Stille, Darlene R. *Matter: See It, Touch It, Taste It, Smell It* (Amazing Science). Picture Window Books, 2004.

Trumbauer, Lisa. *What Are Atoms?* (Rookie Read-About Science). Children's Press, 2005.

Walker, Sally M. *Matter* (Early Bird Energy). Lerner, 2005.

Lesson A-5

Distinguishing Materials

Overview:

The number of different things humans make is without limit. But the basic materials we use to make them are few: metal, plastic, wood, glass, rubber, and clay/stone cover most of them. This lesson will bring children to recognize and distinguish among these basic materials. In this process they will gain the vocabulary for describing characteristics. Further, they will learn why one material and not another is preferred for a given use. All of this will provide a foundation for future studies concerning where and how raw materials/resources are obtained and how various things are made.

Time Required:

> Part 1. Categories of Materials (activity plus discussion, 30-40 minutes)
> Part 2. Characteristics of Different Materials (activity plus discussion, 30-40 minutes)

However, this should be considered as just the beginning of an ongoing study that, for some students, may lead into life careers. Therefore, it will be a topic that you should plan on revisiting from time to time and adding more as occasions arise.

Objectives: Through this exercise, students will be able to:

1. In addition to words they already know, recognize and use the following words in their proper context: METAL, PLASTIC, RUBBER, CERAMIC, TEXTILES, FIBER, FABRIC, TEXTURE, ELASTIC, BRITTLE, FLEXIBLE, MOLDABLE, TRANSPARENT, TRANSLUCENT, OPAQUE.

2. Identify a the material from which various items or parts of items are made.

3. Discuss how they identify various materials, i.e., by look, feel, relative weight, sound, etc.

4. Describe the characteristics of various materials, e.g., light, heavy, hard, soft, brittle, strong, flexible, elastic, etc.

5. Separate items according to the material from which they are made and tell how they distinguish one material from another.

6. Discuss why certain materials are used for given functions.

Required Background:

Lesson A/B-1, Organizing Things Into Categories
Lesson B-2, Distinguishing Living, Natural Nonliving, and Human-made Things
Lesson A-2, Solids, Liquids, and Gases

Materials:

Wood pencils (one for each student)
Plastic spoons
Metal spoons
Wooden spoons
Magnet

Beyond these, no special items or equipment are required. The lesson will involve simply looking at and considering common items such as the furniture, cooking/eating ware, and other things that are present in their surroundings.

Teachable Moments:

Whenever students are looking at or examining a given item, pose the question, "What is it made of?"

Methods and Procedures:

Part 1. Categories of Materials

Have children get out (or pass out) wooden pencils. They will anticipate a project of some sort coming. But ask them to examine their pencils. Then ask, "What is it (at least the stem portion) made of?" Wood! "What other things in the room are made of wood?" Let children get up and explore the room, feeling as well as looking at things. Let them call out things they think are wood and make a list on the board. The list will probably include items of furniture, door and window frames, the doors on cabinets, bookshelves, the broom handle, and so on. (Save paper being made from wood for later.) When they have found what they can, go over the list. If there are mistakes, have them check it again and coach them in correcting the error.

On the same or successive occasions, do the same sort of activity making metal, plastic, ceramic, or another material the focal point. Metal makes an exceptionally fun part of the exercise as children discover (perhaps with some hints) belt buckles, clasps, and zippers in addition to coins, paper clips, doorknobs, hinges, etc.

As children become familiar with different materials, take a few moments to point at or name familiar items and ask what they are made of. The categories listed below will cover most materials. It is obvious that each of the categories has many subdivisions, i.e., different kinds of metals, plastics, etc. Don't push for making these distinctions at first. These will come in time or they can be the subjects of further lessons. The major categories of materials are:

> Metal (includes iron/steel, aluminum, copper, chrome, lead)
> Plastic (all types)
> Wood and wood products including paper of all sorts
> Clay/stone (includes, ceramics, cement, bricks)
> Glass
> Rubber
> Leather
> Fibers (thread, string, fabrics and textiles of all sorts)

Things not included in these categories will usually be self-evident, an ice sculpture for example, or they may be beyond the scope of this lesson.

Part 2. Characteristics of Different Materials

As students learn to identify different materials, they cannot help but become experientially familiar with the properties and characteristics that distinguish them. However, if you ask them, "How do you know this is made of _____," they will probably not be able to answer. We ourselves might be hard pressed to give such answers. We have learned to make distinctions by experience, and the factors of such experiential learning are often subtle, and below the level of conscious recognition. However, bringing students to be cognizant of these factors lays the foundation for greater understanding and more sophisticated methods of analysis that will come later.

A way to proceed is to pass out items such as a metal spoons, plastic spoons, and wooden spoons. From the above, students will recognize them as metal, plastic, and wood. But then ask, "How do you tell these different materials apart?" Coach kids as necessary in describing the differences that they note. These will likely be differences in look, feel, relative weight, hardness, and other characteristics. It is evident that you will be adding to your students' vocabulary at the same time.

The characteristics we commonly use in identifying different materials are listed below. This list is definitely not for memorization. It is to guide students experientially in how they may use a combination of different senses and features in identifying materials. There should be repeated occasions where you challenge students to identify the material from which an item is made and have them discuss the characteristics that make their identification possible. The list will include:

> Look: Materials have looks that often distinguish them. Of course, modern technology makes plastics look like wood and other materials; hence, we may be deceived by looks alone.

> Texture: It may be smooth, grainy, rough, fibrous, etc. Materials have a distinctive texture that is evident by both look and feel.

Feel of surface: In rubbing the surface, materials have distinctive feels. Plastics usually have a waxy feel quite different from that of wood, metals, or other materials.

Sensation of temperature: Metals are markedly cold to the touch, wood is relatively warm, and plastics are intermediate.

Feel of hardness: Just feeling or especially pressing your fingernail into a material gives an impression of its hardness. Stone and metals are especially hard; wood and plastics, less so.

Relative weight: Heft an item of metal or stone in your hand and it feels quite heavy. An equally-sized piece of wood or plastic is much lighter. A simple test is, Will it sink or float? (Technically this is the property of density, the subject of a later lesson.)

Sound: Each material gives its distinctive sound as you tap it with your fingernail.

Strength: To what degree will it stand up to pressures or forces tending to bend, stretch, squash, or break it. Metals, especially iron/steel, are used where great strength is required.

Flexibility: Materials have distinctive degrees of flexibility, the degree to which they can be bent without breaking. Glass, stone, and ceramics are BRITTLE; they break rather than bend. Metals are BENDABLE; they bend without breaking.

Elasticity: Being ELASTIC is being able to return its original shape after bending, compressing (squeezing), or stretching. Rubber is renowned for its elasticity, but note metal springs. Metals, wood, plastic, and even glass are elastic to the degree that they can be bent and will snap back.

Malleable or Moldable: This is the ability to be molded. Moist clay is highly MALLEABLE. It can readily be molded into any desired shape. After drying, of course, it becomes brittle. Metals are also highly malleable, but it takes extreme pressure such as beating with a hammer and usually high temperatures to mold them. Witness blacksmithing.

Transparency: Glass and some plastics are readily recognized by being TRANSPARENT. Many plastics are TRANSLUCENT; they allow one to see light through them but not images. Most other materials are OPAQUE; they do not permit the passage of light.

Magnetic: Will a magnet be attracted to it? Iron and nickel, yes; other metals, no. This test is commonly used to separate iron (and nickel) from other metals. (The topic of magnets and magnetism is exceedingly important and will be investigated further in Lesson A-5A.)

Additional tests that may be used to distinguish different materials, but which are generally reserved for higher grades, include:

Conductivity: Will it conduct electricity? All metals will; other materials (with few exceptions) don't. Also, consider conductivity of heat.

Resistance to elevated temperatures: Metals are very resistant to heat; hence they are used for cooking utensils. Plastics generally melt and/or burn at high temperatures. Wood and wood products burn readily.

Again, there is no way you should even think of covering this list in a single or even a few lessons. Rather, interject a bit here and there. For example, in playing with a rubber ball, you can discuss its being ELASTIC. That is what makes it bounce. Contrast this with what happens if a glass is accidentally dropped. It breaks because it is BRITTLE.

It goes without saying that even with all these tests, we may still be fooled. We have developed the technological skill of making plastics that so resemble other materials that it is difficult to tell the difference. Consider, for example, how closely synthetic leather (a sort of plastic) resembles real leather, and synthetic fibers resemble natural fibers. Also, we may be fooled by coatings of various sorts. Finally, there are composite materials, such as wood chips pressed together with resins ("glues") that are now common in building materials. These "problem items," however, can be treated as extensions of the lesson. They should not detract form the core of the lesson itself.

As the lesson proceeds, integrate discussion of why a given material is used for a certain item. For example: Why is metal used for door hinges and locks? Why not use wood or plastic? Why is rubber used for play balls? Why not use metal or wood? Why is wood most commonly used material for broom handles? Why not use solid metal bars or stone? Why is cement (stone) used for roadways and sidewalks? Why not use plastic, metal, or wood? Note that in some cases the answer will involve cost. For example, innumerable toys are made of plastic, not because it is better than wood and/or metal but because it is much less expensive to make the toys from plastic.

Finally, relate this lesson to the concept of "organizing things into categories." Students have learned (Lesson A-2) that one category of matter is solids. Here we are learning that solids fall into several categories of materials as discussed. In turn, students may go on to learn that each of these kinds of materials can be sorted again into more specific groups, i.e., different kinds of metals, plastics, woods, etc. Thus, categorizing things according to the material(s) from which they are made serves many useful purposes.

Questions/Discussion/Activities to Review, Reinforce, Expand and Assess Learning:

Make a book illustrating the ways in which they would distinguish a metal block from a wooden block of similar size and shape. Do the same for items made of different materials.

Set up an activity center with an assortment of items made from different materials where children can handle and sort them as they choose.

Periodically ask children to identify the material used in making an item they are playing or otherwise working with. Likewise ask, "How can you tell? Help them recall and review the words that apply.

As children become familiar with distinguishing different materials, you may invite them in small groups to a challenge game of sorting a pile of miscellaneous items according to the material from which each is made.

Challenge children to identify materials with their eyes closed or blindfolded, that is, by using only non-visual senses.

You may also introduce a historical perspective. Development of new materials and means of fabricating them into devices was and remains central to advances in civilization.

In small groups, pose and discuss questions such as:

> Pass around an item that students have not previously examined. Ask them to identify the material from which it is made, and discuss why that material is used in preference to another.
> Further cite familiar things such as playground equipment and ask students to identify the material from which it is made, and discuss why that material is used in preference to another.
> Prepare and challenge students to fill in a matrix listing materials across the top and items down the side. The question to be considered in each blank is: Would the material work for the given item/use?
> Challenge students to design something/anything; let them use the freedom of their imaginations. Then ask them to specify the material(s) they would use for it or its various parts. Have other students critique if given materials would work for their specified use. Some students may come up with the need for materials with properties that are not met by existing materials. Point out that this is the ongoing challenge of scientists. Much of modern technology has depended on the development of new materials with particular properties.

To Parents and Others Providing Support:

Take advantage of the countless opportunities in the kitchen and around the house and yard to mention and talk about the material that is used in making given items: What is this ____ made from? Why do you think ____ is used for this purpose? What are the characteristics of ____ that make it good for this purpose? Why would ____ not be used instead? For example, in a break from yard work, you might look at a rake and ask: What are the tines made from? What is the handle made from? Why do you think metal is used for the tines? ... wood for the handle? Why would we not use plastic for the tines? ... metal for the handle?

Bernard J. Nebel, Ph.D.

A fun party game is to blindfold children in turn and have them identify items and materials by feel and other non-visual senses.

Connections to Other Topics and Follow-Up to Higher Levels:

This lesson will naturally lead into discussions/lessons regarding RESOURCES and TECHNOLOGIES.

Resources: What are the raw materials or resources from which basic materials are obtained? Where are they obtained?

Technology: What steps are involved in obtaining the raw material, making it into the basic material, and finally manufacturing the finished product? In the case of metals, for example, ores must be mined and smelted to obtain the metal and finally fabricated into the finished product.

Geography: Where do we obtain those raw materials/resources?

Ecology: Are those resources renewable or non-renewable? What about final disposal? Can they be recycled? What are the implications for pollution?

Energy: What are the energy resources required?

Food and water resources: Where would be without these?

Re: National Science Education (NSE) Standards

This lesson is a steppingstone toward developing student's understanding and abilities aligned with NSE, K-4:

Unifying Concepts and Processes
  • Systems, order and organization
  • Form and function

Content Standard A, Science as Inquiry
  • Abilities necessary to do scientific inquiry
  • Understanding about scientific inquiry

Content Standard B, Physical Science
  • Properties of objects and materials

Content Standard D, Earth and Space Science
  • Properties of Earth materials

Content Standard E, Science and Technology
  • Abilities of technological design

Content Standard F, Science in Personal and Social Perspectives
  • Types of resources

Content Standard G, History and Nature of Science
  • Science as a human endeavor

Books for Correlated Reading:

Numerous titles in the "Start to Finish" series, Lerner Publishing Group.

Numerous titles in the "Welcome books: How Things Are Made" series, Children's Press.

Lesson A-5A

Magnets and Magnetic Fields

Overview:

Through guided exploration, students will discover the properties of magnets and magnetic fields including the magnetic field of the Earth. This will provide the underpinnings for understanding huge portions of modern technology since magnetism is intricately entwined with electricity and thus underlies the functioning of electric motors, generators, recording and playback systems, and innumerable other sorts of electrical devices.

Time Required:

Part 1. Properties of Magnets (guided exploration, 20-30 minutes, plus interpretive discussion, 25-35 minutes, on two occasions)

Part 2. Magnetic Fields (demonstration,15-20 minutes, plus interpretive discussion,15-20 minutes, plus, explorative activity, 20-30 minutes, followed by interpretive discussion, 25-35 minutes)

Part 3. The Earth's Magnetic Field (demonstration, 15-20 minutes, plus interpretive discussion, 15-20 minutes, plus explorative activity and interpretive discussion, 30-40 minutes)

Objectives: Through this exercise, students will be able to:

1. Demonstrate and describe properties of magnets, namely:
   a) how magnets only act on iron or iron-containing materials[11]
   b) how every magnet has two ends or poles
   c) how poles behave toward one another (unlike poles attracting and like poles repelling)

2. Explain and demonstrate how a magnetic field can be seen with a magnet and iron filings.

3. Describe how the strength of a magnetic field changes with distance from the magnet.

---

[11] Nickel exhibits magnetic properties as well as iron. However, nickel is seldom encountered in everyday materials. Therefore, it is omitted from consideration here.

4. Demonstrate that a magnetic field goes through all non-iron materials.

5. Demonstrate how a magnetic field is conducted through iron.

6. Demonstrate how a bar magnet that is free to turn will orient itself in a north-south direction; explain why it does so.

7. Explain the basic operation of a directional compass.

8. Explain how/why we designate the poles of a magnet as we do.

Required Background:

Lesson A-5, Distinguishing Materials
Lesson D-3A, North, East, South, and West

Materials:

Part 1. Properties of Magnets
A variety of magnets including horseshoe, bar, and button types; ideally enough so that each small group can have a set
A variety of small, iron or iron-containing items, such as paper clips, small nails, etc.
A variety of items made from metals other than iron, such as: aluminum soda cans, pennies (copper), silver or gold jewelry, a lead fishing weight, etc.

Part 2. Magnetic Fields
Horseshoe magnet
Piece of poster paper six to eight inches square
Iron filings (may be purchased or obtained free from any machine shop)
Small sheets of paper, cloth, plastic, and aluminum foil

Part 3. The Earth's Magnetic Field
Bar magnet
Tape
Thread
Directional compass
Large nail

Teachable Moments:

Kids love to play with magnets. Allowing them to do so will create ample teachable moments.

Bernard J. Nebel, Ph.D.

Methods and Procedures:

Part 1. Properties of Magnets

Pass out magnets and invite students to test various things and discover what they are and are not attracted to. Suggest and allow them to test personal effects, items about their desks, and around the room. You may elect to have them make tables listing the items tested down the left-hand side and note "Attraction" or "No Attraction" down the right, leaving space in the middle. CAUTION! Admonish students to NOT test computers, audio or video tapes, digital photo cards, or other such things. The information on them is imprinted magnetically and may be destroyed by a magnet held close. After 15 minutes or so, reassemble students for discussion regarding their findings. What is a magnet attracted to? What is it not attracted to?

Children will most likely name specific objects, but after they have done this, ask them to consider the material from which each item is made (Lesson A-5), and have them add this to their lists after the name of each item. Reexamining their lists from this light should reveal that magnets were only attracted to metal. They were not attracted to cloth, paper, wood, plastic, glass, leather, skin, or other nonmetallic material. Returning to metal, ask them, "Can we even say that magnets attract all metal?" What about an aluminum soda can? What about pennies and other coins? ... lead? ... gold or silver jewery? Invite students to retest anything where there is doubt.

Observation and instruction obviously will occur regarding different metals. Of all the common metals, magnets only attract iron and steel, which is mostly iron. Said another way: If a magnet attracts it, it is iron; a metal that is not attracted to a magnet is not iron. Indeed, this is a way that is commonly used to distinguish and separate iron from other metals. There may be some confusion caused by paint, enamel, or other coatings. Refrigerators are a case at point. The fact that magnets are attracted to refrigerators shows that the outside is sheet metal (iron) under the enamel coating.

Next, invite students in small groups to play with an assortment of magnets and discover what they can about them. Allow them the freedom to play/experiment as they wish, putting them together and taking them apart in various ways. (Many refrigerator magnets seem to be a "wafer" of plastic. They are actually very fine iron filings imbedded in plastic. The magnetized iron filings act together as a magnet.)

After 15 minutes or so of such play, bring students to talk about what they have observed/discovered about magnets. Do they always attract one another regardless of how they put them together? Most children will have discovered that, put together in one way, two magnets attract each other. However, when they try to put them together another way, they actually push apart.

Explain that this is a basic property of magnets. Regardless of shape or size, every magnet has two "ends." These ends are called the NORTH-SEEKING POLE, and the SOUTH-SEEKING POLE. (The reason for using these names and how we distinguish them will come in Part 3 of this lesson.) Opposite poles (north and south) attract. But, like poles (north and north or south and south) REPEL. (Introduce "repel" as the word we use to describe the pushing apart.) If any students didn't observe this, invite them to reexamine the behavior of their magnets. They will find that they can actually push one magnet around with another without touching it by bringing like poles close together. Finally, call students' attention to the fact that both poles of a magnet are attracted to iron that is not a magnet. Only when the iron is the like pole of another magnet does one observe the repulsion.

Part 2. Magnetic Fields

Conduct the following demonstration with children crowded around so that they can see what happens. Spread iron filings on a piece of poster paper. You can spread them around with your finger like finger paint. Carefully hold the paper level and bring a horseshoe magnet up underneath to touch the paper. Students will see the filings jump atop one another and line up forming a display of two asterisk-like arrangements over the poles of the magnet. Also, have children note the arching lines of filings between the two poles.

(Practice this beforehand so that you can optimize the amount of iron filings to give the best effect. Of course, you may have students do this also, but if you do, be prepared to have a mess of iron filings scattered about and especially stuck to the magnet. Filings can be mostly removed from a magnet by wiping and pinching them off with a piece of tissue.)

What does this demonstration show? Let students ponder this for a minute or so, but don't expect more than puzzled looks. Then, instruct them that this shows what we call a MAGNETIC FIELD around the poles of the magnet. Each pole has a magnetic force field of attraction that affects iron particles. That magnetic field is always there, although we can't see it. The iron filings only make its presence visible.

What else can we say about the magnetic field? Does it extend out indefinitely? Is it the same strength for a given distance and then cut off sharply, or does it decrease gradually with distance from the pole? Coach kids in reasoning how their observation shows that the field is strongest at the poles and fades gradually with distance. This is evidenced by the fact that filings climb atop one another and line up most markedly at the poles. This effect diminishes gradually with distance from the pole. By more than an inch or so from the poles, it is not seen at all.

This effect may be seen even more clearly by slowly moving the magnet around under the bed of filings on the paper. Students will see filings begin to line up as the field approaches and fall down and "relax" as the field moves away. (In the center, many filings will be dragged along by the field, but keep the focus on the edges of the field.)

Have children relate this effect to their exploration with magnets. What did they experience concerning the attractive (or repulsive) force with distance? Did the attraction of the magnet for an iron object or another magnet get stronger as the magnet got closer? Let students test this again if they are hesitant about the answer. They will experience that the closer the magnet comes, the greater the attractive force until it touches. Similarly, they will experience that the repelling force becomes greater as like poles of two magnets are pushed together. Thus, bring students to conclude that the magnetic field that surrounds the poles is strongest at the poles and tapers off rapidly with distance from the poles.

What else? What can we say about the ability of the magnetic field to pass through other materials? It is evident in the demonstration that the magnetic field passed through the paper and affected iron filings on the other side. What about other materials? Let's do the tests.

A simple way of testing is to see if a magnet on one side of the material will attract a small iron item such as a paperclip on the other side, but allow children to experiment as they may wish. You may have them make another table and record results. Remind them of the factor of distance. If there is no apparent attraction, is it because the magnetic field is not penetrating, or is it because the distance is simply too great? Therefore, in doing this testing they must use thin slips of material. Paper of various sorts, cloth, plastic, aluminum foil, and coins are ready materials, but glass, wood, ceramics, and other materials may be tested as well.

In every case, students will discover that the magnet on one side of the material will attract an iron item on the other side. Thus, they can conclude that magnetic fields can and do go through all materials. There is one notable exception, however. Demonstrate with an empty tin can, which is mostly iron, that a magnet stuck on the outside will not attract a paperclip or other iron item on the inside. Magnets are attracted to iron; why doesn't the field go through? This is another case where you should expect only puzzled looks.

Explain that the iron effectively captures and holds the magnetic field. Most of the magnetic field goes through the iron from one pole to the other and little comes out the other side. Another way of seeing this is in the fact that a magnet will pick up a whole chain of paperclips, small nails, or other small iron items. Kids invariably have fun testing how long a chain of paperclips, for example, they can suspend from a magnet as each one acts as another magnet to hold the next. Allow them to experiment with this. Then have them note that the last paperclip in the chain is held at a distance that would be impossible without the other clips in between.

What does this mean? Explain that the paperclips capture the magnetic field and conduct it along their length. This extends the effective distance of the field from the magnet. Importantly, have students note that the magnet and, hence, its total magnetic force is not changed in the process. It is only that its magnetic field becomes aligned along the chain of iron items.

Part 3. The Earth's Magnetic Field

Have children recall or retest how magnets put together in one way attract, but repel when put together the other way about. Review how we described this as unlike poles (north and south seeking) attracting and like poles (south and south, or north and north) repelling. But why are the poles referred to as north and south seeking, and how does one tell which is which?

While students are pondering this question, bring out a demonstration, prepared ahead, of a BAR magnet hanging horizontally from a two to three-foot piece of thread. (Wrap a strip of tape around the center of the magnet and loop the thread through the tape at the edge of the bar. If you simply tie a loop around the magnet, it is likely to tip into the vertical position.) Attach the thread to a stationary support so that the magnet can swing and turn freely. Hanging it over the edge of a table works well. Give the magnet a slight nudge so that it will turn back and forth but not swing.

Have children watch and note its orientation as it comes to rest. You may try it again hanging it in different locations, but each time students will note that the magnet comes to rest pointing the same direction. (Be sure the magnet is not hanging close to a metal object that will attract it.) What direction is that? From Lesson D-3A students should note that is to the north and south. Label the end of the bar pointing north as "north-seeking" and the other end "south-seeking."

Explain that Earth itself acts as a huge, but relatively weak, magnet. The magnetic north pole of Earth is in the area of (but not exactly at) the geographic North Pole. Likewise the magnetic south pole is near the geographic South Pole. Thus, there is a magnetic field around Earth arching between the poles. Have students recall the lines of iron filings arching between the poles in the preceding demonstration. The field is not strong enough to have a noticeable effect in attracting iron to one pole or the other, but magnets that are free to turn will orient themselves parallel to this field.

Further explain that this is the basis of a directional compass. A compass is really nothing more than a small bar magnet balanced on a needle so that it may turn freely. Allow students to examine a compass. Bring a nail close to it and have children observe how the needle is attracted to the nail just like would be expected of a magnet. Thus, the compass needle orients itself in Earth's magnetic field just like the bar magnet in the demonstration. Now, the end of the magnet that points north is logically designated as the NORTH-SEEKING POLE; the other end is the SOUTH-SEEKING POLE.

Pose the question, "How do we know that unlike ends of magnets attract and like ends repel?" Bring the end of the bar magnet that you have just labeled "north-seeking" up to the compass. Students will observe that the north end of the compass turns away and the south end is attracted. You may choose to suspend another bar magnet from a thread as you did the first and determine its north-seeking end. With the two bar magnets you can demonstrate clearly that it is the north and south ends of the two magnets that attract; like ends repel.

Must you test every magnet in this manner to determine which end is north seeking? Guide students to reason that the answer is no! Once you have determined the north-seeking end of one magnet, you can use it to determine the poles of any other magnet. The compass needle is a convenient magnet with known ends. If a pole of a magnet attracts the north end of a compass needle, which pole is it? Students should reason that must be the south-seeking pole because it is unlike poles that attract. You may follow with an activity of having students test and label the ends of various magnets with this technique.

Questions/Discussion/Activities to Review, Reinforce, Expand, and Assess Learning:

Make books illustrating one or more aspects of the lesson.

Set up an activity center where children can independently repeat the activities described and/or conduct further tests as they may choose. Tests/experiments with iron filings may be allowed but closely supervised to avoid messes.

Will magnets work under water? Test it and see.

How did the discovery and invention of compasses aid early exploration?

In small groups, pose and discuss questions such as:

What are the attributes of magnets?
How do two magnets act toward one another?
How do they interact with other materials?
How would you sort a pile of junk into iron and non-iron metals?
How might you quickly separate needles and pins from buttons in a sewing basket?
How might you quickly separate aluminum cans from tin/iron cans?
What are the attributes of magnetic fields?
How can you show that magnetic fields go through other materials?
If you suspend a bar magnet horizontally on a string, what will it do?
Why/how do compasses point north?
Why do we refer to the poles of a magnet as north-seeking and south-seeking?
How can you determine which is which?

To Parents and Others Providing Support:

Facilitate your children in playing/experimenting with magnets.

Discuss the attributes of magnets. How they behave toward each other? How they behave toward other materials?

Supervise and help your children test the magnetic fields of various magnets using iron filings as described.

Discuss the attributes of magnetic fields.

On an outing, use a directional compass and discuss why it points north.

Connections to Other Topics and Follow-Up to Higher Levels:

Magnets and electricity
Electromagnets
Motors
Generators
Recording and playing information on discs
Other uses of magnetic fields in technology and medicine

Re: National Science Education (NSE) Standards

This lesson is a steppingstone toward developing students' understanding and abilities aligned with NSE, K-4:

Unifying Concepts and Processes
• Systems, order and organization
• Evidence, models, and explanation

Content Standard A, Science as Inquiry
• Abilities necessary to do scientific inquiry
• Understanding about scientific inquiry

Content Standard B, Physical Science
• Properties of objects and materials
• Light, heat, electricity, and magnetism

Content Standard D, Earth and Space Science
• Properties of Earth materials

Content Standard G, History and Nature of Science
• Science as a human endeavor

Books for Correlated Reading:

Branley, Franklyn. *What Makes a Magnet?* (Let's-Read-and-Find-Out Science, Stage 2). HarperTrophy, 1996.

Fowler, Allan. *What Magnets Can Do* (Rookie Read-About Science). Children's Press, 1995.

Olien, Becky. *Magnets* (Our Physical World). Capstone, 2003.

Rosinsky, Natalie M. *Magnets: Pulling Together, Pushing Apart* (Amazing Science). Picture Window Books, 2002.

Walker, Sally. *Magnetism* (Early Bird Energy). Lerner, 2005.

Lesson A-6

Matter II: Air Pressure, Vacuums, and the Earth's Atmosphere

Overview:

In this lesson, students will relate their comprehension of the particulate nature of matter to understanding air pressure. They will come to see air pressure and vacuums as a situation of how closely particles are pushed together. In turn, this will provide the foundation for understanding the Earth's atmosphere as an envelope of air that decreases in pressure with increasing altitude until there is no air in outer space. They will understand wind in terms of high and low-pressure regions.

Time Required:

Part 1. Compressing Air and Air Pressure (activities plus discussion, 30-40 minutes)
Part 2. Air Pressure Inside And Outside (activity plus discussion, 20-30 minutes)
Part 3. Vacuums (demonstration plus discussion, 15-20 minutes)
Part 4. The Earth's Atmosphere (showing pictures plus discussion, 25-35 minutes)
Part 5. Wind (showing pictures plus discussion, 25-35 minutes)

Objectives: Through this exercise, students will be able to:

1. Understand and use the following words in their proper context: COMPRESS, PRESSURE, VACUUM, ATMOSHPERE.

2. Describe COMPRESSING (putting air under pressure) in terms of forcing the particles of air closer together.

3. Describe a vacuum in terms of particles of air being further apart.

4. Describe sucking in terms of reducing air pressure at one location and higher-pressure air pushing toward the lower pressure.

5. Define the Earth's atmosphere.

6. Describe how and why air pressure in the Earth's atmosphere decreases as you go up, until there is no air in outer space, i.e., outer space is essentially a vacuum.

7. Describe movements of air, including wind, in terms of locations of differing air pressures.

Required Background:

Lesson A-3, Air is a Substance
Lesson A-4, Matter I: Its Particulate Nature
Lesson D-1, Gravity I

Materials:

Rubber party balloons
Pictures/videos of astronauts working in outer space
A household vacuum cleaner
An inflatable tire or ball and a pump (desirable, but may be omitted)
Video or pictures of astronauts working in space (desirable, but may be omitted)
Weather maps showing high and low-pressure areas

Teachable Moments:

Teachable moments will include times when you are using a vacuum cleaner, blowing up a balloon or pumping up a tire or ball, driving up or down a long hill and feeling your ears pop, sucking a drink through a straw, and other such occasions.

Methods and Procedures:

Part 1. Compressing Air and Air Pressure

Review Lesson A-4, "Matter I: Its Particulate Nature." Be sure that students have the concept that there is no significant attraction between the particles of a gas, air in this case. Each particle is continually in motion, bumping into other particles, walls, etc., but there is empty space between the particles.

Then commence blowing up a balloon or pumping up a tire or ball, and ask. "What do you think we are doing to particles of air as we do this?" With Q and A discussion, students should recognize that the process is pushing air particles closer together. Forcing the particles closer together increases the PRESSURE, pressure being a measure of how hard the air is pushing on the sides of the balloon, tire, or ball. We refer to it as AIR PRESSURE. (If students are familiar with pounds and square inches, they may note that air pressure is commonly measured in pounds per square inch, and pressure gauges read accordingly.)

The following visualization may help students master this concept. Have them imagine a large empty room with three or four blindfolded children wandering around aimlessly. Occasionally they will bump into one another and the walls, but with no attraction they will "bounce" off and continue their wandering. But now, imagine more and more children in the room, likewise blindfolded and wandering aimlessly. Now the bumping against each other and the walls becomes more and more frequent. The amount of bumping creates a certain pushing or PRESSURE against each other and the walls. Depending on the walls, they may be pushed outward by the pressure, as a balloon expands. Or, the pressure may simply increase, as in a tire, which does not expand significantly.

This is the situation with particles of air. As we pump up a tire or ball, we are pushing more and more air particles into the same space. They are being forced closer and closer together. Their bumping against each other and the walls is seen and measured as increasing PRESSURE.

Go on to introduce the following new words. As we force air particles closer together we say we are COMPRESSING the air. Or, the air under pressure is COMPRESSED air. In blowing up a balloon, we compress air in our lungs with our chest muscles and force it into the balloon. A pump compresses air even more, and forces it into the tire/ball giving greater pressure than we can exert with blowing. The inflated tire/ball/balloon then contains COMPRESSED air. Note that the term "compressed air" by itself does not specify the pressure. The air may be just slightly compressed (low pressure) or it may be highly compressed (high pressure). One really needs a pressure gauge to determine how compressed the air is.

What happens when you release an inflated balloon or open the valve of a tire? Air rushes out! Use Q and A discussion to bring students to describe this in terms of the movement of particles. The air particles are pushed out by the higher pressure inside toward the lower pressure outside. (Admonish children that letting air out of a tire is a very bad thing to do unless they are prepared with a pump to pump it back up. Doing so may cause great inconvenience and cost.)

A generalization of great importance should be emphasized here. Compressed air, air under higher pressure, always flows toward lower pressure. Air by itself cannot and will not flow toward higher pressure. Balloons will not inflate by themselves regardless of how much we might like them to. You may get kids to recite: *water always flows downhill, AND air always flows toward lower pressure.*

For the last segment of this part, have children observe that air and other gases are easily compressed. That is, one can force more and more gas into the same space. Or, the volume of a gas, the space it takes up, is easily reduced by added pressure from the outside. Liquids and solids, on the other hand, do not compress. Pressure on a liquid or solid may increase, but they are not squashed down into a smaller volume as gases are.

Students may see an exception with sponge balls and some other materials. Point out that when you squeeze a sponge ball, you are not compressing the solid portion; you are squeezing out or compressing air within the sponge. Or, it seems easy to squeeze down on a lump of moist clay. But, point out that the volume of the clay is not really reduced; in being squeezed down in one location, the clay bulges out at another. The same is true for solid rubber.

Now, direct students' attention to the question: How is it that gases are easily compressed and liquids and solids are not? Students should be able to reason out the answer from their knowledge of the particulate nature of solids and liquids (Lesson A-4). In solids, the particles being tightly bonded together are already as close together as they can get. They can't be pushed closer. The same is true of liquids; while the particles are loosely attracted such that they slip and side around one another, they are still just as close as they can get. Only in gases is there no attraction; hence, there is empty space between the particles. Thus, the particles can be pushed closer together, that is compressed.

Some students may ask if there is a limit to how much a gas can be compressed. Yes, at enough pressure (or low enough temperature, which slows down the movement of particles and allows them to come closer together) gases actually become liquid. Students may have seen or heard of liquid air, liquid nitrogen, or liquid oxygen.

Lastly have students consider just how fortunate it is that solids do not compress. They will enjoy imagining what would happen to a tall building if the bricks or other supports on the lower levels were squeezed down and compressed by the weight of those above.

Part 2. Air Pressure Inside and Outside

A silly "game" that kids get a kick from is blowing their cheeks out and alternately sucking them in. But this silliness can become an instructive lesson. What are you doing with air particles as you puff out your cheeks? Use Q and A discussion to bring kids to recognize that they are using pressure exerted with their chests to push more air into their mouth. The increased pressure in their mouth pushes their cheeks out.

And what about sucking the cheeks in? In sucking, they decrease the pressure in the mouth and cheeks curve inward. (Sucking is actually an action of filling the mouth cavity with the tongue and then pulling back with the tongue creating a vacuum in the mouth.) But here is a very important point that requires special attention. Have students consider that in normal air around us, air particles are also bouncing around and exerting a certain amount of pressure. We usually don't feel this pressure because it is the same on all sides. Specifically, with mouths open, air particles are pushing equally on both sides of our cheeks. The pressure being equal on both sides, we don't feel it at all.

What happens with sucking in? Use Q and A discussion to bring children to reason it out as far as possible. The conclusion to be reached is that in sucking, they are making the air pressure inside LESS THAN the pressure outside. It is the GREATER PRESSURE OUTSIDE that pushes their cheeks in.

Now, turn to sucking a liquid through a straw. Use Q and A discussion as far as possible to bring kids to reason out what is really happening. In sucking, they are simply reducing the pressure at the top end of the straw. It is actually outside air pressure that is pushing the liquid up the straw.

Bernard J. Nebel, Ph.D.

Many children gain the naïve impression that is retained through adulthood that in sucking, we are actively pulling the liquid up the straw. This erroneous idea can lead to much confused thinking regarding the real world. Therefore, it is important to get children on the right track early on. As an aid, have kids imagine trying to drink soda through a straw in the vacuum of outer space. Not only would they not be able to suck anything up; their innards would be sucked (pushed) down the straw into the vacuum.

This may be used as a lead into discussing vacuums.

Part 3. Vacuums

The word "VACUUM" may or may not be new to students at this level. In any case it should be clarified. When we suck or otherwise remove air from a container, we are creating a VACUUM. Just as air pressure may range from slight to great, vacuums also may be of different degrees. They range from slightly lower presser (removal of some air particles) to total removal of all the air particles. But, vacuums are never active in pulling (sucking) air in; it is always higher pressure on the outside that pushes air (or liquid) into the vacuum. *Air will only flow from higher pressure toward lower pressure.*

Again, have kids consider trying to drink soda through a straw if the soda were in a vacuum.

Children will be familiar with vacuum cleaners. Ask, "How do they come to be called VACUUM cleaners?" Have kids examine a vacuum cleaner and note how a blower forces air out of the machine creating a vacuum (reduced air pressure) at the hose/sweeper end. Outside air rushes toward the vacuum (lower pressure) carrying dust with it. The dust is caught in the filter bag while the air goes through the pores and exits through the blower.

Part 4. The Earth's Atmosphere

On gaining a comprehension of air pressure and vacuums in terms of the relative closeness of air particles, understanding the basic nature of the Earth's atmosphere becomes a simple matter.

The first step is the definition. The Earth's ATMOSPHERE refers to all the air that surrounds the Earth.

From various sources, children are probably aware that there is essentially no air in outer space. Outer space is a total vacuum. If a person were to step into this vacuum without a space suit, she/he would literally explode as air (pressure) in the body pushed outward into the vacuum. Thus, astronauts wear spacesuits that maintain an atmosphere of suitable air pressure as well as supplying oxygen necessary for breathing. Use pictures or videos of astronauts working in space.

Now pose the question, "If outer space is a vacuum, why doesn't the Earth's atmosphere push out into that vacuum?" Let students ponder this question for a few moments. They may not come up with the answer. The answer is gravity! Recall that air does have weight (Lesson A-3). The gravity of the earth literally holds the air down around the Earth and prevents it from blowing out into space.

Now, take students through the following reasoning: If air has weight and it is easily compressed, what will happen in a tall "stack" of air? You may provide a hint by asking, "What is your experience when a bunch of you pile on top of each other?" The ones on the bottom feel squashed by the weight of those above. The one on top are fine. Point out that the same is true for the column of air. The air at the bottom is squashed down or compressed by the weight of the air above. As you go up, there is less and less weight of air above, hence, less and less compression (pressure). The air particles are further and further apart until there are essentially none, and one is in the total vacuum of outer space. Note that there is no distinct boundary between the atmosphere and outer space. There is only the gradual thinning of the atmosphere until there is none present.

Some students may note that there is no atmosphere on the moon and ask, "Why not?" The moon's gravity is not strong enough to hold air. Any air present does blow out into the vacuum of outer space.

The decrease in air pressure as you go up in height and the increase as you go down is commonly experienced as your ears "popping." What causes the popping? Diagram that the inner chamber of your ear has the structure of a bottle with a long neck opening into the back of your throat. Air must come through this long "bottleneck" to equalize the pressure inside and outside your eardrum. (Great pain is experienced when pressures are unequal.) As you go up into decreasing air pressure, air must come out of the "bottle;" as you go down into increasing air pressure, more air must come into the "bottle." But the long "bottle neck" is rather tight; therefore air tends to come in or out in bubbles that you experience as hearing and feeling your ears "pop."

Part 5. Wind

The last point that should be integrated into this lesson is a consideration of wind. It follows very naturally from a discussion of different air pressures.

Pose the question, "What makes the wind blow?" Use Q and A discussion to take students through the following steps of reasoning. We have seen that air moves from a place of higher pressure to one of lower pressure. If there is no pressure difference, will air move? No. The only thing that causes a mass of air to move is a pressure difference. Are there differences in air pressure from one location to another? Show students weather maps and/or guide them to pay special attention to weather reports. The weather map or weatherperson will point out regions of high pressure and regions of low pressure and how they are influencing the weather. Without going into detail you can have students reason that air will flow toward low-presser regions from higher-pressure regions, and the result will be experienced as wind.[12] It will also be evident from weather reports that the

---

[12] Children may note that air/wind does not flow straight toward the low-pressure area. Instead it circles around. This is the same effect as water going down a drain. Instead

high and low-pressure regions come and go and move about. Thus, winds will vary accordingly.

The weatherperson will also usually give BAROMETRIC pressure, or say, "The BAR0METER is at so many inches and rising/falling/holding steady." You can simply tell kids that a barometer is an instrument that measures air pressure; the measurement is given in inches. The only significant point at this stage is to have students note that air pressure in a given region does change from day to day. The causes for these fluctuations will be the subject of later lessons.

Questions/Discussion/Activities to Review, Reinforce, Expand, and Assess Learning:

Make books illustrating a container with air particles under high pressure, low pressure, and with a vacuum. (Children will be apt to show a difference in the particles in the different situations. Correct them on this. Illustrations should show only a difference in the relative number of particles in the container. Size and other features of particles should be the same.)

Set up an activity center with pumps, inflatable toys, and/or other devices so that students can test and gain experience with compressing air and creating vacuums.

Assign students to be on the watch for and report examples where they see air pressure or vacuums playing a role.

A fun activity is to take a soda bottle and fill it 100 percent full of water. Place a small scrap of paper over the mouth so that there is no air between the rim of the opening and the paper. Holding the paper in place invert the bottle, then remove the finger holding the paper. The paper holds the water in the bottle. How does this work? Would this work in outer space? (Air pressure is holding the paper against the rim of the bottle. Water will not come out unless air can get in. It is the air pressure of the atmosphere that holds the paper in place.)

Relate breathing to changes in pressure. For inhaling, one expands the volume of the chest. This deceases the pressure within the lungs and air from the outside pushes in. The reverse occurs when exhaling.

In small groups, pose and discuss questions such as:

What are you doing to air particles as you blow up a balloon or pump up a tire or ball?
Is compressed air a different kind of air? What is different about it?
Why don't liquids and solids compress?
How does a vacuum cleaner work?
Why does the wind blow?

---

of going straight down, it swirls around. The reasons for this will be addressed in another lesson.

What is happening as you drink through a straw? Is liquid being actually pulled or pushed up the straw? Could you drink a soda through a straw in outer space? Why not?

What makes your ears pop as you drive up or down a long steep hill?

How does air pressure change as you go up?

Why is air pressure less as you go up?

What is meant by air becoming "thinner" as you go up?

Why do astronauts have to wear spacesuits?

Air will always flow in what direction (in terms of pressures)?

To Parents and Others Providing Support:

Read, review, and discuss all parts of the lesson with your children.

Take advantage of teachable moments such as blowing up a balloon or ball, seeing a SCUBA diver with compressed air tanks, or seeing astronauts in their spacesuits to talk about air particles and how they are packed tighter, spread out, or absent.

On a windy day, pose questions such as: Why is the wind blowing? Can you point to where there must be high pressure? … low pressure?

When cleaning the house with a vacuum cleaner, ask and discuss how the vacuum cleaner works in terms of moving air and picking up dust?

In drinking through a straw, ask: "What is really moving the liquid up the straw? What would happen if the drink were in a vacuum?"

Discuss the action of pumps and other devices that involve air pressure or vacuums as you come across them in everyday routines.

In looking at weather reports, draw your child's attention to high- and low-pressure regions and barometric pressures. Discuss related wind.

Connections to Other Topics and Follow-Up to Higher Levels:

Any and all aspects of atmospheric science, weather, and climates

Volume/density changes with temperature

Rising and descending air currents

Atmospheres on other planets

Re: National Science Education (NSE) Standards

This lesson is a steppingstone toward developing students' understanding and abilities aligned with NSE, K-4:

Unifying Concepts and Processes
  • Evidence, models, and explanation

Bernard J. Nebel, Ph.D.

• Constancy, change, and measurement

Content Standard A, Science as Inquiry
•Abilities necessary to do scientific inquiry

Content Standard B, Physical Science
• Properties of objects and materials
• Position and motion of objects

Content Standard D, Earth and Space Science
• Properties of Earth materials
• Changes in Earth and sky

Books for Correlated Reading:

Branley, Franklyn M. *Oxygen Keeps You Alive.* HarperCollins, 1985.

Lauber, Patricia. *You're Aboard Spaceship Earth* (Let's-Read-and-Find-Out Science, Stage 2). HarperTrophy, 1996.

Parker, Vic. *Air* (What Living Things Need). Heinemann, 2006.

Stewart, Melissa. *Air is Everywhere.* Compass Point Books, 2004.

Lesson A-7

Air: A Mixture of Gases
(Mixtures and Chemical Reactions)

Overview:

Students have learned that air is made up of particles (Lesson A-4). In this lesson, they will discover that air is actually a mixture three important gases: oxygen, nitrogen, and carbon dioxide. They will observe that oxygen is essential for any fire to burn and they will relate this to the necessity of oxygen for the respiration of all animals. Similarly, they will learn that carbon dioxide is a waste product of both processes. Finally, they will learn that oxygen and carbon dioxide are held constant in the atmosphere by green plants, which consume carbon dioxide and emit oxygen. This lesson will also introduce the concept of chemical reactions.

Time Required:

  Part 1. Defining a Mixture (discussion plus demonstrations, 20-30 minutes)
  Part 2. Burning Requires Oxygen (demonstration plus interpretive discussion, 30-40 minutes)
  Part 3. Burning Releases Carbon Dioxide (demonstration plus interpretive discussion, 30-40 minutes)
  Part 4. Mixtures and Chemical Reactions (discussion, 15-20 minutes)

Objectives: Through this exercise, students will be able to:

1. Describe how a mixture is different than a single material.

2. Demonstrate how a flame goes out when deprived of air; state how this is because of a lack of OXYGEN gas.

3. Explain that burning involves consumption of both fuel and oxygen.

4. Contrast and compare a flame's need for fuel and oxygen with our (and all other animal's) need for food and breathing (obtaining oxygen).

5. Describe how the demonstration (2 above) shows that air is not all oxygen; only about one fifth is oxygen and the other four fifths is mostly NITROGEN gas.

6. Demonstrate how the gas produced from mixing baking soda and vinegar, when

"poured" on a flame, will put it out; identify this gas as CARBON DIOXIDE.

7. Recognize describe the distinction between a mixture and a chemical reaction.

8. State that carbon dioxide is a by-product of burning and respiration.

9. Conclude from these activities that air is actually a mixture of oxygen, nitrogen, and a small amount of carbon dioxide.

10. State the exchange of gases that occurs in burning and respiration.

11. Describe the exchange of gases that occurs in photosynthesis.

12. Describe the role of green plants in maintaining the balance of gases in the atmosphere.

Required Background:

Lesson A-2, Solids, Liquids, and Gases
Lesson A-3, Air Is a Substance
Lesson A-4, Matter I: Its Particulate Nature
Lesson A-5, Distinguishing Materials
Lesson B-3, Distinguishing Plants and Animals
Lesson C-3, Energy II, Kinetic and Potential Energy

Materials:

Candle stub 2-3 inches or birthday candles
Matches or lighter
Soup bowl
Water
A clear drinking glass
Empty jar (about 1 pint size)
Baking soda
Vinegar

Teachable Moments:

In the course of an activity/game reviewing how they distinguish materials (Lesson A-5), pose the question, "Is air all one thing, or might it be a mixture of different gases?" Invite children to watch some demonstrations that will answer this question.

Methods and Procedures:

Part 1. Defining a Mixture

Start by making sure that children understand what is meant by a MIXTURE.

Children's experience is such that you will only need to apply the word to familiar situations to convey the meaning. Explain that whenever we MIX or stir two or more things together, we will have a MIXTURE. Give common examples. A handful of loose change is a mixture of pennies, nickels, dimes, and quarters. Vegetable soup is a mixture of ___. (Have kids name various vegetables that may be present.) Invite children to cite additional examples of mixtures.

Students may give some instances, such as stirring sugar into water, where the mixing results on one substance dissolving in the other. Here, the result is properly called a solution. However, note that solutions are a special class of mixtures. Any attempt to make a technical distinction at this stage will result in unnecessary confusion. Therefore, it is practical to accept them as examples of mixtures. It is only significant for students note that in some mixtures the different components are still visible to the unaided eye as in the mixture of coins and vegetable soup. In other cases, they break up into particles that are too small to see, but each is still there.

The distinctions between mixtures and solutions will be addressed further in Lesson A-9. Of course, when we mix some things, baking soda and vinegar for example, there is a chemical reaction. Save this for later in the lesson, however. Be sure that students have a firm understanding of a mixture first.

Moving on, it will be readily apparent to children that you may have a mixture of various solid items. You may demonstrate this by "stirring" together any combination of items from your desk and saying, "Here we have a mixture of paperclips, staples, pencils, and whatever." Similarly, children will appreciate that a liquid can be a mixture of two or more liquids such as milk and chocolate syrup, or solid and a liquid such as lemon juice and sugar. Invite students to give further examples that they are familiar with.

Finally ask, "There are different kinds of gases just like there are different kinds of solids and different kinds of liquids. Do you think that air might be a mixture of different gases, or is it all the same gas?" That is, does air consist of particles that are all exactly alike, or are there different sorts of particles representing different gases? Let students ponder this a few moments and give their thoughts, as they will.

This provides another opportunity to point out that answers in science are not given by majority opinions or votes. They are given by finding evidence. We are going to proceed to find evidence for one or the other.

Part 2. Burning Requires Oxygen

Children get a kick from the following demonstration. Mount a candle stub 2 to 3 inches long in a soup bowl by dripping some wax into a "puddle" and inserting the base of the candle into the puddle. Pour about half a cup of water into the bowl around the candle. Light the candle, and after it is burning steadily, invert a clear glass tumbler over the candle into the water. Students should be able to watch the candle flame burning through the inverted glass. After a few seconds, the candle flame abruptly dims and goes out. After a few seconds more, water starts being drawn up inside the glass and will be pulled up to fill about one fifth of the glass. A variation on this demonstration is to mount a short piece of birthday candle in a walnut-shell boat. The boat then rises up in the glass after the flame has gone out.

Children will invariably want to see this demonstration repeated two or more times. Do so at your discretion, then proceed into interpretive Q and A discussion. Why does the candle go out? Guide children to recognize that some part of the air is required for burning as well as the "fuel," wax in this case, that is being burned. (Many children are apt to think that it is just the wick of the candle that is burning. Just as in an oil lamp, the wick only serves to spread the melted wax into a thin film that burns more readily. The wax is the fuel.)

Is all of the air equally supportive of burning? If so, why does the flame go out when the largest portion of air still remains? Guide students in reasoning how the demonstration shows that only a portion of the air supports burning. When that portion is used up, the flame goes out. Finally, tell students that the portion of the air that supports burning is called OXYGEN, more specifically oxygen gas. Most of the air remaining in under the glass is another gas called NITROGEN. Thus, air is a mixture of oxygen gas, nitrogen gas, and another very important gas we will come to in a moment.[13]

In discussion, make a connection here with Lesson C-3, "Kinetic and Potential Energy." Review and emphasize that fuel (wax in this case) is a form of potential energy. In burning, that potential energy is released as kinetic energy, namely heat and light. But now we see that releasing the potential energy requires OXYGEN in addition to the fuel. We also learned that food is potential energy for our bodies (in addition to being material for actual growth and maintenance).

Why do you think we breathe, and need to keep breathing all the time? Guide children in reasoning out the answer. The release of the energy from food requires OXYGEN just like the candle flame. Our breathing brings oxygen into our bodies. A short activity that helps make this point is to have everyone do vigorous jumping jacks for a full minute. At the end, everyone will be breathing hard. Why? Use Q and A discussion to guide students in reasoning out the answer. More exercise requires more energy. That requires the burning of more "fuel," which requires more oxygen to burn it, hence, deeper, faster breathing.

---

[13] It must be admitted that we have fudged significantly in this demonstration. The flame does go out as a result of depletion of oxygen, but the actual change in volume of gas is minimal because the oxygen consumed is replaced by carbon dioxide, the byproduct of combustion. The change in volume we observe is because in placing the glass over a burning candle, we place it over hot air. As the hot air cools after the candle goes out, it decreases in volume, and this is what actually pulls the water up. Nevertheless, this demonstration makes the point so elegantly that forgoing the fuller explanation seems forgivable at this stage. We will come it in a later lesson when students have mastered more background.

We obviously don't go up in flame. The "burning" of food in our bodies takes place at modest temperature in a number of small steps that release the energy in a way that runs all the functions of our bodies. But, the overall requirements and effect is the same. Our bodies require both food and a continual supply of oxygen to "burn" it.

You may choose to insert some discussion regarding safety here. Anything that cuts off access to oxygen (choking, drowning, being closed in a small space, etc.) leads to rapid death. Why? Just like the candle flame, life is quickly extinguished by lack of oxygen. You may continue and expand on this discussion as you see fit.

There may well be questions, because it appears that breathing simply pumps air in and out. The air we exhale seems to be the same as the air we inhale. If students don't bring up the question, pose it yourself, "Is the air you exhale the same as the air you inhale?" Guide students in reasoning that the answer must be NO. Air that is exhaled will have less oxygen because some has been taken into the body. There is no change in the nitrogen however; that portion, nearly four fifths of the air, does just go in and out without change. (Of course, there is also a change in carbon dioxide content, but save this to the next section.)

Kids have undoubtedly seen and may even know persons who wear breathing tubes. What is the purpose of such tubes? With Q and A discussion guide students in reasoning: If air is only one fifth oxygen people with respiratory difficulties are aided by increasing the oxygen content of the air they are breathing. The extra oxygen is supplied to the person's nostrils through tubes from tanks of pure oxygen.

Part 3. Burning Releases Carbon Dioxide

There is one more gas to consider: CARBON DIOXIDE. A demonstration that students get a kick out of is the following.

Set up a lighted candle in a location without drafts. In the same draft-free location, place about a tablespoon of baking soda in an empty mayonnaise or other such jar. Then add about a quarter cup of vinegar. This causes fizzing as the acidic vinegar reacts with the baking soda to produce carbon dioxide.

The "magic" comes when, as the fizzing subsides, you carefully pick up the jar and tip it as though pouring something over the candle flame. (Be careful not to pour out any of the visible remains of the vinegar and soda.) The candle flame will suddenly go out. Indeed, you did pour something on the flame that caused it to be extinguished.

Explain to students that the bubbles of fizz coming from the vinegar and soda are CARBON DIOXIDE. Carbon dioxide is a gas that is heavier than oxygen and nitrogen. Therefore, as it is produced by the reaction, it accumulates in the jar and pushes the oxygen and nitrogen out. As the fizzing subsides, you have a jar full of carbon dioxide. Being heavier than the rest of the air you can pour it over the candle flame as you would water. The carbon dioxide blocks the flame's access to oxygen and smothers it. (You may note that carbon dioxide is commonly used in commercial fire extinguishers.) Kids will certainly want to perform this demonstration on their own. Supervising their doing so will be excellent reinforcement.

Go on to explain that carbon dioxide is not produced just with baking soda and vinegar. It is also a byproduct of burning. As a flame consumes oxygen and fuel, it releases carbon dioxide into the air. The amount produced is the same as the amount of oxygen consumed.

The same is true of the "burning" of food in our bodies. As our bodies consume oxygen, they also produce carbon dioxide, which exits through the lungs. The air we inhale has plenty of oxygen and little carbon dioxide. The air we exhale has less oxygen and more carbon dioxide. (The carbon dioxide content of our breath is not so great, however, that we can put a flame by gently exhaling on it. Blowing out a candle is a matter of cooling the wax on the wick below a temperature at which it will burn.) Add that the same is true for all members of the animal kingdom. They all get their energy from the consumption of food and "burning" it with oxygen. Fish and other aquatic organisms are no exception. Through their gills, they are able to absorb oxygen that is dissolved (mixed) in the water.

Conclusion

Now for the conclusion: Is air all one gas, or is it a mixture of different gases? Use Q and A discussion to guide kids in reviewing the demonstrations and reaching the conclusion. They have seen evidence for three different gases: oxygen, nitrogen, and carbon dioxide. If they have gotten into fractions you may add the proportions as well: air is about 1/5 oxygen, about 4/5 nitrogen, and a relatively tiny amount of carbon dioxide.[14]

Keeping The Atmosphere Constant

Finally, some students may observe that if all burning of fuels and all animal respiration consume oxygen and release carbon dioxide, it follows that the atmosphere should be depleted of oxygen, and carbon dioxide should accumulate to the point of suffocating all fire and life. What prevents this from happening? If kids don't raise this question, I recommend letting it ride for a time until later reviews show that they have fully absorbed what has been presented thus far. Then bring it up and let them ponder the problem.

The revelation is green plants and photosynthesis. Using the energy of sunlight, green plants take in carbon dioxide from the air and construct it, along with water, into the tissues of their bodies. In turn, plants become food (fuel) for animals and humans. In photosynthesis, plants also release oxygen gas as a by-product. In short, green plants with

---

[14] You may note that this is not the total content of air. Additionally, there is almost always a certain amount of water vapor present. There may also be dust particles and various pollutants. Then, there are small amounts of other gases that that have no biological significance. But let's not overload children with too much detail. These may be added later as occasions warrant.

the energy of sunlight are what prevent us from all suffocating. While we use oxygen and give off carbon dioxide, green plants do the reverse; they use carbon dioxide and give off oxygen. The balance between plant and animal life on Earth keeps the balance between oxygen and carbon dioxide in the atmosphere relatively constant.

Part 4. Mixtures and Chemical Reactions

Return to the concept of a mixture. Review that a mixture means that two or more things are simply mixed or stirred together. Each ingredient remains the same despite being mixed with other things. Thus, we have seen that air is a mixture of oxygen, nitrogen, and carbon dioxide. This is to say, air is individual particles of oxygen, nitrogen, and carbon dioxide mixed together, but not affecting one another.

But sometimes when things are mixed together, they do "get into a fight," "tear each other apart," and reform into quite different particles. When this occurs, we call it a CHEMICAL REACTION. Mixing baking soda and vinegar was an example. Particles of baking soda and particles of vinegar actually come apart and form particles of carbon dioxide, which are different from either the baking soda or the vinegar. The fizzing was evidence of this CHEMICAL REACTION.

Burning is also an example of a CHEMICAL REACTION. In the candle flame, there is more going on than a simple mixing of oxygen and wax. Both wax and oxygen gas particles are broken apart and reform into particles of carbon dioxide (and water vapor, but omit this complexity for now). The heat and light of the flame are evidence of this chemical reaction.

In summary, intense fizzing or release of intense heat (and perhaps light) are indicators of a chemical reaction. When nothing particular occurs on mixing, beyond perhaps a change in color, it indicates no chemical reaction; you have only a mixture of the ingredients. (This distinction will not hold in every case, but it works as a starter.)

To be sure, this discussion is likely to boggle many of your children's minds. Don't worry about it, and don't dwell on it. It is sufficient, here, to simply "sow the seed." It will be revisited and reinforced on many occasions in future lessons, especially as we begin to describe substances in terms of atoms and molecules (grades 3-5).

Questions/Discussion/Activities to Review, Reinforce, Expand, and Assess Learning:

Make books illustrating that air is a mixture of gases, that burning consumes both fuel and oxygen, and/or other aspects of the lesson.

Draw children's attention to how their rate of breathing varies with activity. Have them explain why this is so. Discuss, why we, or any other animal, cannot live without breathing, or otherwise obtaining oxygen.

Discuss the crucial role of plants on Earth. Could animals exist without plants? Why not?

Bernard J. Nebel, Ph.D.

In small groups, pose and discuss questions such as:

> What is air? Is it all one kind of particle, or is it a mixture of different kinds of particles? What is the evidence for your answer? (How did you demonstrate the existence of each?)
>
> Why does a flame go out when a glass is inverted over it? What gas does burning depend on? If air is about one fifth oxygen, what gas makes up most of the remainder?
>
> How is burning related to our eating and breathing?
>
> Why does vigorous exercise cause us to breathe harder?
>
> How are eating and breathing connected?
>
> Could you survive in an atmosphere of carbon dioxide?
>
> How does the air you inhale differ from the air you exhale?
>
> Both burning and animal respiration consume oxygen and produce carbon dioxide. What would be the consequence if these were the only processes occurring on Earth? What prevents this consequence?
>
> Compare plant photosynthesis, animal respiration, and burning in terms of what gas is consumed and what gas is released as a by-product in each case.
>
> How does the amount of oxygen and carbon dioxide in the atmosphere remain relatively constant?
>
> When two things are mixed, what happens to the particles of each if they simply form a mixture? ... undergo a chemical reaction? How can you tell which occurs?

To Parents and Others Providing Support:

Read, review, and discuss all parts of the lesson with your children.

You may repeat the demonstrations, perhaps, with close supervision, allowing your children . Discuss what each demonstration shows.

While eating, ask: "What is it that food and breathing are providing for your body?"

After vigorous exercise, your child will be breathing hard. Ask, "Why should exercise make you breathe hard?"

When working in the kitchen and mixing ingredients, ask, "Is this just making a mixture or is it causing a chemical reaction."

Connections to other topics and follow up to higher levels:

Photosynthesis in more detail
Respiration in more detail
Ecology
Chemistry/chemical reactions

Re: National Science Education (NSE) Standards

This lesson is a steppingstone toward developing students' understanding and abilities aligned with NSE, K-4:

Unifying Concepts and Processes
• Evidence, models, and explanation

Content Standard A, Science as Inquiry
• Abilities necessary to do scientific inquiry
• Understanding about scientific inquiry

Content Standard B, Physical Science
• Properties of objects and materials
• Light, heat, electricity, electricity and magnetism

Content Standard C, Life Science
• Characteristics of organisms

Content Standard D, Earth and Space Science
• Properties of Earth materials

Content Standard F, Science in Personal and Social Perspectives
• Personal health

Books for Correlated Reading:

See also list under A-6 above.

Pipe, Jim. *Changing Materials: Fire and Ice.* Stargazer Books, 2005.

Tocce, Salvatore. *Oxygen* (True books), Children's Press, 2004.

Young, June. *Look How It Changes!* (Rookie Read-About Science). Children's Press, 2006.

Lesson A-8

Matter III: Evaporation and Condensation

Overview:

In this lesson students will learn to describe their experience of things drying (evaporation) and water droplets forming on a cold surface (condensation) in terms of water particles going into or coming out of the air. This will serve to both reinforce their understanding and appreciation of the particulate nature of matter and will lay the foundation for understanding one of the most important parameters of nature, the water cycle.

Time Required:

Part 1. Evaporation (activities plus interpretive discussion, 40-50 minutes)
Part 2. Condensation (activities plus interpretive discussion, 40-50 minutes)

Objectives: Through this exercise, students will be able to:

1. Understand and use the following words in their proper context: EVAPORATION, CONDENSATION, WATER VAPOR, HUMIDITY.

2. Describe EVAPORATION in terms of particles of water leaving the liquid state and entering the air in the gas state, namely, WATER VAPOR.

3. Present evidence that supports number 2 above.

4. Describe CONDENSATION in terms of particles of water in the gas state (water vapor) coming together to reform liquid on a cold surface.

5. Distinguish between dew and rain.

6. State how temperature influences evaporation. … condensation.

Required Background:

Lesson A-2, Solids, Liquids, and Gases
Lesson A-4, Matter I: Its Particulate Nature
Lesson A-6, Matter II: Air Pressure, Vacuums, and the Earth's Atmosphere

Materials:

> Part 1. Evaporation
> A covered container
> Two wet paper towels or similar "rags" of equal size
> Dinner plate
> Soda bottle

> Part 2. Condensation
> Ice
> A drinking glass or glass jar
> Salt

Teachable Moments:

Teachable moments will occur when hanging out wet clothing or other items to dry, using a laundry or hair dryer, seeing moisture condense on a cold glass, and on other occasions when evaporation or condensation is observed.

Methods and Procedures:

Part 1. Evaporation

On an occasion of children coming in from a rainy day and putting things up to dry (or you may create the situation by actually spilling water, mopping it up with a towel, and hanging the towel up to dry) pose the question, "Where does the water go as things dry?"

Some students may know the word EVAPORATE, and declare that the water evaporates. Answer, "Yes, but what does that mean?" Review as necessary that water is made up of particles (Lesson A-4). Where do the particles of water actually go? Many children will have the notion that they simply disappear. If this is the case, ask questions such as: When you can't find a certain toy or your hat, do you think it simply disappeared? Or, do you think you left it somewhere? If it simply disappeared, would you even bother looking for it? Let kids give examples of losing things and finding them somewhere.

Continue this Q and A discussion far enough to bring students to the generalization/conclusion that things don't just disappear or vanish. They go somewhere. Our task is to search and find them (or discover who has taken them). Emphasize that this is a basic truism for all of nature, science, and human society. Nothing just disappears; everything goes somewhere.

So, as wet things dry, where do the particles of water actually go? Ask leading questions, such as: Why do things dry faster when they are spread out and hung up? Why does a "wind" of warm air from a hairdryer make your hair dry faster? Will things dry if they are in a closed container? With such Q and A discussion, bring students to reason that particles of water must be going off into the air among the particles of air.

Continue this discussion guiding students to give evidence that supports the idea that water particles go off in the air? Evidence should include items such as: When we put things in a container or otherwise prevent contact with the air they stay wet. Spreading things out so that they are more exposure to the air speeds up the rate of drying. Likewise blowing air over them speeds up the rate of drying.

Students may elect to set up an experiment. Wet two paper towels or similar rags; place one on a line and place the other in a closed container. When the towel hanging out is dry, check the one in the container. It will be as wet as when you started. Similarly, you may pour a quarter cup of water on a dinner plate and the same amount into an empty soda bottle. Set the two side by side out of way. Kids may observe the two from time to time during the day. They will note the water on the dish diminishes first. Why?

Have children note and discuss how these results are consistent with water particles going into the air. The closed container or neck of the bottle prevents or impedes water particles from going into the air with the result that drying does not occur or occurs more slowly. If the particles simply disappeared, they should disappear just as readily from the closed container, should they not?

Introduce how we use the words EVAPORATE and EVAPORATION in talking about the above. Then, we refer to water particles in the gas state as WATER VAPOR. Add that other liquids such as alcohol and gasoline also evaporate. Thus, EVAPORATION refers to any substance going from a liquid to a gas or vapor state. In every case, it is a matter of particles breaking free from their loose attachment to other particles in the liquid and floating freely as a gas into the air. You may have children model this with an activity similar to the "solid-liquid-gas" game described in Lesson A-4. Here, children will move from the loose-hand-holds, which models the liquid state, to going off independently to model the vapor (gas) state. They should go from their "liquid formation" to the "vapor formation" one at a time to model the fact that evaporation occurs gradually, not all at once.

Students may elect to test whether various liquids will or won't evaporate. They may simply pour a small amount on a dish or paper towel; set it aside, and see if it dries.

Children may be surprised and even resistant to the concept that there can be water particles among the air particles but, indeed, there are. Students have probably heard the terms dry air, moist air, low humidity and high humidity. Explain that these words refer to the relative number of water particles (amount of water vapor) in the air. Explain that HIGH HUMDITY or moist air refers to relatively large amounts of water vapor in the air. Dry air or LOW HUMIDITY refers to relatively small amounts.

Lastly, point out that a final proof of water particles going into the air should be in observing that they will come out of the air again. Again, things don't disappear; they go somewhere. It follows that they can be gotten back.

Part 2. Condensation

Condensation is regularly observed as water droplets forming on a cold, ice-filled glass on a warm, humid day. (If you don't have such conditions you can add a couple tablespoons of salt to the glass filled with ice and water. The salt forces the ice to melt faster and, consequently, lowers the temperature still further. The colder temperature will cause condensation even in relatively low-humidity conditions.) Or, simply exhale your breath on a cold glass surface, as people do for cleaning their eyeglasses. Exhaled breath invariably has high humidity, since it has picked up water vapor from the moist surfaces of the lungs. Children love to fog windowpanes with their breath. You may also cite bathroom mirrors being covered with water droplets after a warm bath or shower, an experience children will be familiar with.

In any case, calling students attention to one or more of these instances, ask, "Where are the droplets of water coming from?" In the case of the glass of ice water, they may think the glass is leaking; but then, why doesn't a glass of room-temperature water leak also? And, if the glass is leaking, why doesn't the water level go down?

Have students consider that if water particles go into the air, as they concluded occurred in evaporation, shouldn't they be able to come out of the air again? Explain that this, indeed, is exactly what is happening. Water vapor, the particles of water in the air, are settling on the cool surface. As more and more settle and group together, they are seen as droplets of water on the cooler surface. We speak of this as water vapor CONDENSING.

Discuss their experiences of dew. As surfaces cool overnight, water vapor commonly condenses from the air producing dew. Frost is exactly the same idea with one difference. When condensation occurs at temperatures below freezing, the water particles form directly into ice crystals, which are seen as frost.

Some students may put evaporation and condensation together and derive the basic idea of the water cycle. Evaporation occurs from wet areas; then, high in the air, condensation occurs producing rain, or snow if temperatures are below freezing. Indeed, you may choose to sow this seed, but we are going to leave the bulk of this for another lesson.

Questions/Discussion/Activities To Review, Reinforce, Expand, And Assess Lesson:

Make books illustrating what water particles are doing in evaporation, in their vapor state in the air, and in condensation.

Facilitate students setting up additional experiments that will test how the rate of evaporation increases with air movement (wind from a fan), with increased temperature, or other variables as students may choose. In each case, guide students in reasoning how results may be interpreted as enhancing (or impeding) the movement of water particles from the liquid to the vapor state.

Have children play-act evaporation and condensation as describe.

In small groups, pose and discuss questions such as:

> Ask students to cite examples of evaporation and/or condensation that they have recently experienced. Ask them to describe what was happening to water particles in each instance? (Where were they going from? ...coming from?)
>
> How does temperature affect evaporation? ...condensation?
>
> What is meant by high humidity? ... low humidity? ...dry air?
>
> Why do we use hair dryers? ... clothes dryers? How/why do they speed evaporation?
>
> Why don't wet cloths dry when you leave them in a heap on the floor?
>
> If you wish to prevent something from drying out, what should you do?
>
> Does rain/snowfall have something to do with evaporation and condensation? How so?
>
> Why should moving air or wind increase the rate of evaporation?
>
> Why should heat increase the rate of evaporation?
>
> Rain/snow comes from high up. How do water particles, which evaporate at ground-level, get high up where condensation occurs.

To Parents and Others Providing Support:

Call your child's attention to instances of evaporation, such as wet pavements drying up, and ask/discuss: Where is the water going? What is the process called?

Do the same regarding instances of condensation, including seeing dew or frost.

Call you child's attention to factors affecting evaporation and condensation: temperature, air movement, humidity, and sunshine.

When you experience the bathroom mirror fogged up after a hot shower, have your kids explain it in terms of evaporation and condensation.

A simple activity is to hold a glass of ice 4 to 6 inches above a dishpan of hot water. You will notice water droplets forming on the glass. Have your kids explain it in terms of evaporation and condensation.

Connections to Other Topics and Follow-Up to Higher Levels:

The water cycle

Factors that cause precipitation

Why some regions are wet and others are dry, and how this determines the type ecosystem present: desert, grassland, forest.

Re: National Science Education (NSE) Standards

This lesson is a steppingstone toward developing students' understanding and abilities aligned with NSE, K-4:

Unifying Concepts and Processes
  • Evidence, models, and explanation

Content Standard B, Physical Science
  • Properties of objects and materials

Content Standard D, Earth and Space Science
  • Properties of Earth materials

Books for Correlated Reading

Bailey, Jacqui. *A Drop in the Ocean: The Story of Water* (Science Works). Picture Window Books, 2004.

Cobb, Vicki. *I Get Wet*. HarperCollins, 2002.

Curry, Don L and Gail Saunders-Smith. *The Water Cycle*. Capstone Press, 2000.

Dorros, Arthur. *Follow the Water From Brook to Ocean.* (Let's-Read-and-Find-Out Science, Stage 2). HarperTrophy, 1993.

Fowler, Allan. *It Could Still Be Water* (Rookie Read-About Science). Children's Press, 1993.
_____. *What Do You See in a Cloud?* (Rookie Read-About Science). Children's Press, 1996.

Frost, Helen, Numerous titles. Capstone, 2000.

Kerr, Kimberly. *The Raindrops' Adventure*. K. Kerr Press, 1999.

Morrison, Gordon. *A Drop of Water.* Houghton Mifflin, 2006.

Rau, Dana Meachen. *Fluffy, Flat, and Wet: A Book About Clouds* (Amazing Science). Picture Window Books, 2006.

Rauzon, Mark J. and Cynthia Overbeck Bix. *Water, Water Everywhere* (Reading Rainbow Book). Sierra Club Books for Children, 1995.

Robinson, Fay. *Where Do Puddles Go?* (Rookie Read-About Science). Children's Press, 1995.

Rosinsky, Natalie M. *Water: Up, Down, and All Around* (Amazing Science). Picture Window Books, 2003.

Sherman, Josepha. *Flakes and Flurries: A Book About Snow* (Amazing Science). Picture Window Books, 2004.

Bernard J. Nebel, Ph.D.

Simon, Seymour and Nicole Fauteux. *Let's Try It Out in the Water: Hands-On Early-Learning Science Activities.* Simon & Schuster, 2001.

Waldman, Neil. *The Snowflake: A Water Cycle Story.* Millbook Press, 2003.

Lesson A-9

Matter IV:
Dissolving, Solutions, and Crystallization

Overview:

    In this exercise, students will extend their understanding of the particulate nature of matter to interpreting how some substances dissolve and form solutions. Similarly, they will interpret the process of crystallization. Further, they will learn to distinguish solutions from mixtures. This will lay the foundation for comprehending basic concepts underlying chemistry, biology, and ecology.

Time Required:

    Part 1. Some Things Dissolve: Solutions and Mixtures (activity plus interpretive discussion, 45-60 minutes)
    Part 2. Soluble and Insoluble, (activity plus interpretive discussion, 30-40 minutes)
    Part 3. Crystallization (10-15 minutes to set up followed by periodic observations)

Objectives: Through this exercise, students will be able to:

1. Recognize and use the following words in their proper context: DISSOLVE, SOLUTION, POLLUTION, SOLUBLE, INSOLUBLE, CRYSTAL, CRYSTALLIZATION.

2. Explain the process of a substance dissolving in terms of its breaking up into fundamental particles.

3. Some substances dissolve, others don't. Give the terms that apply to each and the reasoning as to why this occurs.

4. Distinguish between a solution and a mixture.

5. Conduct an activity demonstrating crystallization of a substance from water solution, and explain the process in terms of the particles involved.

6. Cite an observation that demonstrates how evaporating water leaves impurities behind. Give the implications of this for water purification and the Earth's water cycle.

7. Recognize and describe the hazard of drinking water from untested sources.

8. Explain what is meant by polluted water.

Required Background:

Lesson A-4, Matter I: Its Particulate Nature
Lesson A-8, Matter III, Evaporation And Condensation
The concept of mixtures, which was addressed in Lesson A-7

Materials:

Clear glass or plastic tumblers with water
Packets of powdered beverage (optional)
Sugar
Salt
Flour or cornstarch
Spoon
Small jar with lid
Cooking oil
Detergent
Dark colored plates
Drinking straw
Paper towels

Teachable Moments:

Create an occasion where children are mixing a beverage powder into water to prepare a drink. Invite them to examine the granules more closely as they dissolve.

Methods and Procedures:

Part 1. Some Things Dissolve: Solutions and Mixtures

Invite children to make a drink for themselves in which they start with a beverage powder. Rather than just pouring the whole amount into a glass of water and stirring, however, ask them to add the powder just a tiny pinch at time to a glass of still water and watch carefully what happens to the particles as they fall through the water.

They will note that colored particles leave wavy trails of color and get smaller and may disappear they descend. If a pile of the granules ends up on the bottom they may see wavy lines of light and shadow come from it into the water. (This does require close observation. It can be easily done or repeated or repeated with just sugar granules. The powdered beverage is only to help focus their attention. The observations will be the same except for the color.)

Explain that the word we use to describe something apparently disappearing into water is DISSOLVE. The ___ has DISSOLVED. Move to Q and A discussion to bring students to interpret what is happening in terms of particles. Review as necessary that even a small granule of sugar, for example, is actually a clump of many individual sugar particles (Lesson A-4). The basic particles of a substance are too small to be seen. With this review, guide students to reason that DISSOLVING is a process of the substance coming apart into its basic particles, which then "swim" in among the particles of water. If some students claim that the granules just disappear, ask, "How is it that you can taste their presence even though you can't see them?" (SAFETY PRECAUTION: Stress to children that taste should NOT be used as a general test to determine if something is dissolved in water. Many dissolved substances can be harmful.)

Go on to explain that when something dissolves in water, the resulting combination of water and dissolved material is called a SOLUTION. Thus, the beverage they have prepared is a SOLUTION of the ingredients in water. As was describe in Lesson A-7, solutions are a special class of mixtures. The difference is that a mixture refers to any ingredients mixed together. As in vegetable soup, there may be chunks of the different ingredients; they are not broken up into their fundamental particles. Also, a mixture may refer to solids or gases mixed together, as well as liquids.

A SOLUTION refers specifically to a liquid and means that one or more ingredients are dissolved in the liquid; that is the ingredients are broken up into their fundamental particles which are too small to be seen with the unaided eye. Children may visualize the solution as particles of the dissolved material "swimming" among the particles of liquid. Most commonly, a solution refers to ingredients dissolved in water, but you can also have solutions of things dissolved in alcohol, oil, or other liquids.

Moving on, have students stir a teaspoon or so of flour (or cornstarch) into a glass of water and contrast the result with a sugar-water solution. They will observe that flour-in-water is cloudy, whereas the sugar-water is clear. Should we call the flour-in-water a mixture or a solution? Explain that another fact about solutions is that they always appear clear, although they may have color. The cloudiness says that the particles, although are too small to be seen individually, are not fully dissolved. Thus, flour-in-water remains a mixture; it is not a true solution. Students may distinguish various mixtures from solutions on this basis of clearness versus cloudiness.

Ask students about their favorite drinks: milk, sodas, fruit juices, and so on. Are they solutions or mixtures? The key is holding it up to the light and testing: Is it clear or cloudy? Kids usually get a kick out of recognizing that most drinks, including sodas, are really solutions of various things in water. Milk and orange juice with pulp are exceptions. Have children note that there may be a number of different things simultaneously dissolved in water. It is a solution nonetheless, just a more complex solution.

Guide children to note that ingredients in solution (dissolved in the liquid) may be gases or liquids, as well as solids. For example, the bubbles that come out of sodas are from dissolved gas (carbon dioxide). Under pressure, the gas remains dissolved; when the pressure is released, the gas starts coming out of the water. Fish and other aquatic organisms depend on dissolved oxygen (oxygen gas being dissolved in the water). Alcohol is a liquid that readily dissolves in water, thus various alcoholic beverages.

When water contains one or more materials that make it unfit for a given use, we speak of it as POLLUTED. It will bear stressing that there is no change in the fundamental water particles themselves. It is only that there are other harmful things "swimming" in among them. Thus, water can be PURIFIED by removing the polluting particles or otherwise rendering them harmless. There are a number of different techniques for doing this that will be addressed in another lesson.

It will be well to add a SAFETY lesson here. Point out that it is possible for harmful substances to be dissolved in water, and sometimes such substances are colorless, odorless, and tasteless. Thus, water may look clear but can still contain hazardous substances. In developed countries, at least, tap water is constantly tested and monitored so that it may be assumed safe unless instructed otherwise. However, they should never drink water from a stream, lake, or other untested/untreated source, even though it may look clear.

Part 2. Soluble and Insoluble

You may invite children to drop other things in water and test whether they dissolve. They will discover that plastic, metals, wood, stones, and many other things don't dissolve. Despite stirring, they will just sit there in water without change.

Ask questions such as: Why don't they dissolve? Is it because they are not made of tiny particles? Emphasize the conclusion from Lesson A-4; everything is made of particles. Things that don't dissolve are no exception. The answer is that their individual particles are too tightly attached together. Water particles can't get in between to separate and bring them into solution. You may introduce two additional words here: SOLUBLE and INSOLUBLE. If a substance will dissolve, it is said to be SOLUBLE. If it doesn't dissolve, it is said to be INSOLUBLE. Thus, sugar and salt are soluble; marbles and coins are insoluble.

You may have children prepare a chart with five columns: Material; soluble (dissolves); insoluble (doesn't dissolve); makes a solution; makes a mixture only. They can test materials of their choosing and fill out their charts accordingly. (Are your bodies soluble? We should hope not! It would be pretty silly, and quite a disaster, if you were!)

The behavior of oil and water is a special case that fascinates most children. Add some cooking oil to water in a jar and shake vigorously. At first it seems as if the oil is dissolving as it breaks up into tiny droplets dispersed in the water. But then, as the jar sits, the droplets rise and reform a layer of oil on top of the water. In short, oil is insoluble in water. It does not dissolve. Why not? The answer is that oil particles repel water particles. They refuse to "swim" in among them. They repel them much like the like poles of a magnet (Lesson A5). Thus, as the mixture of oil and water is left to sit, the oil particles join together again, and being lighter than water, they rise to the top.

But, add some liquid detergent to the oil and water—you will need an amount of detergent about equal to the amount of oil—and shake vigorously as before. Students will observe that now the oil stays mixed in the water, effectively dissolved. How so? Soap/detergent particles attract both water and fat/oil particles. This provides a means for fat/oil to dissolve in water. Kids may reason out how soap/detergent serves to clean their bodies and greasy dishes. (Some detergents are more effective than others in this demonstration. Thus, this activity may be extended to testing the effectiveness of different detergents.)

Part 3. Crystallization

Review with students how they may detect if one or more things are dissolved in water. They should note three ways: color, smell, or taste; but admonish them, again, that it is DANGEROUS TO TEST ANY UNKNOWN SOLUTION BY TASTE.

Tell students that here is another way. Have them pour some salt solution (a tablespoon or so of salt in a cup of water) into a dark-colored plate. (Dark blue plastic-coated "paper" plates work well; salt crystals show up conspicuously on the dark surface.) Set the dishes of salt water aside and observe periodically over the course of the day. Students will discover that as the water evaporates, salt crystals are left behind and remain on the dish. Introduce these new words in context: CRYSTALS have formed. The salt has CRYSTALLIZED. CRYSTALIZATION has occurred.

With Q and A discussion, guide students to interpret the result. Review with them their knowledge from Lesson A-8, that evaporation is a process of water particles leaving the liquid state and going into the air. They should reason that the salt particles did not evaporate; they remained behind and joined together to reform crystals.

Another way to observe this effect is to wrap a straw with a bit of paper towel and place it in a glass of salt solution such that the paper on the straw is in the solution, but also extends up and out of the glass. (The straw is only to give support to the paper to prevent it from flopping over the edge.) As evaporation occurs, students will observe salt crystals forming on the paper, especially at the top end.

Again, use Q and A discussion to guide children to reason out the result. Their conclusion should be to the effect that water with salt particles in solution creep up the paper. Near and at the end of paper, water particles leave and go into the air, i.e., evaporate, but salt particles remain behind. As more and more salt particles accumulate at the end of the paper, they join together and form crystals.

Water purification and the Earth's water cycle

Further, have children reason out what the above means for water purification. Water that evaporates consists of water particles alone—impurities have been left behind. Therefore, as the evaporated water particles come back together in condensation (Lesson A-8) they form pure water. Therefore, the water cycle of the Earth does not just recycle water through evaporation and condensation; it purifies it in the process.

Bernard J. Nebel, Ph.D.

Growing Crystals

Growing crystals is an activity that intrigues many children. When water evaporates very slowly in a still atmosphere, minerals in solution often form quite large crystals. Put some salt solution (as above) on a plate and cover it with another inverted plate. The inverted plate/cover impedes evaporation, but the slight crack between the rims allows it to proceed slowly. In making periodic observations, be very careful not to shake or jar the dish of salt solution. It may take two weeks or more, but large crystals usually form as water evaporates slowly.

Questions/Discussion/Activities To Review, Reinforce, Expand And Assess Learning:

Make books illustrating a substance dissolving. …crystallizing.

Set up an activity center where students can continue observing: dissolving, crystallization, and testing the solubility of materials as they choose. Have a chart at the center where students can record the solubility or insolubility of materials that they test. (Note that solubility is another means by which materials may be distinguished, Lesson A-5.)

As you and/or students mix paints and other things in the course of art projects, take the opportunity to discuss and contrast mixtures, solutions, dissolving, and crystallization.

As children wash up from projects, have children observe and note whether the "dirt" going into the water makes a solution or a mixture. Would the term "polluted" apply to the dirty wash water? How so?

In small groups, pose and discuss questions such as:

Without speaking, dissolve something in water and ask students to tell what is going on both naming the process and describing what is happening in terms of particles.
Have examples of solutions (clear) and mixtures (cloudy) and ask them to tell which are solutions and which are mixtures. How can they tell?
What is meant by a substance being soluble? …insoluble?
What are four ways in which you might detect if there are materials dissolved in water?
As water with salt in solution evaporates, what happens to the water? …the salt?
Open a soda and ask: Where are the bubbles coming from?
Cite various beverages and ask whether each is a solution or not. How can they tell?

Blood, the transport medium of the body, deserves special mention. Blood is largely water, but countless things are in solution (food particles, oxygen, carbon dioxide, hormones, etc.). They go into the blood at one location and come out at another. Other things, most notably red blood cells, which carrying oxygen and give blood it red color, are carried along in the current. Thus, blood is both a mixture and a solution.

To Parents and Others Providing Support:

Review the lesson, repeat activities, and discuss interpretations as needed.

In working about the kitchen with children, observe, ask, and discuss questions, such as: Is ___ dissolving or just making a mixture? Is ___ a solution or just a mixture? What is happening in terms of particles?

On seeing a "scum" left behind from evaporating water, ask: "Where did that scum come from?" Expect/review an answer to the effect that the scum is stuff that was dissolved/mixed in the water. It was left behind as the water evaporated. Yes, tap water usually contains calcium and perhaps other minerals in solution and will leave a scum as it evaporates.

Kids get a kick out of recognizing that their urine is a solution. It is actually a complex solution containing a wide assortment of waste products from the body. The waste products in urine are those filtered from blood by the kidneys.

See how large a salt crystal you can grow. You may experiment by adjusting the concentration of the starting solution and other factors. Similarly, kids can experiment growing crystals of other substances that will dissolve in water.

On visiting or seeing pictures of limestone caverns, discuss how they were formed. Namely, water seeping through the ground gradually dissolved the limestone, which is slightly soluble in water. Then, the stalactites, stalagmites, and other features form by the recrystallization of limestone. Water with limestone in solution seeps from above, evaporates in the cavern, and leaves a deposit of limestone behind.

If salinization of soil is a familiar problem in your region, it is worth discussing in terms of this lesson. Irrigation water contains traces of salts in solution. As water evaporates, these salts are left behind gradually making the soil too salty to grow crops.

Connections to Other Topics and Follow-Up to Higher Levels:

The Earth's hydrological (water) cycle: salt and fresh water
Water purification
Water pollution
Numerous aspects of basic chemistry
Numerous aspects of basic biology

Re: National Science Education (NSE) Standards

This lesson is a steppingstone toward developing students' understanding and abilities aligned with NSE, K-4:

Unifying Concepts and processes
  • Evidence, models, and explanation

Bernard J. Nebel, Ph.D.

Content Standard A, Science as Inquiry
- Abilities necessary to do scientific inquiry
- Understanding about scientific inquiry

Content Standard B, Physical Science
- Properties of objects and materials

Content Standard D, Earth and Space Science
- Properties of earth materials

Books for Correlated Reading:

Angliss, Sarah. *Matter and Materials* (Hands-On Science). Kingfisher, 2001.

Dayton, Connor. *Crystals*. PowerKids Press, 2007.

Frost, Helen. *Keeping Water Clean*. Capstone, 1999.

Harrison, David L. *Caves: Mysteries Beneath Our Feet*. Boyds Mills Press, 2001.

Kramer, Stephen P. *Caves* (Nature in Action). Carolrhoda Books, 1994.

Maki, Chu. *Snowflakes, Sugar and Salt: Crystals Up Close*. Lerner, 1993.

Selsam, Millicent E. *Greg's Microscope*. HarperTrophy, 1990.

Siebert, Diane. *Cave*. HarperCollins, 2000.

Stewart, Melissa. *Crystals*. Heinemann, 2002.

Wick, Walter. *A Drop of Water: A Book of Science and Wonder*. Scholastic, 1997.

Lesson A-10

Rocks, Minerals, Crystals, Dirt, and Soils

Overview:

The entirety of the Earth's crust is made up of various rocks and minerals. They are the source of all metals and countless other resources. Soils, the base for all crop production, are largely broken down rocks and minerals. But exactly what are rocks and minerals? In this lesson students will learn to distinguish rocks from minerals; they will gain understanding concerning the nature of crystals; they will discover the mineral nature of soil. Finally, they will gain an appreciation for the abundance of resources obtained from the Earth's crust. This lesson will provide a foundation for all further studies in the field of geology and other earth sciences.

Time Required:

    Part 1. Rocks and Minerals (activity plus interpretive discussion, 50-60 minutes)
    Part 2. Crystals (activity plus interpretive discussion, 50-60 minutes)
    Part 3. Dirt and Soil (activity plus interpretive discussion, 40-60 minutes)
    Part 4. Rocks, Minerals, Dirt, and Soil as Resources (periodic discussions, 10-15 min each, as desired)

Objectives: Through this exercise, students will be able to:

1. Describe what is meant by the EARTH'S CRUST.

2. Separate an assortment of rocks and minerals into those which are rocks and those which are minerals, and describe the basis for such separation.

3. Recognize and name certain common rocks and minerals that occur in their region.

4. Describe and/or model with blocks how chemical particles making up minerals pack together in certain ways.

5. Relate the uniform, repeating pattern of packing (4 above) to CRYSTALIZATION, i.e. the formation of crystals

6. Describe the connection between rocks, minerals and dirt; name and describe the

processes that convert rocks and minerals to dirt.

7. State the distinction between dirt and soil.

8. Name the ways in which rocks, minerals, and soil serve as basic resources for humans; particularly cite and discuss the importance of ores to technology and the importance of soils to agriculture.

Required Background:

Lessons A-4, Matter I: Its Particulate Nature
Lesson A-5, Distinguishing Materials
Lesson A-9, Matter IV: Dissolving, Solutions, and Crystallization
Lesson B-2, Distinguishing Living, Natural Nonliving, and Human-made Things

Materials:

Part 1. Rocks and Minerals
An assortment of various rocks and minerals. (You and your students may have collected these yourselves on various excursions, or sets may be purchased. For this part of the lesson, you will need pieces of granite, an igneous rock that shows a conspicuous, speckled, granular structure. (If necessary, ask for help from an Earth science teacher or other professional.) Save sedimentary rocks (sandstone, shale) for later (Lesson, D-8).
Pocket magnifiers

Part 2. Crystals
Dish of salt crystals that students have grown (Lesson A-9)
Photographs of various crystals and actual specimens so far as possible
Cube-shaped, toy building blocks; a set of at least eight equal blocks per student or group
Samples of mica
Shaker of table salt
Video of lava flowing from a volcano (optional)
Photographs of real snowflakes
Samples of various ores (Sets of ores may be purchased.)

Part 3. Dirt and Soil
Soil (Only a small handful will be plenty. Don't use commercial potting soil or compost, which is mostly organic matter, decaying plant material and/or manure. Better, arrange to have children collect a bit of soil/dirt from the schoolyard during a recess.)
Saucers with white, flat bottoms
Water
Pocket magnifiers
Samples of various ores (sets may be borrowed or purchased)

Part 4. Rocks, Minerals, Dirt and Soil as Resources
Nothing additional is required

Teachable Moments:

An excellent prelude to this lesson is a field trip along a streambed facilitating students collecting various stones as they choose. Alternatively, simply allow them to examine an assortment of rocks and minerals that you have obtained.

Methods and Procedures:

Part 1. Rocks and Minerals

Allow children to sort through and examine an assortment of rocks and minerals, as they will. Point out that we refer to these stones as ROCKS and MINERALS, but which is which? What is the distinction between a rock and a mineral?

Have students use a magnifier hand lens to examine a stone of granite, which has a conspicuous, granular, speckled appearance. (Recall that the most effective way to use a magnifierhand lens is to hold it very close to your eye steadying it with your hand against your cheek. Then bring the item to be examined up to the lens until it comes into focus; that will be almost to your nose.) Ask students to discern how many different kinds of material are present? Describe each.

In granite (by definition) students will observe three distinct materials: a clear to whitish glassy material (quartz); a light brown to black material that shows one smooth, shiny surface (mica); and a dull, usually pinkish material (feldspar). You may have students make a chart listing and describing characteristics of each.

Explain that each of the different materials in the stone is called a MINERAL. A MINERAL is a single material of uniform color, texture, luster, and structure. The stone they have been examining is made up of three distinct MINERALS as they have noted. Now, what is the difference between a rock and a mineral? Guide students in formulating an answer to the effect: A mineral is a single material; a ROCK is two or more minerals mixed together, as in the stone they have observed. Thus, that stone may properly be called a ROCK.

Impress upon students the commonness of minerals. Occurring alone or in combination with other minerals to form rocks, they make up the entirety of the Earth's crust—all the rock of mountains, continents, and what underlies oceans and other bodies of water as well. In short, they are what the solid portion of what the Earth is made of.[15]

---

[15] You may wish to add that various lines of evidence show that when the Earth was formed it was entirely molten (melted rock) because of very high temperatures. All the water was in the vapor state in the atmosphere. Now the outer portion has cooled and solidified to form the crust, which can be visualized as rather like the peel of an orange. Below the crust the rock is still molten (magma) and still occasionally erupts

You may add that the particular rock they have been looking at is called granite. It is the most common kind of rock underlying continents.

Challenge students to separate an assortment of stones into those that are rocks and those that are minerals. If the stone has an even color and luster throughout, it will be a mineral. If it has a specked, granular appearance indicating different sorts of materials, it will be a rock. Importantly, point out that in addition to the features already noted, each mineral has a distinct chemical composition, i.e., each mineral is made from specific kinds of chemicals arranged in a specific way.[16] This brings us to the topic of crystals.

Part 2. Crystals

Students will be familiar with the terms crystal and crystallization from Lesson A-9 where they observed the formation of salt crystals forming from evaporating salty water. The objective now is to help them understand the relationship between crystals and minerals. Skipping to the conclusion first, a CRYSTAL is a piece of mineral that exhibits a characteristic shape. Have children re-examine the salt crystals they grew previously (Lesson A-9) and describe their shape. They will note that they all are shaped like boxes or cubes, although of different sizes and many have probably grown together in various ways. Similarly, have students observe a few grains of table salt with a magnifier.hand lens. Again they will observe that each grain has the shape of a tiny cube.

Explain that when we say, "a crystal of table salt," we are referring to a specific single, cube-shaped piece, regardless of its size. However, when we say, "Table salt is a mineral," we are referring to the fact that table salt is solid, hard, natural, earth material with a specific chemical make up. Of course, salt also dissolves readily in water and, hence, is contained in seawater.[17] (Most minerals are essentially insoluble, or dissolve

---

through the surface—this is what volcanoes are—and is seen as lava flows. The pressures deep down toward the center of the Earth are so great that material, while extremely hot, is in a solid state. This forms what is called the CORE of the Earth. Molten magma exists between the core and the crust. The surface of the crust, in turn, has been modified by various physical and biological processes giving rise to soils and other kinds of rocks and minerals, but these are all topics for later lessons. (See later in this lesson and Lesson D-8.)

[16] You may seek the help of a local naturalist to help you and your students learn to recognize and identify the minerals and rocks that occur most commonly in your region. But, recognize that there are hundreds of different minerals. Therefore, unless children are so inclined, it is impractical to burden them with specific names at this time.

[17] Along the way in this discussion, there are likely to be questions regarding salt and other foodstuffs. Take time to make the distinction clear. Emphasize that salt is a mineral, a mineral called halite. It occurs naturally in the Earth's crust like other minerals, and dissolved in seawater. (Most other minerals are quite insoluble.) Chemically, it is sodium chloride. In contrast, food for every member of the animal

116

only slightly.)   In other words, "mineral" refers just to a natural, solid, hard earth material; it says nothing about amount, size, or shape. "Crystal," on the other hand, refers to a piece of a mineral with specific characteristics of shape, color, etc.

This should imply to students that crystals of different minerals will have different shapes, and indeed, this is the case. Show them photographs and actual specimens, so far as possible, of crystals of various minerals. Children are invariably fascinated by the intricate shapes that many crystals have. (Quarts crystals are renowned and readily available; likewise intriguing and available is mica, which flakes readily into thin, semitransparent, cellophane-like crystals.) Children are intrigued to note that some minerals and crystals have characteristics and shapes of such beauty that they are sought as gemstones. Yes, gemstones are minerals of certain kinds. (Unfortunately, the simple cube-shape and other properties of salt crystals provide no such value.)

Children are likely to ask why crystals of different minerals have such different shapes. Explain that it has to do with the different fundamental particles (chemical elements) involved; they differ in their size, their relative attraction for one another, and in other respects. Students might visualize this as follows: If they pack cubic blocks of the same size together, they will get the shape of a larger cube. You may demonstrate this with eight such blocks. If they pack tennis balls together, they will get a different shape. Thus, each mineral, as it crystallizes, will tend to form crystals with a unique shape. As more particles of the given mineral are present, they may form a larger crystal, but the same shape is maintained. Or, they may form many additional crystals, again, with the same shape.

Draw students' attention back to the piece of granite rock that they observed at the outset of this lesson. Why should it be comprised of separate granules, sometimes discernable as partial crystals of the three minerals noted? Explain that granite is an IGNEOUS rock, which means that it formed directly from the cooling of magma (molten rock). In the liquid/magma state, the particles of all the various chemical elements may be evenly mixed. As the magma gradually cools and solidifies, however, the particles of various chemicals tend to sort themselves out and pack together to form crystals of the different minerals.

Have children visualize the different chemical particles as cube-shaped blocks and tennis balls moving over, under and around one another as they would in the liquid state. As their motion slows and they begin to pack together into the solid state, as occurs on cooling, packing will work best if blocks and tennis balls pack together separately. This provides a general conceptual picture of what happens as magma slowly cools. As the fundamental particles have different size and relative attraction for one another, they tend to sort themselves into arrangements that pack most efficiently. The result is grains of

---

kingdom, including humans, is and must be preexisting plant or animal tissue as they learned in Lessons B-3 and B-5. Therefore, salt, being a mineral, cannot be considered a food. We only use it to flavor other foodstuffs, and we also need very small amounts of sodium in our bodies. There is no way that we could eat salt as a food. Indeed, in more than the tiny amounts we use for favoring, salt would be toxic.

different minerals packed together forming rock such as in the piece of granite that was observed at the outset.

(Of course, there is much more to the story. Igneous rocks weather into small grains, are carried about, deposited, and may eventually be compressed into rock again, i.e., sedimentary rocks such as sandstones and shale. These topics are addressed further in Lesson D-8.)

Part 3. Dirt and Soil

We have said that the Earth's crust is entirely comprised of various rocks and minerals. Yet, children experience that much of the material beneath their feet is dirt/soil. How does this fit into the picture?

Have students take a bit of soil, no bigger than a pea, place it on a white saucer, dribble on a bit of water, smoosh the soil in the water, and carefully tip the saucer back and forth. (The effect you want is to see individual specks of soil spreading out against the white surface of the saucer. Experiment with this yourself ahead of time so that you can guide your students in getting the proper proportions of soil and water. If there is too much soil, the effect will be simply muddy water. Again, the best effect is achieved with just a crumb of soil and no more than about 50 drops of water. Don't use commercial potting soil or compost. It is mostly organic matter and will yield only organic "gobs.")

Direct students to examine the specks with a magnifier.hand lens. Observation is facilitated if the dishes are set aside for a time so that the water evaporates and the dishes are observed in the dry state. (An even better experience is obtainable by letting water with soil specks dry on a glass slide and observing them under a microscope if one is available.) What do they see? What about size of the specks? … their color? … their shape?

Soils may differ markedly from location to location. Therefore, the details of what is observed will vary according to the particular sample of soil you have chosen. Differences notwithstanding, there are some generalities. Students will note that there is a variety of sizes ranging from sand grains down through specks that are barely visible, even with a magnifier.hand lens. Point out again, however, that even these tiniest specks are made up of many hundreds of the basic chemical particles we have discussed in Lessons A-4 to A-9. (This is why we are using the word "specks" rather than "particles.") Likewise, students will note that specks are a variety of different colors ranging from black through various shades of brown to some that are clear glass-like. Where shapes can be seen, students will observe that they look like pieces of minerals. Indeed, that is exactly what they are. In topsoil, students may also observe various fragments of partially decayed vegetation, commonly referred to as organic matter or humus. We will hold this for a later lesson (Lesson B-12).

Focusing on the mineral specks, explain that there are a number of forces in nature that cause the gradual breakdown of rocks and minerals into smaller and smaller fragments. The natural forces include freezing and thawing; physical abrasion by water, ice, and/or wind; rocks being tumbled in a stream, and chemical attack by acids: both a slight natural acidity of rainwater and acids produced by plants. The total of all these forces is termed WEATHERING.

In short, rocks and the minerals they contain are gradually WEATHERED into tiny fragments. Then, we refer to any sizable mass of such fragments as dirt. Said the other way around, dirt amounts to a mass of the bits and pieces of weathered rocks and minerals. There may be any number of different minerals present in dirt, depending on the particular rocks and minerals that were the source. Dirt may also include various amounts of decaying organic matter (decaying plant matter and/or manure).

When it comes to growing plants or crops, however, we refer to the dirt base as SOIL. Then all sorts of additional considerations, including how much organic matter is present, come into play in assessing how well suited the soil is for supporting the growth of plants. These will be considered at greater length in Lesson B-12.

Ores

Show and let students examine samples of various ores. By their appearance, students will classify them readily as rocks and/or minerals. Explain that the chemical particles contained in some minerals may be those of iron, copper, aluminum, or other metallic elements. There may be just one kind of metal in a given mineral, or sometimes there are two or more. In turn, the metal-containing mineral is frequently mixed with other minerals in the same rock.

The term ORE refers to any natural, earth material that is mined and processed to obtain a desired metal. For example, iron ore is any sort of rock that is mined and smelted to obtain iron. Copper ore is any rock that is mined and processed to produce copper, and so on.[18] How far you wish to go with where given ores are found and the actual mining and smelting operations will be your option. Videos describing such operations are available. In any case, do impress upon students that rocks, minerals, and dirt/soil provide a vast abundance of resources.

Part 4. Rocks, Minerals, Dirt and Soil as Resources

Review and emphasize the basic point that humans make nothing starting with nothing (Lesson B-2). To make anything requires starting with certain materials. The starting place or resources for all materials is either living things or natural non-living things. Here we have been looking at the natural non-living category, at least the solid portion of that category, which also includes water and gases of the atmosphere.

---

[18] More exactly, the definition of an ore has an economic aspect. It is: An ore is any natural material that can profitably be mined and processed to obtain the desired product. In turn this depends on the value of the product and the costs of mining and processing. Thus, a non-ore rock that contains very small amounts of a desired material may become an ore as the price of the desired material increases and/or technological advances lower the costs of mining and processing.

Bernard J. Nebel, Ph.D.

After this review, use Q and A discussion to consider the tremendous abundance of resources that come from the crust of the Earth. You may name or point to particular things and have kids reflect on where the resources for making it come from. The general categories that come to light should include:

Rock is used directly as building material for walls and monuments.

Sand and gravel is used directly in making cement for pavement, structures, and cinder block.

Clay (the smallest of dirt/soil particles) is used in making bricks, pottery, and ceramic items of all sorts.

Metals all come from ores (minerals of one kind or another).

Glass is from another mineral.

Uranium, the key resource for nuclear power and weaponry, is from a mineral.

Oil (crude) and coal are sometimes classified with mineral resources since they are obtained from the earth, but they are actually derived from ancient plant and animal material.

Plastics, in turn are mostly derived from crude oil.

Growth of land plants, and hence, all land animals, depend on soil itself and certain elements from minerals being present in the soil as will be explored further in Lesson B-12.

Questions/Discussion/Activities To Review, Reinforce, Expand, and Assess Learning:

Make books illustrating the physical structure of rocks, minerals, crystals.

Encourage students to bring in samples of rocks, minerals, or crystals they may have found or acquired, and give show-and-tells during the preclass session.

Set up an activity center where students can continue to examine and learn to identify common rocks and minerals of your region.

Discuss where and how various ores or other minerals are obtained, particularly those in your region.

Engage students in discussion regarding the resources that come from rocks and minerals.

For students who wish, facilitate their making and labeling collections of rocks and minerals found in your region.

Show a video illustrating the processes involved in producing a metal from its respective ore.

Discuss history in terms of how advances in civilization hinged on learning to obtain and use the resources contained in rocks and minerals.

In small groups, pose and discuss questions such as:

Show students examples of a rock, a mineral, and a crystal. Ask them to tell which is which. How can they tell? What are the distinguishing attributes?

What distinguishes a mineral from an ore?

What is the distinction between a mineral and a crystal?

Why do crystals of different minerals have distinctive shapes?

What is dirt?

What is soil?

Where do metals come from? What processes are involved in producing them?

What materials/resources might you obtain from rocks and/or minerals?

As you pursue the chemical nature of minerals further, you and your students will observe that nearly every mineral is made up of at least two chemical elements. For example table salt, which has the mineral name, *halite*, is made of sodium and chlorine particles (atoms) giving it the chemical name, *sodium chloride*. The mineral *quartz* is *silicon dioxide* (silica + oxygen). There are some exceptions to this "rule." Gold, where it occurs at all, may be found in nearly pure grains or nuggets; hence the practice of panning for gold. Sulfur, silver, and a few other elements are also occasionally found by themselves. Still most minerals, including those containing metals, contain two or more elements.

Why should this be the case? It is because the atoms of most elements may acquire an electric charge, either positive or negative. Like the poles of a magnet, like charges repel; unlike charges (+ and -) attract. In a mineral, one of the elements is positive; the other is negative. The attraction between positive and negative is what draws them to pack together into a solid as described. Again, they pack together according to size, but also, they must pack together in a way that balances the electric charge, i.e., the total positive charges must equal the total negative charges.

Let students handle and compare an ore and its respective metal, a sample of the mineral hematite (iron ore) with some iron, large nails for example. The differences between the ore and the metal are many and conspicuous. Instruct students that the process of obtaining a metal from its mineral ore involves a process called smelting. Smelting involves a chemical reaction that removes electric charges from the metal atoms. When this occurs, the metal atoms in the mineral "melt" together to form the metal. Iron comes out of the mineral *hematite*, for example.

We generally think of minerals as being strictly earth materials, but parts of our bodies are mineral as well. Ask students, "Can you guess what parts those are?" Bones and teeth! Bones and teeth are largely made of a mineral, calcium phosphate. This is why bones and teeth are hard. In the growth of bones/teeth, however, more is involved than the simple crystallization of calcium phosphate. The crystallization occurs along a protein matrix that controls the resulting size and shape. Exactly how this works is indeed awesome and not entirely understood.

Also, a number of marine organisms, particularly clams and snails, make shells of calcium carbonate in a similar manner. As the animals die, this shell material is deposited and over millions of years may build up into very thick layers that we now experience and may quarry as limestone.

Bernard J. Nebel, Ph.D.

To Parents and Others Providing Support:

On any outing, facilitate children picking up stones and pebbles and examining them with a magnifier.hand lens. Help them identify whether a given stone is a rock or mineral. Discuss how they can tell from their observations.

Facilitate their making a collection of rocks and minerals as they desire. Help them, or find someone to help them, in identification.

Make a point of viewing rock and mineral collections in museums and nature centers.

Take advantage of a snowfall to view snowflakes with a magnifier. Flakes landing on a furry or fuzzy coat will be supported so that they may be viewed individually. Just guard against melting the flake with your breath as you view it. Likewise, view frost on a windowpane.

Connections to Other Topics and Follow-Up to Higher Levels:

Lesson D-8, "Rocks and Fossils" and Lesson B-12, "Plants, Soil, and Water" should be given in close proximity and integrated.
All areas of geology
Production of metals
Non-renewable resources

Re: National Science Education (NSE) Standards

This lesson is a steppingstone toward developing students' understanding and abilities aligned with NSE, K-4:

Unifying Concepts and Processes
  • Systems, order, and organization
  • Evidence, models, and explanation

Content Standard A, Science as Inquiry
  • Abilities necessary to do scientific inquiry
  • Understanding about scientific inquiry

Content Standard B, Physical Science
  • Properties of objects and materials

Content Standard D, Earth and Space Science
  • Properties of Earth materials

Content Standard F, Science in Personal and Social Perspectives
  • Types of resources

Content Standard G, History and Nature of Science

• Science as a human endeavor

Books for Correlated Reading:

Bailey, Jacqui. *Cracking Up: A Story About Erosion* (Science Works). Picture Window Books, 2006.

Bingham, Caroline. *Rocks and Minerals* (Eye Wonder). DK, 2004.

National Geographic. *My First Pocket Guide: Rocks and Minerals.* National Geographic, 2001.

Prager, Ellen. *Sand* (Jump Into Science). National Geographic Children's Books, 2006.

Riley, Joelle. *Erosion* (Early Bird Earth Science). Lerner, 2007.

Rosinsky, Natalie M. *Dirt: The Scoop on Soil* (Amazing Science). Picture Window Books, 2003.
_____. *Rocks: Hard, Soft, Smooth and Rough* (Amazing Science). Picture Window Books, 2003.

Stewart, Melissa. *Down to Earth* (Investigate Science). Compass Point Books, 2004.
_____. *Minerals.* Heinemann, 2002.
_____.*Soil.* Heinemann, 2002.

Tomecek, Steve. *Dirt* (Jump Into Science). National Geographic Children's Books, 2002.

Walker, Sally M. *Minerals* (Early Bird Earth Science). Lerner, 2006.
_____. *Rocks* (Early Bird Earth Science). Lerner, 2006.
_____. *Soil* (Early Bird Earth Science). Lerner, 2006.

Thread B

Life Science

Pursue Threads A, C, and D in Tandem

For a flowchart of the lessons and an overview of concepts presented see pages 8-13.

(Lesson B-1, see A/B-1, page 41)

Lesson B-2

Distinguishing Living, Natural Nonliving, and Human-Made Things

Overview:

This exercise will stimulate and nurture young children's natural interest in exploring and learning about things in their surroundings. They will exercise organizational skills by learning to assign any item to one of three categories: a) living or biological, b) natural nonliving, and c) human-made. In the process, they will learn to recognize the key attributes that distinguish these categories. This lesson will be the foundation for launching numerous ongoing studies that may well lead all the way to rewarding, life-long avocations, and professional careers.

Time Required:

Part 1. Collecting Critters and Things (Out of school time as necessary to solicit the support of parents and others; short excursions with children around the schoolyard and to local parks as administrative rules/requirements permit)

Part 2. Beginning to Classify Things (game/activity, 30-50 minutes, probably on more than one occasion)

Part 3. How We Make Separations (interpretive discussions, 5-10 minutes on several occasions as necessary)

Part 4. Keeping The Lesson Ongoing

Objectives: Through this exercise, students will be able to:

1. Correctly assign any item to one of the following three categories:
   a) living or biological things
   b) natural nonliving things
   c) human-made things

2. Recognize, name, and describe key attributes that enable the above distinctions.

3. Recognize and use the following words in their proper context: BIOLOGICAL, RESOURCES, RAW MATERIALS, CONSERVATION.

4. Recognize that stories and videos commonly give plants, animals, and nonliving things attributes that they don't have.

Required Background:

Lesson A/B-1, Organizing Things Into Categories
Lesson A-2, Solids, Liquids, and Gases

Materials:

Part 1. Collecting Critters and Things

This exercise hinges on children bringing in shopping bags of items representing the three categories that will be described. Ideally, each child will bring in her/his own bag of items. To facilitate this happening, you will need to contact and solicit the support of parents/caregivers. Appended to the end of this lesson, you will find a suggested letter introducing yourself, asking for their support, and giving specific instructions regarding this lesson. This letter may serve as the beginning of ongoing communications. Note that engaging the active support of parents/caregivers will be of inestimable assistance and value to say nothing of being the greatest step you can take toward the National Science Education Standard of "Develop communities of science learners (Teaching Standard E).

A list of instructions concerning "Dos and Don'ts Regarding Outings and Collecting" is likewise included at the end of this lesson. This information should also be conveyed to parents/caregivers as well as considered for your own use. If you conduct a fieldtrip yourself, each student/group should be equipped with:

shopping bag
miscellaneous unbreakable jars or containers for collecting things
magnifierhand lens (optional)
tongs/gloves
3 x 5 cards and pencil

Part 2. Beginning to Classify Things
Three cardboard boxes (medium size) labeled:
     a) living or biological things
     b) natural, nonliving things
     c) human-made things
Alternatively, you can just mark and label areas on the floor.

Teachable Moments:

On the appointed day children will come in with their bags of critters and things eager to see what they will do with them. Proceed! In the case of conducting your own field collecting trip with children, move into this exercise immediately on returning to the classroom.

Methods and Procedures:

Part 1. Collecting Critters and Things

The crux of this exercise is to have children sort an assortment of critters and things into categories as described below. Consider that "critter" refers to every kind, shape, and size of living organism: plant, animal, insect, worm, etc., domestic or wild; and "things" refer to *every thing* that is not living, natural things such as rocks, water, and air, and all human-made things as well.

To keep this exercise child-centered, I strongly advise your instructing children to bring in a shopping bag of items that they have collected themselves. Of course, this will require some cooperation and aid from parents/caregivers. Therefore, communication with parents/caregivers will be required. You may use the suggested "Letter to Parents/Caregivers" at the end of this exercise to facilitate this process. While burdensome at first, this will be an invaluable first step in bringing parents/caregivers into the community of learners, and will aid immeasurably in fostering the advancement of your students.

In inviting parents/caregivers to facilitate their children in collecting and bringing critters to class for further observation, identification, and study, issues of SAFETY and respect for living things and nature must be addressed. You will find instructions regarding "Dos and Don'ts" also appended at the end of this exercise. You may communicate this information to students and their parents/caregivers as circumstances warrant.

Part 2. Beginning to Classify Things

Have children sit with their bags of items in a circle around three boxes, or areas of the floor labeled:

a) LIVING OR BIOLOGICAL THINGS
b) NATURAL NONLIVING THINGS
c) HUMAN-MADE THINGS

(There may be children who forgot to bring their bag of items. Allow them to sit in the circle and participate in discussion, but, sorry, they will not have anything to place in the boxes. Also, the allotted time will probably expire before everyone gets to place all their items. Finally, there will be additional points of discussion that will/should be brought up, as described in the next section. Therefore, plan on pursuing this activity on consecutive days. This will give another chance to those students who forgot to bring their bags. It will be unlikely that they will forget again.)

In turn, each child will choose one of the items from their bag, say what it is, and place it in the appropriate box. Don't start with a lot of explanation. Let children exercise their own judgment. Even kindergarteners have had enough experience so that, for many items, they will intuitively make the correct placement. But, along the way, there will be questions and hesitation. As these arise, further explanation/discussion will be in order.

Here is the essence of the further explanation that may be added as questions arise.

Living or Biological Things. This category includes actual living things such as a pet mouse, an insect, fresh leaves, and so on. Note cards reporting sightings of wildlife and wildlife photographs may be included as representing the real thing. It also includes things that are not actively living now, but are conspicuously from a living thing: shells, pinecones, a piece of bark, a bone, a feather, etc. Introduce and define the word "biological." The word BIOLOGICAL refers to any "critter" that is currently living, was once living, or is a part of what was once a living thing.

Natural Nonliving Things. This category includes all non-biological things that are not human-made, e.g., rocks and stones, dirt, sand, and clay. Very importantly, it also includes air and water. Young students readily collect stones, but they will invariably overlook air and water as also being major parts of the nonliving environment. They are simply taken for granted. Nevertheless, their collection of stones, which has their attention, provides an entry point for prompting with a line of questioning such as: What nonliving things did we leave out of our collection? Well, what would the world be like without water? Would there be rain? Snow? Streams or rivers? Lakes? Oceans? Would we be able to live? Would plants be able to live? Would anything be able to live?

The same sort of questions may be asked concerning air. Thus, students can be brought to conclude that air and water are parts of the nonliving environment that are equally as important as the stone, rock, and dirt that make up the bulk of the Earth's mass.

Don't miss the opportunity to relate this portion of the discussion to Lesson A-2, "Solids, Liquids, and Gases."

Human-Made Things Or Materials. This category includes metals, plastic, paper, ceramics, rubber, and things made from these materials, as well as things made from wood or other biological material.

Questions Regarding Making Placements

As we have said, children will intuitively make correct placements, especially as they understand the above definitions. However, there are points that may still cause confusion. The following are points that commonly arise. Use these as opportunities for expanding children's understanding.

Are plants living? Students may be disinclined to place a leaf, an insect, and a shell together in the same category with their pet mouse that is obviously a living thing. Therefore, grouping such things leads to discussion concerning: What are the attributes of living things? How do we distinguish living/biological things from natural nonliving things, such as stones?

Have kids recall their experience of seeing plants grow, flower, and form fruits with seeds. In short, plants grow and reproduce, two primary aspects of living things. Thus, have kids recognize that there are two major groups of living things, PLANTS and ANIMALS. In turn, there are several major categories of animals. However, save getting into a prolonged discussion regarding the distinction between plants and animals and different categories of animals for later lessons (Lessons B-3 and B-7).

In doing this activity on a subsequent occasion, you may use two boxes for "Living or Biological," one for plant material and the other for animal material.

Where should very simple human-made things go? Confusion and hesitation commonly occurs when it comes to items such as a simple wooden block. Should it be placed in the biological category—it is a biological material—or should it be in the human-made category because it has been shaped by human hands?

The point to bring out and emphasize is that we humans make nothing starting from nothing. To make anything, we must start with some material(s). Introduce the terms RAW MATERIALS and RESOURCES. The starting materials for making things are called RAW MATERIALS or RESOURCES. Without exception, the raw material or resource for making an item comes from either the "Biological" category or the "Natural Nonliving" category. Allow kids to ponder this concept and try to think of any exceptions. Use Q and A discussion to bring them to recognize that any exception that they might propose is false.

Discuss whatever examples may come up, but recognize that this will be the tip of an iceberg. Raw materials and how we make things from them is a fascinating study that can be pursued to any length and depth including life careers. It is probably most evident that all wood and paper products come from the wood of trees. The raw material for nearly all plastics is crude oil or natural gas, which are derived from ancient life on Earth. The raw material for all metals is one or another mineral ore.

In some cases, the human input is great: the mining, smelting of minerals to extract the metals, fabricating parts from those metals, and finally assembling the parts to make the finished product. In other cases, the human input is quite minor, just shaping a piece of wood, for example.

In pursuing this activity, have children say words to the effect that the raw material for the wooden block was a living tree, but it was cut and shaped by humans. Therefore, I will place it in the box for human-made things. Similarly, in the case of a polished stone, they might say that the raw material or resource was a natural stone, but since it was polished by humans it should go in the box of human-made things.

Along the way in this discussion, introduce the concept of CONSERVATION. Emphasize to children that ALL resources are more or less limited in quantity, and invariably they cost money. Therefore, we all should try use as little material as we can in making something or doing a job. The watch words should always be DO NOT WASTE MATERIALS and EXERCISE CONSERVATION.

Where should foodstuffs go? Foodstuffs are a category that also deserves special attention. Whole fruits and vegetables are clearly from living things and belong in the biological category. But, what about highly processed foods such a breakfast cereals, bread, and pasta? From the above reasoning, it would be logical to put them into the human-made category. Here, I find it best to keep the emphasis on the fact that all foodstuffs are derived from living plants or animals. There is no way that humans can make food without starting with living, growing plants and animals. Everything we eat is biological plant or animal material. There is no way we can make palatable food from plastic, for example.

Hence, I keep all foodstuffs in the biological category. Ensuing discussion and lessons may delve into how we process wheat into bread, obtain sugar from sugarcane, and all other aspects of the agriculture and food industries. (Table salt is the one exception; it is a mineral.)

Part 3. How We Make Separations

As children demonstrate proficiency in making proper placements for all or at least most of their items, shift discussion by asking, "How can/do you tell that _____ is a living or biological thing, a natural nonliving, or a human-made thing?" As we have said, children's experience is such that they will intuitively make correct placements for the most part. Still, it is instructive to bring out the actual factors on which such intuitive judgments are based.

There is no way that such discourse can be scripted. It will follow the course that it does. Also, don't try to cover all the points in one lesson; it may be addressed in bits and pieces. Thus, take the following as general points to be brought out in Q and A discussion regarding each category as occasions arise.

Regarding Living or Biological Things. In our own schooling, we have probably learned that the attributes of living things are growth, reproduction, need for air, water, food/nutrients, metabolism, cells, and other such features. That is all very well, but note that none of these attributes are actually observed in many of the biological things we have collected, much less in pictures. Additionally, children will invariably cite movement or response to being touched or poked as an attribute of living things. Have them note that this is typical of the animals alone, not plants. Likewise, we don't see this in biological items such as pinecones or shells, even less in pictures.

Yet, when we see a picture of a living thing, or even a biological item, we will usually identify it as such even though we have not observed its growth, reproduction, need for food/water, etc. Children (and adults) will say that it simply looks like a living/biological thing. So, What are the "looks" that distinguish living/biological things? There are four aspects that you should bring out and discuss with children:

a) Orientation: Living things have a distinct orientation. Have kids note that there is a head end and tail end, a top side and bottom side, a right-side-up and an upside-down. Contrast this with a stone, for example. Note that it has no particular orientation—unless that is, we paint eyes on it.

b) Symmetry: Living things have a pronounced symmetry, a right side being a close mirror image of the left side (bilateral symmetry) or similar parts arranged in a circular pattern like a starfish or a pinecone (radial symmetry). Natural nonliving things, such as stones, rarely show such symmetry.

Taken together, orientation and symmetry provide a form and shape that is unique to living things. Water, clay, and air, in contrast have no particular form or shape (unless it is imposed by humans, but then it becomes human-made, as discussed above).

c) Fine structure and detail: Close examination of living/biological things reveals an abundance of fine detail and structure, almost always in an even, repeating pattern. Consider for example: the veins of a leaf, the ridges on a clamshell, the segments of a feather, the scales of a fish, the segments of an earthworm, the gills of a mushroom, and countess other things.

If at all possible, have enough pocket magnifiers available for students to share and instruct them in their use—most use them improperly. The best view is attained when the magnifier lens is held very close to the eye—so close that the eyelashes are essentially brushing the lens. Then bring the item to be viewed up to the magnifier lens until it comes into focus. It will be only about one to three inches from the lens.

The additional clarity of fine detail that can be seen with a 10x or even a 5x magnifier will be amazing. It is this fine structure and detail that enables us to distinguish a biological product from a human-made product, wood from plastic for example. Of course, fine, detailed, repeating structure is even more glaring as one uses a microscope for examination of cellular structure, but it is advisable to save this until children are more mature (third grade and up). Younger children rarely have the fine motor skills needed for using a microscope effectively and avoiding its damage

In conclusion, our intuitive recognition of living things is not on the basis of growth, reproduction, and metabolism, although these things may be inferred; it is on seeing ORIENTATION, SYMMETRY AND FINE STRUCTURE.

The fourth attribute that often comes into children's experience and is worthy of discussion is:

d) Tenuous Quality: Simply put, this means that living things are prone to die. At least, we need to maintain certain conditions (temperature, light, etc.) and provide water and food/nutrients to maintain a living thing. Even then, any individual will progress through a cycle of aging and death. Life is only maintained through a cycle of reproduction and growth. (See Lesson B-4, Life Cycles.)

Regarding Human-Made Things. Human-made things frequently have orientation and symmetry, but they lack the other attributes of living things. Children get a kick out of noting that things such as bottle caps don't have babies. Nor do human-made things have the fine structure and detail of living things, and they lack the tenuous quality. Further, they are often made from materials such as metals, paper, and plastic that do not occur in nature except as litter. Also, they are made for a specific function, use, or purpose. Finally, they may require fuel or power to run and they may break, but this is conspicuously different from the needs of living things.

Regarding Natural Nonliving Things. Naturally occurring nonliving things are rocks and minerals of all sorts and sizes and, very importantly, air and water. Again, we distinguish them from living things simply by the absence of those features of living things noted above. We note that they are naturally occurring by their existence in nature apart from human activities.

Stories and Reality

Throughout this activity, there will be many instances and opportunities where you may contrast fiction and reality. Use them! From countless stories, videos, and their own imaginations, many children will have the actual belief that nonliving things, especially their stuffed animals, actually do talk, move, and think in one way or another. One may hate to burst their bubbles of fantasy, but it is important to do so—gently and tactfully, of course. A major step in learning at this age (K to grade 1) is to have kids separate fantasy from reality and to distinguish what is really observed from what is only imagined or pretended.

Part 4. Keeping The Lesson Ongoing

You will not be able to cover nor will students be able to absorb all the points of this lesson in a single sitting. Therefore, follow-up sessions are in order. This is true not only for this lesson but for all other lessons as well. We strongly recommend establishing a routine of devoting the first 30 minutes or so of each day (or at least two to three times a week) to such follow-up sessions. The numerous advantages of such preclass sessions were described in Chapter 1 (page 19).

In this case, invite students to continue bringing in individual items that are of particular interest to them, give a show-and-tell, and classify it. Activities and/or discussion appropriate to future lessons will be evident as you come to them. In every case, these preclass sessions serve for extended review and reinforcement, and they provide opportunities to identify and clear up misunderstandings. Even more, they create a setting and teachable moments that you can use to fill in additional points and progress seamlessly into further lessons.

Cleaning up

Cleaning up at the conclusion of this activity may be a challenge. Be sure that actual living critters are given respect and properly cared for or taken back to be released where they were found. Still, there will be much miscellany. What to do with it, of course, will be your option, but recognize that it may be the "raw material" for a number of future exercises concerning further classification and discussion concerning attributes of each of the three categories. Therefore, it will be to your advantage to create a means of storing items in respective categories. Cardboard boxes neatly labeled and stacked may suffice.

Questions/Discussion/Activities to Review, Reinforce, Expand, and Assess Learning:

Continue to encourage parents and other caregivers to pursue the habit of taking children on outings. A discussion entitled "Dos and Don'ts Regarding Outings and Collecting" is included at the end of this lesson. Encourage children to bring in their "discoveries" and do a brief show-and-tell at the beginning of class. At the end of their show-and-tell, have them assign it to its proper category. Note that categories may extend to higher levels of classification as students advance.

Make one or more books depicting the three categories discussed and the points we use to distinguish them.

In small groups pose and discuss questions such as:

Give each child a particular item and ask her/him which category it should be put in and why. What are the features that indicate that it belongs in that category?
What do all items in category _____ have in common?
Can humans make anything starting from nothing? What is the resource/raw material that was used in making _____?

Extend discussion into where/how various resources are obtained and how given things are made as students' questions and interest dictate.

To Parents and Others Providing Support:

You will note that the essence of this lesson hinges on your guiding your child to put together a shopping bag of miscellaneous stuff covering the three categories discussed. More specific instructions are in the letter that you should have received separately (also appended at the end of this lesson).

Nothing stimulates, enhances, and preserves a child's interest and joy in learning more than giving them freedom to explore their natural surroundings, discover whatever they do, and make collections as they may wish (within the limits of safety, of course). Much of your children's ongoing learning will hinge on experiences of what they see and observe in their natural surroundings. Therefore, most important will be to establish the habit (try for at least once a week) of taking your kids on "outings" and facilitating their exploration and potential collection of "critters and things." A discussion of "Dos and Don'ts Regarding Outings and Collecting" is included at the end of this lesson.

In addition to seeing what is present at the moment, weekly outings should call children's attention to seasonal changes in plant and animal behaviors, weather conditions, changing day-length, etc., as well as learning to identify both common plants and animals including birds, certain insects, and so on (see Lessons B-4, B-4A and B-4B). Keep in contact with your children's teachers so that you can keep the focus of outings in harmony with lessons being addressed.

For the current lesson, ask, discuss, and review with your children the category to which various items belong and how they can tell that it belongs there.

Connections to Other Topics and Follow-Up to Higher Levels:

This lesson sets the stage for any number of ongoing studies concerning all of the three categories we have discussed. "Living or Biological Things" leads naturally into all further studies concerning life sciences. "Natural Nonliving Things" leads to further study of rocks, minerals, and geology in general. "Human-Made Things" is a launching point for further inquiry into how things are made from available resources.

Weather: One can't take nature walks without also noting the weather. This can foster questions regarding what causes the weather to change from day to day and also seasonally. Numerous additional lessons may follow accordingly.

Re: National Science Education (NSE) Standards

This lesson will be instrumental in promoting Teaching Standard E: "developing communities of science learners" and the following content standards:

Unifying Concepts and Processes
• Systems, order, and organization
• Form and function

Content Standard A, Science as Inquiry
• Abilities necessary to do scientific inquiry

Content Standard B, Physical Science
• Properties of objects and materials

Content Standard C, Life Science
• Characteristics of organisms

Content Standard E, Science and Technology
• Abilities to distinguish between natural objects and objects made by humans

Content Standard F, Science in Personal and Social Perspectives
• Types of resources

Books for Correlated Reading:

Kalman, Bobbie. What Is a Living Thing? (Science of Living Things). Crabtree, 1998.

Royston, Angela. Living and Nonliving (My World of Science). Heinemann, 2003.
_____. Materials (My World of Science). Heinemann, 2001.
_____. Natural and Man-made (My World of Science). Heinemann, 2004.

Twist, Clint. Materials (Check It Out!). Bearport, 2006.

Zoehfeld, Kathleen Weidner. What's Alive? (Let's-Read-and-Find-Out Science, Stage 1). HarperTrophy, 1995.

Suggested Letter to Parents/Caregivers Asking for Their Help

Date
Name
Phone/email/other

Hello _____(parent(s)/support givers of _____ _____(student),

I am _____ (your name), _____ (child's name) teacher for the coming school year. I am looking forward with great eagerness and enthusiasm to this role in aiding _____'s advancement of knowledge and skills.

Most of all, for _____'s benefit, I am looking forward to developing a working partnership with you and having your support in this progression. Therefore, this will be the first of many communications in which I will ask for specific aid in helping _____(child) prepare for and master particular lessons.

Specifically, on _____(day, date) we are planning to do Lesson B-2: Distinguishing Living, Natural Nonliving, and Human-Made Things. If you have not received your e-mail or printed copy of this lesson, the idea is as follows:

Students will assign various items to one of three categories:

1. Living or Biological—any living thing or part of a once-living thing. A picture of a plant/animal or the name of an animal that was seen may suffice as the living thing

2. Natural Nonliving—sand, stones, rocks of any sort, water, air

3. Human-Made Things or Materials—metals, plastic, paper, ceramics, rubber, and things made from these materials, as well as things made from wood or other biological material

Bernard J. Nebel, Ph.D.

As well as the basic skill of organizing things into categories, students will be learning attributes that distinguish these categories, the concept of resources and making things from various resources, and other pertinent things.

For this exercise, I am asking children to bring in a shopping bag of 5 to 10 items of their choosing on _____ (day). I am asking for your assistance in seeing that they do so. Please do not select items and put them in a bag for your child. An important aspect of children's learning is for them to look around the house and out-of-doors and make selections of their choosing. Only guide your child toward items that will give them one or more items representing each category. Still, let them make the decision to include it in their bag or not even if this results in items of all one category.

Please contact me if you have questions. Otherwise, I will look forward to seeing _____ on _____ (day) with his/her bag of items.

Sincerely yours,

_____

Dos and Don'ts Regarding Outings and Collecting

1. Don't underestimate the importance of this activity.

On a par with reading, writing, and math, development of observational, analytical, and organizational skills are paramount in elementary education. Every field of science and many other professions as well is constructed on a foundation of observing, analyzing, and interpreting—whether experimentation is involved or not.

Like every skill, these traits are only developed through practice and use. There is no better place to develop and use these skills than in the identification and classification of plants and animals. It requires discerning observation to note similarities and differences, and analyzing how these similarities and differences lend to organization of living things into taxonomic groupings.

Just as importantly, virtually every field of life science—zoology, botany, ecology, and all specialized studies of particular groups such as birds, reptiles, insects, etc.—begin with gaining a familiarity with the diversity of species present in natural surroundings.

Likewise, efforts toward conservation grow out of a respect and love for nature that begins with becoming acquainted with the abundance and diversity of living things in our surroundings. On a purely aesthetic level, becoming able to recognize common plants and animals of your region is, to me, rather like getting to know and feel at home with neighbors versus forever being in a world of strangers. The use of plants and animals in art and design should not escape notice.

There is no better place to start this development than in exploring the out-of-doors and finding, considering, and examining whatever is there to be seen. Happily, it is greatly aided by the fact that young children have a strong natural urge to explore their surroundings if given opportunities to do so. Whatever is found in explorations provides the pieces for building knowledge and skills that may lead all the way to and through professional careers.

In short, in collecting, examining, and studying critters, children are learning much more than the names of species. They are developing a repertoire of understanding and experience that they can draw on throughout their lives in their own creative endeavors, as well as simply coming to appreciate the intricacy and wonder of creation.

In traditional school settings, opportunities to take children into the out-of-doors are conspicuously limited. Therefore, it is most important to communicate this need to parents/caregivers and gain their support in facilitating this as far as possible. (Note that this is entirely in keeping with the National Science Education Teaching Standard E: Develop a Community of Science Learners.)

2. Don't make this activity more complex than it needs to be.

When we ask parents to take their children on outings where they can explore nature, we are not implying that they need to take their children into "wilderness" areas. Simple ambles about the yard, the neighborhood, or to local parks may well suffice, particularly at the early stages. Such outings may well be combined with walking to a friend's house or running errands. At the very least, there will be various trees and smaller plants, flowers, birds, insects, small mammals such as squirrels, and many other living things that may be encountered and noted along the way. Of course, if there are opportunities for more extensive forays into natural surroundings, so much the better. Here again, the nature study activity can be added onto a picnic, camping, or other sort of trip.

3. Do address issues of safety.

All regions have their particular poisonous plants, biting/stinging insects, and perhaps other hazardous "critters." Do not let these deter you from taking children on outings, but take appropriate precautions. Call on a local naturalist to help you learn to recognize potential hazards, poison ivy for example. Likewise learn what precautions to take to avoid contact or getting bitten, and actions to take if contact, bites, or stings do occur. (Note that this also is in keeping with the National Science Education Teaching Standard E: Develop a Community of Science Learners.)

Consider that the hazards present in nature are not unlike those confronted in the kitchen: cuts, burns, spoiled food, etc. To reduce these risks by staying out of the kitchen would be foolish. One learns to recognize the hazards and handle implements and food accordingly. The same principle applies to nature; learn the hazards and act according. To stay away from nature because of hazards would be the same as never entering a kitchen, but worse; you would cut yourself and your children off from even more learning and experience. Even, the hazardous "critters" are a part of nature and may well be studied in their own right.

4. Do promote respect for living things.

There is fantastic knowledge, appreciation, and joy to be gained by observing and studying both living and nonliving things seen in nature. Keep the emphasis on this fact. Many kids seem naturally inclined to chase, throw stones at, and otherwise harass birds, squirrels, and other wildlife. Discourage this sort of activity and repeatedly emphasize that our role should be to simply watch and see how animals behave and carry on their lives in their natural habitats, how a bird builds its nest or a squirrel gathers and stores nuts, for example. Every plant or animal has its role to play in the scheme of nature and that role should be observed and respected. (Of course, there are exceptions when it comes to weeds, mosquitoes, and other pests, but even here there is much to be learned by observation and study. Issues of pest control and wildlife management are topics for future study.)

5. Consider whether or not to catch and hold "critters" in captivity.

There is hardly anything that excites young children more than capturing a "critter," and there is much to be gained in doing so. Many of the finer features, essential for identification, can only be seen by holding the sample close and using a magnifier. Then, the keeping of a living thing provides an unparalleled opportunity for observing its behavior and discussing its needs, i.e., air, water, proper food, and suitable environment. Have students consider the specific environment (wet, dry, light, dark, physical surroundings, etc.) in which they found the animal and recognize that these conditions must be duplicated in the captive situation as far as possible. Keeping any animal without regard to its needs, it may be added, is a form of cruelty, and usually leads to the early demise of the organism. Students will note that the needs for other living things are not basically different from their own needs.

How can we suggest both respecting nature by leaving plants and animals alone in their natural habitats and catching and bringing them to class for further study? The two are compatible by noting the following. For larger mammals and birds there is no real possibility of children catching them. Just watch them so far as interest is maintained and let the name and perhaps other notes written on a card suffice as part of the collection. But what about a toad, turtle, salamander, insect, snail, or other such critter that can be readily caught and do not present a danger?

Emphasize to children that collecting them is going against our general rule of respect and leaving them in their natural habitat. Therefore catching "critters" simply for the sake of catching/collecting them should be discouraged. In parks it is also generally illegal. However, it is worth picking up a critter to examine it more closely, if it can be done without injuring it in the process or hazarding a bite or sting. Instruction and guidance regarding safe but gentle handling is in order. After examination, it may be returned to its "home," or there is the further decision to keep it for class sharing. In deciding to keep it for class sharing, there are a number of additional considerations.

First and foremost, emphasize to children that in order for a "critter" to survive without injury, the conditions of its natural environment must be duplicated and maintained to a high degree. Being able to provide suitable food is perhaps most evident, but maintaining suitable temperature, light, moisture/dryness and materials for shelter may also be critical. In view of this:

- Do we have a container in which it can be kept and transported without injury?
- When we get it home, do we have a cage or container in which we can provide its needs for a few days?
- Can we return it back to where we found it in a few days?
- Is the critter in question abundant (e.g., insects, snails, spiders, tadpoles), so that, if our care fails, we are not going to be depleting the population significantly?

If the answer to each of these questions is positive, capture of the critter may proceed. If not, the surrogate collection of an identification card in the bag should substitute.

Even if conditions and materials for keeping a critter in captivity are present. there are several reasons for not keeping captured animals such as turtles, frogs, snakes, or small mammals for more than a few days. First, even with the best of intentions, one often falls short of providing for all of the animal's needs and it ends up suffering and dying. Second, even when maintained in proper conditions, an animal kept in captivity is removed from its native breeding population, and as far as propagating its population in the wild is concerned, it is the same as if it had been killed. Many species of wild plants and animals from around the world are being threatened with extinction because of being taken and sold for the exotic pet market. Thus, making permanent pets of animals taken from the wild should be discouraged. After everyone has had the opportunity to observe and study it, an animal should be returned to where it was found. A card with its name, where and when it was found, and any further descriptions can be kept as a permanent collection.

Finally, putting a time limit on how long an animal may be kept forces students to make the best of it. They can make their observations and gain the desired knowledge, experience, and memory in that time. Beyond that time, boredom and forgetfulness set in, and the animal ends up suffering and often dying. Acceptable pets are those that are raised specifically to be pets: dogs, cats, gerbils, guinea pigs, canaries, etc., assuming responsible care is maintained.

Zoos may seem to violate these rules. Point out that large, modern zoos are now making every attempt to maintain wild animals so that they will reproduce. Indeed, having some wild animals reproduce in the safety of zoo environments may be the only way of protecting them from extinction.

There are exceptions to the general rule that living things should be returned to their natural environment within a few days. For example, a large caterpillar found migrating across the sidewalk or up a tree in autumn is probably on its way to spin a cocoon and/or pupate. Put it in a large jar with a twig on which it can climb and watch it periodically. If it begins to spin a cocoon, it is worth stopping anything else and allowing children to

Bernard J. Nebel, Ph.D.

watch this process. Continuing, keep the cocoon in a clear jar, covered with onion bag mesh for good air circulation, and children may also observe the emergence of a moth. Everyone I have talked to who witnessed this event as a child still has the memory fresh in their minds. Is there any other lesson we might conceive of that children will remember more vividly for the rest of their lives?

The same kind of fascination and lifelong memory can be produced by watching a spider spin a web and seeing what happens as a fly blunders into that web (catch a fly and throw it into the web); witnessing a praying mantis capture and feed on a grasshopper (put a grasshopper into a small cage with the praying mantis); watching tadpoles develop over the course of several weeks; and other miracles of nature.

6. Consider whether or not to make a collection.

The educational benefit of making and keeping collections is great. First, there is the exercise of those observational and organizational skills that go into making collections. Then the collections provide everyone with an immediate overview of the members of the group and their diversity. They provide ready material for further observation and analysis. They may be used for review and assessment. Collections representing every living thing found have been made and are carefully stored in various museums and universities for such study proposes. Of course, making wide collections of specimens in an elementary school setting is neither possible nor practical. Also, every child collecting everything they came across could be highly damaging to nature—hardly in keeping with respecting nature. There are two areas, however, where the benefits of making collections may be attained with little work or expense and without impacting nature: leaf collections and insect collections.

Leaf collections

Trees and shrubs larger than small saplings are not harmed by taking a few leaves or the end of a twig bearing a few leaves. (The twig with leaves is desirable because placement of the leaves on the twig and buds are sometimes necessary in identification.) Place them in a plastic bag that may be sealed to maintain moisture. They may be kept this way until you are ready to press and dry them—up to a day or longer if they are refrigerated. Identification may be conducted before or after pressing.

For pressing and drying, spread out the leaves or twig with leaves between pages of newspaper interspersed with layers of corrugated cardboard to allow air circulation. Add weight to the top of the stack and leave them for several days to dry. The amount of drying time required will depend on the humidity of the environment.

After pressing and drying, leaves may be fastened to a sheet of typing paper (or larger poster paper if you desire) with tape or paste, labeled with name, location where found, and date. There may well be leaves that you are unable to identify. Press and mount these along with the others; only put a question mark in the place of the name. Along the way, you will likely find someone who can help you identify these unknowns.

You can easily store the mounted leaves in file folders on an open shelf if a cabinet is not available and use them as desired for review in identification, classification, assessment, and so on. Art projects should not be overlooked.

Learning to recognize common trees and shrubs by their leaves through making a leaf collection should be an ongoing project. It will be natural for students not to repeatedly collect leaves that they learn to recognize, but they should remain on the outlook for species they don't recognize, identify these, and add them to their collection(s).

Insect collections

Insects are of prime economic importance, as certain ones are the most troublesome of agricultural and garden pests. Others are significant disease vectors. Some are beneficial in eating other insects; others are valuable in pollination. All have ecological significance, as they are a major food source for many birds, reptiles, amphibians, and others. This is to say that the study of insects (and related organisms) has all kinds of practical importance and can lead into any number of fascinating careers. Inspiration for art projects should not go unnoticed.

In addition to the educational benefits already mentioned, insect collections are fantastic for illustrating how living creatures fall naturally into subcategories of orders, families, genera, and species. Like leaves, they are very easy to collect, preserve, and mount, and with the exception of a few rare butterflies and moths in certain regions, their collection will not negatively impact nature. Finally, guided to examine insects with a magnifier, I have not found a youngster who is not fascinated by the exquisite detail and intricacies of antenna, jointed appendages, veined wings, compound eyes, and other features revealed.

A butterfly net is useful in capturing insects. To kill them, they may be placed in any suitably sized jar or container with a tight fitting lid and a cotton ball wetted with rubbing alcohol. Larger insects are mounted by inserting an insect pin through the thorax (middle section). (You will have to purchase pins but they are not expensive.) Tiny insects, if you get into these, are mounted by placing the insect in a bit of transparent glue on the tip of a tiny "arrowhead" of paper. The pin is previously inserted through the wider part of the arrowhead.

Clear plastic clamshell containers, which are used for salads and other produce in grocery stores, are great for display. Cut a piece of corrugated cardboard to fit the bottom of the container, insert the pins with insects into the cardboard and add labels with names and other information. Additional insects may be added to collections and they may be rearranged as studies proceed. A primary observation will be noting that insects are distinguished by having six-legs and three distinct body sections: head, thorax, and abdomen. Then, variations on this basic design lead to their being placed in different orders: beetles, flies, grasshoppers, butterflies and moths, etc., each with many families, genera, and species. Identification of species will often be impossible, but learning to recognizing the major orders is relatively easy. For information regarding starting an insect collection, google "starting an insect collection."

7. Do keep the activity child-centered.

It is all too easy to make the discovery and collection of items into an adult dominated exercise. Come look at this; Don't bother with that; that's yucky. Isn't this pretty, and so on. Note how such statements will strike the child as constant commands ordering them to do or not do. Many children become highly resistant to such commanding and may become turned off to the entire process. Keep in mind that it is the child's own natural inclination to explore and discover that we want to nurture. Therefore, allow, follow, and facilitate them in doing their own exploration (watching out for their safety, of course), and take an interest in whatever it is that they find. In addition to more conspicuous things, allow and facilitate their poking and scratching in the ground among dead leaves or under decaying logs. Equip them with a small trowel and gloves if you feel the need to do so. There are all kinds of critters to be found there: worms, beetles, crickets, spiders, millipedes, etc. Your words should be to the effect of: What are you seeing/finding? Show me. Tell me about it. Let them make their own judgment concerning its beauty or ugliness.

Of course, you will want to draw their attention to things you see, but try to do this in a manner that makes you a role model rather than a "commander." "Oh, wow, I see/found something that I think is interesting. Do you want to see it too?" The distinction may seem trivial, but to the child's mind, it is vast.

8. Safety!!

Do Not have food or drinks out at or during the time you and/or children are handling animals, and/or other natural materials. Do wash hands thoroughly after handling animals and other natural materials.

Lesson B-3

Distinguishing Between Plants And Animals:
The Plant and Animal Kingdoms

Overview:

In the previous lesson (B-2), students learned that both plants and animals are living things, despite their conspicuous differences in appearance and habits. Here, students will learn that such differences lead scientists to divide living things into two major groups, the PLANT KINGDOM and the ANIMAL KINGDOM and how (most) living things can be assigned to one or the other of these groups. (The assignment of fungi, protozoa, and bacteria to additional kingdoms will be the subject of future lessons.) Furthermore, students will learn that the most basic distinction between plants and animals lies in how the organism obtains its source of energy: plants from sunlight and animals from food. This understanding is a cornerstone for all of biology and all of ecology. It is also central to integrating basic concepts of physics, chemistry, and biology.

Time required:

Part 1. The Plant and Animal Kingdoms (game/activity, 40-50 minutes)
Part 2. The Distinction Between Plants and Animals (Q and A discussion, 30-50 minutes)

Objectives: Through this exercise, students will be able to:

1. Assign any living thing or biological item to one of two categories, plant or animal.

2. Explain how all living things can be/are divided into two categories known as the PLANT KINGDOM and the ANIMAL KINGDOM. (There are three other kingdoms (fungi, bacteria, and protozoa) but save these for another lesson.)

3. State the fundamental feature that distinguishes members of the plant kingdom from those of the animal kingdom, namely:
   a) tell how all living things require a source of energy
   b) tell how plants get their energy from sunlight
   c) tell how animals get their energy, as well as essential nutrients, from food

4. Point out features of plants that enable them to get energy from sunlight.

5. Point out features of animals that enable them to find and capture food.

Required Background:

Lesson A/B-1, Organizing Things Into Categories
Lesson B-2, Distinguishing Living, Natural Nonliving, and Human-Made Things
Lesson C-1, Concepts of Energy I: Energy Is What Makes Things Go
Lesson C-3, Concepts of Energy II: Kinetic and Potential Energy

Materials:

Two medium-sized cardboard boxes labeled Plant Kingdom and Animal Kingdom
Cards with the names of a wide variety of familiar living things, including both plants and animals
Pictures of different plants and animals to supplement the above
Actual plants and animals and/or biological materials to supplement the above insofar as it is convenient and practical
Well-dried peelings of oranges, potatoes, or other vegetables
A bit of dirt
Cards with certain animal characteristics written on one side; similar cards with plant characteristics
A small package of "plant food"

Teachable Moments:

Part 1. The Plant and Animal Kingdoms. This activity dovetails with and builds on Lesson B-2, "Distinguishing Living, Natural Nonliving, and Human-Made Things." Invite children to play this "game" but with a new slant. Assign all their living or biological material to the plant kingdom or the animal kingdom.

Part 2. The Distinction Between Plants and Animals. Review and invite students to ponder that plants and animals are both living things; yet they seem so different in every respect. Why should this be so? What is the "secret" behind this "mystery"? Let's unravel this mystery.

This lesson may also be the focus on a nature walk or a discussion while viewing pictures of animals in their natural surroundings, which will invariably include plants as well.

Methods and Procedures:

Part 1. The Plant and Animal Kingdoms

Have children sit in a circle around two boxes labeled PLANT KINGDOM and ANIMAL KINGDOM. Each child should have an assortment of items representing living or biological things, both plant and animal. These items may be cards with names of familiar living things, pictures of living things, and/or actual specimens that they have collected. In turn, each student will take one of their items, state what it is, and place it in one of the two boxes.

From everyday language, children will have gained the concept that the word ANIMAL refers to horses, bears, elephants, lions, and other MAMMALS. They think of other "critters" as just what they are: birds, fish, insects, spiders, clams, frogs, worms or whatever. Similarly, children will likely have the notion that the word PLANT refers only to house and garden plants. They will likely think of trees, grass, moss, etc. as simply what they are.

Consequently, along the way you will have to discuss and emphasize that we (scientists) divide living/biological things into two major categories: plants and animals. These two major categories of living things are referred to as the PLANT KINGDOM and the ANIMAL KINGDOM. (To be sure, there are three additional kingdoms for fungi, bacteria, and protozoa. However, these are best saved for later lessons because the factors delineating them require more understanding concerning cell structure and metabolism. If representative "critters" should come up at all, they can temporarily be assigned to plant or animal kingdoms as they were until the middle of the 1900s.)

Pursue this game/discussion until students gain the understanding that all birds, frogs, turtles, insects, crabs, worms, clams, etc. are members of the ANIMAL KINGDOM and can correctly be called animals, along with horses, bears, and other mammals. Hasten to add that the animal kingdom is, in turn, divided into the various kinds of animals, as will be addressed in Lesson B-7. For now, however, we wish to emphasize that they are all animals, that is, members of the animal kingdom. Similarly, guide students to understand that trees, grass, moss, ferns, seaweed, etc., are members of the Plant Kingdom and are properly referred to as plants along with familiar house and garden plants. As with the animal kingdom, the plant kingdom will be divided into various categories of plants.

Where we humans fit may be an issue of special discussion. Clearly, we are living things, and considering our nature, we logically fit in the animal kingdom. Have students cite the many characteristics we share with other animals, mammals in particular. To be sure, we are very special animals, with unique attributes and abilities, but in terms of classification, we are animals nevertheless.

Finally, emphasize that making categories, the plant and animal kingdoms or any others, is a human construct. In nature, we find whatever we find. Sorting them into categories is a means toward observing similarities and differences and aiding our understanding of them (review Lesson A/B-1).

Students will demonstrate their mastery of this concept by becoming able to sort their living/biological items into plant and animal kingdoms without hesitation. As this occurs, bring their attention to the following question/problem:

Plants and animals are both living things. Yet, we can readily separate them because they are so different. Let's consider the differences and see if we can condense all of them into one special, key difference.

Bernard J. Nebel, Ph.D.

Part 2. The Distinction Between Plants and Animals

Invite students to cite and discuss characteristics of animals and plants that enable their placement in one kingdom or the other. Allow this discussion to go as it will, listing characteristics mentioned in two columns on the board: Plant Characteristics and Animal Characteristics. As this discussion winds down, challenge students of identify one single trait that all animals have, but which no plants have. In other words, find one single trait that will distinguish all animals from all plants.

Some students will point at eyes and other sense organs. Point out that worms, clams, and a number of other members of the animal kingdom don't have eyes. Movement will be another candidate. Point out that not all animals move around. Oysters, for example, stay put, and so on down the list. You can find counter examples that eliminate characteristics from being the key candidate until it comes to this, and this may not have made it onto the list. It concerns where the organism (plant or animal) obtains its source of energy.

Have students recall from Lesson C-1 that everything needs a source of energy to make it go, work, move, or change. Review this as necessary and emphasize that living things are no exception to this rule. Living things also need a source of energy to live, grow, move, and otherwise function.

Where do we and all other members of the animal kingdom obtain energy? Guide children to recall that the answer is FOOD. Food is fuel, the source of energy (as well as body-building nutrients) for all members of the animal kingdom. Conversely, we can assign all organisms (living things) that eat and consume food as their source of energy to the animal kingdom.

Further, have children ponder what food is. What is it that we or other animals eat? Have children cite items and generate a list on the board. The list will be lengthy and diverse, and students will undoubtedly bring up the fact that various animals eat different things. With Q and A discussion, however, guide children to recognize that in every case food is preexisting plant or animal material (the living or biological category). (Earthworms, if they should be mentioned, are not an exception. They actually feed on dead and decaying plant material in or on the soil surface.)

In summary, what can we say about all members of the ANIMAL KINGDOM? They are all living things that get their energy from where? By eating something. And that something is what? It is always plant and/or animal material.

It follows that all animals have some form of feeding behavior, and usually, a conspicuous means of moving from place to place to find and capture more food. Looking at the distinctive mouth parts, feeding behaviors, and means of locomotion of various types of animals is a fascinating follow-up study and will be pursued in Lesson B-5.

148

Turning to Plants

Conduct similar Q and A discussion with reference to plants. What is the source of energy for plants? Where/how do they obtain energy? Many students will have the notion that plants feed on soil for sustenance. Two lines of discussion/reasoning will dispel this misconception.

First, bring students to reason: If a large tree had been feeding on soil for its sustenance, we should find it growing in a hole where it has eaten up the soil under it, should we not? Students will reflect that this is not actually the case. They may even check to make sure.

Second, have students reflect on the fact that an animal's food is always other plant and/or animal material. Such matter has considerable potential energy/fuel value, whereas soil does not. With due precautions, conduct a demonstration of this. Dried orange or other peelings burn readily, showing that they have high potential energy (Lesson C-3). Dried soil or stones, on the other hand, will not burn, showing that their energy value is nil.

Eliminating soil as a source of energy, let students ponder the question some more. If necessary, provide the hints by asking, "Is light a form of energy?" (With knowledge from Lesson C-1, students should answer yes.) "Are plants designed to capture light energy?" If there is hesitation, ask, "Why is it that nearly all plants have broad, flat, green leaves?" Those that don't still have green or deeply colored surfaces. Further, what happens to plants if they don't get light? Can you keep houseplants in a dark closet and expect them to survive? The conclusion that students should derive is that plants are able to use light as the energy for "making them grow." They do get certain nutrients from the soil, but the mass of such nutrients is trivial in comparison to the mass of the plant.

Most of a plant's mass is made from water and a component of the air, carbon dioxide. Importantly, review and emphasize again that light is NOT converted into plant material. Light is only the energy source that runs the chemical "machinery" of photosynthesis. That machinery makes sugar from carbon dioxide and water. Review the similarity with making cookies. Electricity is not turned into cookie batter; it only runs the mixer that churns the ingredients into batter.

(There may be some confusion in that plant fertilizers are frequently referred to and even labeled as "plant food." Show and let students feel such "plant food." They will find that it has a granular texture much like salt. In fact, it belongs to the natural nonliving category of things, whereas animal food must be from the living biological category. Finally, you may demonstrate trying to burn plant food. It will not ignite, showing its nil energy potential. Thus, "plant food" is not a source of energy; it is simply a mixture of a few natural chemicals that plants require, as we require calcium for our bones and iron for our blood.)

Finally, have students observe that plants, by conducting photosynthesis, make their own bodies (roots, stems, leaves) from water, a component from the air, and a few nutrients from the soil. Then animals consume plants as food. In short, PLANTS MAKE FOOD; ANIMALS CONSUME FOOD. With the understanding behind this statement, students can memorize it and it becomes a pillar supporting many subsequent studies.

Seeing this difference, it may seem that the needs of plants are totally different from those of animals. However, they do have one very significant need in common. Ask students if they can think of what this is. You may need to give hints. What do we need to give houseplants every few days? What do you need to do to flowers after picking them to keep them looking good? What do you need to give your pet dog/cat besides food? Children will readily conclude, WATER.

Go on to emphasize that water is absolutely crucial for all living things including humans. Some children may object that they rarely drink water; they drink milk, sodas, or juices of various sorts. Point out that all such drinks are really water. There are just other things mixed into it.

You may go on to point out that water is so crucial to all living things that scientists cannot even imagine anything living in its absence. Therefore, in the exploration of Mars and other planets or moons they are particularly interested in the question: Is there water, or did water ever exist on the planet/moon. If there is evidence for water, there is also the potential for living things, and this would be most fascinating. If there is no evidence for water, however, further searching for living things, they believe, would be fruitless.

Some students may note that all living things require air as well. Point out that air is a mixture of gases as will be addressed in Lesson A-7. Different organisms require different gases. There is no one gas that all require. Therefore, water remains the prime thing that all living things require.

Questions/Discussion/Activities to Review, Reinforce, Expand, and Assess Learning:

Make a book illustrating animals getting energy from food and plants getting energy from light.

Have students assign various living/biological things to the plant or animal kingdoms. Have them identify the key features that lead them to make each placement. (Utilize the preclass show-and-tell/discussion period described in Chapter 1, page 19.)

In small groups, pose and discuss questions such as:

We observe and know that in order to keep any animal alive, we must supply it with ample food (as well as with air and water). Why is this so? Why is food so necessary? What does an animal obtain from its food?
Why do plants nearly always have leaves? What function do they serve?
What is a key difference between animal food and what is called "plant food."
What is one thing that all living things require in addition to energy?

Have a stack of cards with a single animal characteristic such as teeth, eyes, or mouth written on one side and similar cards with plant characteristics such as leaves, green color, stems, or roots. Let students draw a card from the shuffled deck, read the item, identify it as a plant or animal trait, and explain why. What function/purpose does it provide for the plant or animal?

To Parents and Others Providing Support:

On any outing, draw your child's attention to one or another living/biological thing and ask questions such as: "Does it belong in the animal kingdom or the plant kingdom? What makes it a plant? What makes it an animal? How is it getting the energy it needs?

To a large degree, physical and behavioral features of every animal center around its obtaining food in one way or another. As you look at any given animal, discuss how this is the case. Ask and discuss why this is so important.

Similarly, observe and discuss why obtaining light is so important for plants.

Connections to Other Topics and Follow Up to Higher Levels:

Looking at the distinctive feeding behaviors and means of locomotion of various types of animals can be a fascinating follow-up study.

Further identification and classification of plants and animals into their various groups and subgroups: phylum, class, order, family, genus, and species.

The process of photosynthesis in more detail

Food metabolism in animals in more detail

Plant and animal anatomy and physiology

Basic ecology:

Food chains

Food webs

Energy flow

Nutrient recycling

Re: National Science Education (NSE) Standards

This lesson is a steppingstone toward developing student's understanding and abilities aligned with NSE, K-4

Unifying Concepts and Processes
• Systems, order, and organization
• Form and function

Content Standard A, Science as Inquiry
• Abilities necessary to do scientific inquiry

Content Standard C, Life Science:
• Characteristics of organisms

Content Standard F, Science in Personal and Social Perspectives
• Types of resources

Content Standard G, History and Nature of Science
• Science as a human endeavor

Books for Correlated Reading:

See also list under B-4.

Numerous titles in "The Science of Living Things" and "What Kind of Animal Is It" series. Crabtree Publishing.

Numerous titles in the "Clasifying Animals" series. Raintree.

Numerous titles in "The Animal Kingdom" series. Capstone Press.

Numerous titles in "Rookie Read-About Science" series. Children's Press.

For invertebrates, see numerous titles in the "Bugs! Bugs! Bugs!" series. Capstone Press.

Krebs, Laurie. *The Beeman.* National Geographic Children's Books, 2002.

Rockwell, Anne. *Bugs Are Insects* (Let's-Read-and-Find-Out Science, Stage 1). HarperTrophy, 2005.

_____. *Honey in a Hive* (Let's-Read-and-Find-Out Science, Stage 2). HarperTrophy, 2005.

Selsam, Millicent. *Benny's Animals and How He Put Them Together.* Harper & Row, 1966.

Stewart, Melissa. *A Place for Butterflies.* Peachtree, 2006.

Lesson B-4

Life Cycles

Overview:

Understanding the life cycle of a living thing is central to all other studies concerning its existence on Earth. Whether one wishes to preserve an endangered species, control a pest, or simply know it better, understanding its life cycle is the starting point. In this lesson, students will gain the concept of life cycles and they will learn the major stages in the life cycles of mammals, birds, frogs and toads, flies and butterflies, and flowering plants. This will be a cornerstone in the foundation for further studies in the life sciences.

Time Required:

Part 1. The Concept of Life Cycles (Q and A discussion, 25-35 minutes)
Part 2. Examples of Life Cycles (observing and discussing examples as they are encountered, 10-20 minutes each). Rearing fruit flies, butterflies, and or other organisms may be conducted as desired.

Objectives: Through this exercise, students will be able to:

1. Recognize and use the following words in their proper context: GENERATIONS, LIFE CYCLE, MALE, FEMALE, FERTILIZATION, REPRODUCTION, LARVA, PUPA, PUPAE, METAMORPHOSIS, DORMANT.

2. State how no individual of any kind of living thing may be expected to live indefinitely.

3. Describe how individuals of every kind of living thing follow a course from conception through growth and development to maturation, aging, and eventual death.

4. Tell how the continuing existence of every kind of living thing depends on individuals reproducing.

5. Define a life cycle as all the events that occur between the adults of one generation and the adults of the next generation.

6. Identify the major stages in typical life cycles of mammals, frogs, fruit flies and butterflies.

Required Background:

Lesson B-2, Distinguishing Living, Natural Nonliving, and Human-Made Things
Lesson B-3, Distinguishing Between Plants and Animals: The Plant and Animal Kingdoms
Lessons B-4A, Identification of Living Things, and B-4B, What Is a Species? may be conducted concurrently and will provide a synergy

Materials:

Part 1. No special materials are required, but take advantage of your students seeing baby animals suckling, birds feeding their young, finding tadpoles in a pond, flowers blooming and forming fruits and seeds, and other examples of reproduction.

Part 2. A culture of fruit flies (may be purchased or captured as described here)
Materials for rearing butterflies and/or other "critters" as desired

Teachable Moments:

Any time students witness animals mating, a mother caring for her babies, birds sitting on eggs and/or rearing young, bees pollinating flowers, seeds within a fruit, or any other aspects of life associated with reproduction may be seized as a teachable moment. It is most instructive to see these things in real-life, but pictures and real nature stories may be used also. You might also make arrangements to visit a farm or zoo at times when baby animals may be viewed.

Methods and Procedures:

Part 1. The Concept of Life Cycles

In their preschool years, children will have gained the experience of animals having babies, babies growing up, and individuals passing away from one cause or another. Thus, it is only necessary to introduce new vocabulary words that we use in describing this experience. It will be well to introduce these words on successive occasions, so as not to overwhelm students in one sitting. Nevertheless, we put them together here for efficiency of description.

In viewing any situation that illustrates offspring being born or eggs being laid, explain that this is REPRODUCTION. Scientists prefer to use the word "reproduction" because "having babies" refers mainly to humans and other mammals, laying eggs refers mainly to birds, and so on. REPRODUCTION is a term that is used in connection with any kind of living thing, plants as well animals. It simply refers to the process of the plant or animal producing offspring, something that every kind of living thing does. Thus, we can say that every kind of living thing has the ability to and, at some point in its life, does REPRODUCE.

Explain further that fathers, boy offspring, and men are referred to as MALE; mothers, girl offspring, and women are referred to as FEMALE. Scientists prefer to use the terms male and female because they refer to the sex difference alone; they do not carry any of the other implications of mother, father, boy, or girl. Additionally, male and female are used in connection with parts of flowers, whereas mother, father, boy or girl would be absurd.

By definition, the member that produces eggs or has the babies is termed the FEMALE. But, in every case, a MALE is also on the scene. Children will almost invariably ask, "What does he do?" For five and six-year-olds, it will generally suffice to simply say that in order for the eggs to develop, FERTILIZATION is necessary. The male FERTILIZES the eggs. (In every kind of living thing, the female reproductive cell is called an EGG. Differences arise in whether eggs are laid and development occurs apart from the female's body, as in birds, or whether the eggs are held internally and initial development occurs inside the female, as in humans and other mammals.)

Likewise, when dogs, birds, and other critters are seen mating and children ask what they are doing, it will generally suffice to just say that they are MATING; the male is fertilizing the female. If children do ask for more details, you can provide them as you see fit. It is considered best to provide children with simple, straightforward, truthful answers but only to the extent that answers their questions. There is plenty of time and occasions for details later.

When students are familiar with the above terms, bring them to focus on their ancestors, various famous persons, and/or others who have passed away. Likewise, have them consider pets and other experiences of animals and also plants that have died. Pose the question, "Does any living thing live forever?" With Q and A discussion, guide students to the conclusion that the answer is NO. Every member of every kind of living thing goes through a series of stages from conception (an egg being fertilized) through growth, development, and maturation to finally aging, and eventually passing away. (If any children are upset by the eventuality of their own death, assure them that they have the prospect of living at least ten times longer than they have lived thus far. There is so much they can learn, discover, create, and enjoy in that time.)

Then pose the question, "If every member of every living thing is destined to eventually die, how does any living thing maintain its existence on Earth over a long stretch of time?" With a little pondering, children will recognize that it must be through reproduction. Every kind of living thing reproduces; thus, new offspring grow up and take the place of older members as they die. Children may recognize their own roles in this process. Have them reason: Without reproduction any/every kind of living thing would indeed pass out of existence. It would become extinct, like the dinosaurs.

As students digest this concept, introduce the words that define it. All the members that are bearing offspring are spoken of as one GENERATION. All their offspring are spoken of as the next generation. As they grow up, they will reproduce and form another generation. Have your students formulate examples. Their parents are one generation; they themselves are a second generation; when they grow up and have children, those children will be a third generation; and so on. Students may draw additional examples from pets, farm animals, and so on. Note that generations are usually numbered; first, second, third, and so on. However, where one starts the counting is arbitrary and chosen to suit one's purpose. If one is descended from a famous person, one might start with her/him and say that I am ___ generations removed. If one is looking at hereditary traits, one usually designates the parents showing the trait as the first generation and examines second and third generation individuals (children and grandchildren) for those traits.

Finally, explain that all the stages of growth and development that take place from the adults of one generation to adults of the next generation are referred to as the LIFE CYCLE. Every kind of animal, plant, fungus, protozoan, and bacterium has its respective life cycle. In some cases, the life cycle is very fast. For example, fruit flies, which we will examine later, go from adults of one generation to adults of the next generation completing a life cycle in a matter of two weeks or even less. Also, "critters" may pass through a number of different forms in the course of their life cycle. Butterflies and moths, which have the caterpillar stage, are classic.

In every life cycle, however, there is a stage and time in which REPRODUCTION occurs; that is, when eggs are laid, babies are born, or in plants, seeds are set. Thus, reproduction is a central feature of every life cycle. Then, in every case, of course, the life cycle is repeated again and again. This is why it is called a "cycle"; it goes around again and again like a turning wheel.

Have children note that having a life cycle is absolutely unique to living things. Children invariably get a kick out of imagining the absurdity of bottle caps, or any other human-made thing, mating and having babies. Likewise, natural nonliving things, such as rocks, do not have babies that grow up into big rocks again.

After students have gained the general concept of a life cycle, proceed at later times to consider the life cycles of various organisms as they as they may come to their attention. That is, so far as possible, keep this connected to children's real-world experience.

Part 2. Examples of Life Cycles

Humans and Other Mammals: For humans and other mammals, the life cycle is relatively simple; adults have babies, babies grow up and mature, females mate with males, and babies are born, starting the next generation. (You may insert a morals/ethics lesson here concerning the special status of humans, as you see fit. Humans should not just mate and have babies as wild animals do. Our life cycle includes many rules and responsibilities surrounding men and women coming together and bearing and raising children.)

Frogs and Toads: In the early spring it is common to hear frogs and toads singing in ponds and wetlands. Going out in the evening just after dark with a strong flashlight to spot them swelling their throats with their shrill call is a great adventure that all will enjoy. You may spot pairs mating, the male sitting on top of and grasping the female as she lays her eggs. Later, during daytime, you may go out and find the eggs, little black balls about the size of peppercorns imbedded in a mass or string of jelly-like substance. Then, a few days later, you are likely to find the pond teaming with tadpoles. Do what you can to encourage parents and other caregivers to give their children, and themselves, this exciting experience.

It is possible to collect a few eggs, maintain them, and observe their development into tadpoles and further development into tiny frogs or toads, but this does take some special care. Seek help as necessary in doing this. (google, "raising frogs").

As students see these events, you can discuss them in terms of the life cycle of the frog or toad. Different from birds, the life cycle of frogs/toads has the additional tadpole stage. The female's eggs, fertilized by the male, develop first into the totally aquatic tadpoles, which breathe with gills. Then, after a few weeks of growth, the tadpole rather rapidly "sprouts" legs, develops lungs, and becomes the baby frog/toad. These, of course, grow up to become adults and repeat the cycle.

Insects: Moths and butterflies are renowned for their life cycle that includes a caterpillar turning into a beautiful moth or butterfly. This type of life cycle is shared by flies, bees and wasps, and certain other (but not all) insects. One can purchase caterpillars from butterfly farms (google "butterfly farms") and watch this fascinating process with your students. You can also conveniently observe the life cycle in its entirety using fruit flies.

Fruit flies may be purchased (google, "raising fruit flies") or you may capture them as follows. Any time from midsummer through fall, put a piece of ripe banana in an open jar and set the jar outdoors. In a day or so, it will collect fruit flies. It will undoubtedly collect other bugs as well, but you can recognize the fruit flies. They are the tiny flies (about 2 millimeters) with blood-red eyes.

On the banana, fruit flies lay eggs that hatch and develop more rapidly than those of other bugs present. Hence, after a day in the presence of fruit flies, remove the piece of banana without other bugs and place it in another jar covered with a piece of cloth held with a rubber band. (The banana has sufficient moisture; no additional water is needed.) After a few days at 70–90°F (25–30°C), children may observe the fruit fly LARVA, tiny worm-like things, crawling over and feeding on the banana, which may become more and more soupy, but that is okay. In a week or so, some of the larva will crawl up the sides of the jar and PUPATE; that is, their skin hardens down to encase the larva in an egg-like shell. This "shell" with the lava inside is called the PUPA (plural, pupas or pupae).

Within the pupa, one of the most remarkable processes of all biology occurs: METAMORPHOSIS. The body of the larva is totally rearranged and restructured so that, in another few days (about two weeks from the start), the new generation of fruit flies begins to emerge. Within hours the new fruit flies will mate and the female will begin

157

laying eggs that hatch into larva and another cycle is underway.

Summarizing, the life cycle of the fruit fly may be described as:

ADULTS → EGGS → LARVAE → PUPA →(metamorphosis)→ ADULTS

If some of your students wish, a culture of fruit flies may be kept and propagated indefinitely. Just lay the jar with flies on its side, mouth-to-mouth with another jar containing a fresh piece of ripe banana. Some of the fruit flies will soon migrate to the new jar. Cover the new jar with cloth and keep it as you did the first jar. (You may add that the ease of raising fruit flies and their rapid life cycle make them a favorite experimental "animal" for use in studies of heredity and genetics.)

Plants: All plants have their respective life cycles as well. Seeds in flowering plants, or spores in the case of mosses, ferns, fungi, and other non-flowing plants, are the "babies," so to speak. A seed is more than just a baby plant, however. It is the plant embryo plus a supply of stored food that will nourish the embryo through its germination. It is also significant to note that the embryo is "asleep," or DORMANT, and may remain so for a matter of years. Given suitable conditions of temperature and moisture, however, it "wakes up" and recommences growth, and germination results. Spores are the equivalent of seeds, but are single cells surrounded by a resistant cell wall.

Plant reproduction will be covered in greater detail in Lesson B-10, Plant Science I, Basic Plant Structure. Full details generally await high school or college courses.

Summary Comments

This lesson, as it is directed at five to six-year-olds is only intended to give students the general concept of life cycles and the key points of some of the most classic ones. But this provides a foundation upon which details and variations can be meaningfully added. Concerning details, note that the life cycle of animals also includes such things as finding/attracting mates, mating behaviors, building nests, involvement in rearing young, migration or hibernation, etc., as well as details of reproduction and development. In the plant kingdom, there are agents of pollination, mechanisms of seed dispersal, factors that maintain or break seed dormancy, and so on.

Further, the examples given here by no means cover the variations on the theme. Many parasites, for example, go through a number of different stages in/on different hosts. Mosses, ferns and other non-flowering plants have life cycles that are quite different from those of flowering plants.

Whenever a scientist takes an interest in any given kind of plant or animal, discovering and describing all the stages and details of its life cycle will be the focus of his/her attention. Emphasize that we know what we do about the life cycles of various animals and plants because people have taken the pains to observe them closely over long periods of time and report their findings. Such studies are not only central to biology. They are also central to the studies in ecology, wildlife management, pest management, parasitic diseases, and many other fields. Interest in this area can well lead into any number of rewarding life careers.

Questions/Discussion/Activities to Review, Reinforce, Expand and Assess the Lesson:

Make books illustrating the life cycle of various animals or plants.

So far as possible and practical, provide materials and time for students to rear fruit flies, butterflies, and/or other "critters" as they may choose. Have them discuss their observations/results in terms of life cycles and generations.

With your children, read natural history stories depicting any given animal or plant and discuss it in terms of its life cycle.

In small discussion groups pose and discuss questions such as:

Focus attention on a certain kind of animal and ask students to describe its life cycle.
What does every kind of living thing have the capacity to do? What do we mean by "reproduce"?
What is the role of the female? ... the male?
Every living thing, in order to maintain its existence on Earth, must do what? Why?
How many generations are living in your household? Describe them.
How many generations are you removed from your great-grandfather?

To Parents and Others Providing Support:

In addition to discussion and activities suggested above:

Take advantage of opportunities to visit farms and/or other sites to enable your children to witness births, eggs hatching, and parent's involvement (if any) in rearing young. Discuss observations in terms of the life cycle of the given animal.

With your children, read natural history stories regarding any particular plant or animal and discuss it in terms of its life cycle.

When you spot a caterpillar, acorn, bird's nest, tadpoles, flowers, or any number of other things, ask and talk about it in terms of its life cycle. What stage of its life cycle does it represent? What came before? What will come after?

Connections to Other Topics and Follow-Up to Higher Levels:

Details of life cycles
Population dynamics
Plant and animal breeding
Genetics
Ecology
Wildlife management, including protecting endangered species

Bernard J. Nebel, Ph.D.

Pest and disease control

Re: National Science Education (NSE) Standards

This lesson is a steppingstone toward developing students' understanding and abilities aligned with NSE, K-4:

Unifying Concepts and Processes
  • Systems, order, and organization
  • Constancy, change, and measurement

Content Standard A, Science as Inquiry
  • Abilities necessary to do scientific inquiry
  • Understanding about scientific inquiry

Content Standard C, Life Science
  • Characteristics of organisms
  • Life cycles of organisms

Content Standard G, History and the Nature of Science
  • Science as a human endeavor

Books for Correlated Reading:

Ehlert, Lois. *Waiting for Wings.* Harcourt, 2001.

Gibbons, Gail. *Monarch Butterfly.* Holiday House, 1989.

Heiligman, Deborah. *From Caterpillar to Butterfly* (Let's-Read-and-Find-Out Science, Stage 1). HarperTrophy, 1996.

Pfeffer, Wendy. *From Tadpole to Frog* (Let's-Read-and-Find-Out Science, Stage 1). HarperTrophy, 1994.

Posada, Mia. *Ladybugs: Red, Fiery, and Bright.* Carolrhoda Books, 2002.

Rockwell, Anne. *Becoming Butterflies.* Walker Books for Young Readers, 2002.

Ryder, Joanne. *Where Butterflies Grow.* Puffin, 1996.

Zemlicka, Shannon. *From Egg to Butterfly* (Start to Finish). Lerner, 2003.
_____. *From Tadpole to Frog* (Start to Finish). Lerner, 2003.

Numerous titles in "The Life Cycle" series. Crabtree Publishing.

Lesson B-4A

Identification of Living Things
(Discerning Similarities and Differences)

Overview:

Central to all learning is the skill of discerning similarities and differences. Nothing provides better practice in developing this skill than the identification of living things. Furthermore, such identification is the starting point for all life sciences. For some, it will become the launching point for professional careers. For many, it will become a life-long avocation of interest, joy, and reward. For all, it will create a certain sense of "feeling at home" with nature.

Additionally, this exercise will draw you, your students, their caregivers, your colleagues, and outside persons into a community of learners, a prime objective of the National Science Education Standards.

Time Required:

This is not a lesson to be conducted and completed in a set time. It is an ongoing study taking a few minutes here and there as opportunities and interests dictate. Utilize the preclass show-and-tell/discussion period described in Chapter 1, page 19.

Objectives: Through this exercise, students will be able to:

1. Recognize and identify on sight common plants and animals (members of plant and animal kingdoms) of their region.

2. Use field guides for the identification of common plants and animals.

Required Background:

Lesson B-2, Distinguishing Living, Natural Nonliving, and Human-Made Things
Lesson B-3, Distinction Between Plants and Animals

Materials:

The common plants and animals of your region as they are encountered
Field guides covering the common trees, flowers, mammals, birds, and insects of your
    region (Used copies of field guides are generally available, and you can share a

set of guides with your colleagues.)
Materials for making collections of leaves, insects, and other specimens as desired

Teachable Moments:

When any plant or animal is at hand, guide children in identifying it to the extent possible. This should often involve children following your using a field guide. On any outing, draw children's attention to given trees, flowers, birds, animals, insects, etc. and identify them to the extent possible.

Methods and Procedures:

It will be natural to, and by all means you should, proceed from Lessons B-2 and B-3 into more specific identification of the plants and animals (members of plant and animal kingdoms) of your region. This task may sound horrendous, since many teachers are totally unfamiliar with them.

In a way of alleviating this fear, recognize that children have no such hesitation. Learning the common names of plants and animals growing naturally in your region is a simple extension of learning to identify horses, sheep, tulips, and other farm and garden animals and plants. Also recognize that everyone who has the mental capacity to distinguish among and learn the names of different people can certainly do the same for different kinds of trees, birds, etc. Indeed the differences in the leaves between various species of trees, for example, is much greater than the differences in the faces of the many people you know.

Therefore, if you are among those who feel they know nothing about and can't identify the common plants or animals in your region, the recommendation is simply to start. The only wrong way to do it is not to start. Don't be embarrassed to be a learner along with your students. In fact, being a fellow learner is the way to be the best of all role models. Further, get together with your colleagues with a set of field guides and make identification a group effort; share your knowledge and experience. Call in a local naturalist from a nature center, zoo, or museum. You may have such a person among the parents/caregivers of your students. In short, make yourselves into a community of learners.

There are so many kinds of living things that it should be conspicuous that this activity will be ongoing and extending through all the grades. It is not unlike the experience of continually meeting and learning the names of new people you meet. Encourage children to bring in their new "discoveries" (actual or sightings of) and utilize the preclass period for show-and-tells and identification.

Identifying trees by their leaves is a very convenient place to start. Trees, at least those of any size, are not harmed by cutting off a small twig with a few leaves attached. Thus, you may ask parents to facilitate their children in bringing in a sample from one or more trees in their yard or neighborhood. Second, tree leaves are relatively easy to identify by matching them to pictures in a field guide. To be sure, more sophisticated identification involves technical words defining overall shape, margins, and other detailed characteristics. This is a splendid turn-off for many children and adults alike. Therefore, just matching your leaf to a picture and finding the common name is fully sufficient at this stage and most children enjoy the process. Finally, children enjoy the activity of pressing leaves and mounting them on paper such that they may be stored for later reviews, testing, art projects, etc. (Google "pressing mounting leaves" for more information.)

There will be cases where you cannot find a suitable match of your leaf in the field guide. Press, mount, and save it along with the others; just label it with a question mark. Bring it to the attention of other people and eventually you will meet someone who can identify it for you and your class. Likewise, you and your students may misidentify some specimens. On occasion, have someone with more knowledge review your collection to catch such errors.

There should be no embarrassment in the discovery of errors. Indeed, you can make this into an important lesson in the scientific process. Scientists always have others review their work at one stage or another to pick up possible errors. It is this process of review by others that keeps science on a truthful and honest course.

Birds are another group that lends itself especially well to identification. Enlist help as necessary in setting up a bird feeder in a location where it can be seen from your classroom window and keeping it stocked. Again, visiting birds can be identified by matching them with pictures in a field guide. The same species will tend to visit the feeder over and over, and students can learn to recognize and identify them accordingly. Then, more rare visitors will stand out and attract the extra attention they deserve.

Likewise, students should learn to recognize and identify the common flowers of gardens. Here, there may be difficulty, because there are so many varieties of each garden flower that picture matching becomes difficult. Using garden catalogues for field guides may help. But there is also a benefit in the difficulty. By learning the various varieties of roses, for example, students will learn to recognize the common characteristics shared by all roses, the variations on the theme, so to speak. Recognizing that an unknown specimen must belong to a certain group on the basis of certain characteristics is a huge advance in learning.

How this activity will be ongoing, always extending in depth and breadth, should be abundantly clear. Most instructive and beneficial will be to set aside a few minutes at the start of each day to allow one or more students to do a brief show-and-tell regarding a new discovery they have made in the world of living things. Likewise, allow and encourage children to work on identifications and collections during free/recreation times, coaching them as necessary. Encourage them in the habit of self-testing and testing one another.

Some students will become particularly intrigued with one or another group of organisms. Facilitate their further independent work so far as you can.

Bernard J. Nebel, Ph.D.

As students become able to recognize and name the common plants and animals of their region, they can extend their interest into observing and describing additional aspects. For example, each kind of animal has its own requirements for food and shelter and exhibits its own unique behaviors. Each kind of plant favors certain conditions of temperature, sunlight, and moisture and blooms at a given time of year. We will address this somewhat further in Lesson B-5, "Food Chains and Adaptations." So far as possible, however, encourage this sort of information coming from, or at least being reinforced by, students' own observations, as opposed to "book learning."

Learning to recognize and identify the common flora and fauna of your region becomes the underpinning for later studies in ecology, wildlife management, anatomy and physiology, and other life sciences. As you conduct such studies with your students, you cannot help but become increasingly knowledgeable yourself. Gradually, you will become the expert and newer teachers will be seeking your knowledge—not a bad reward.

Questions/Discussion/Activities to Review, Reinforce, Expand, and Assess Learning:

The ongoing nature of this lesson/activity provides its own review, reinforcement, expansion, and assessment.

Have students practice their reading skills on natural history books/stories of appropriate level. Focus on stores/books that present factual information. For animal stories that are fanciful, discuss what aspects may be factual and what is make believe or simply not true.

Facilitate students or groups of students in making collections of leaves, seeds, insects, and/or other such things as they may desire.

In small groups, pose and discuss questions such as:

Have students describe how they recognize and distinguish one species from another
Describe/discuss needs of a particular animal or plant in as much detail as possible, i.e., kind of food, amount of moisture, etc. Encourage this being based on their own observations as well as from "book or lecture learning."
Observe and discuss how plants and animals generally do not occur as single unique species but occur in groups along with similar species.

Note: If you use pressed, mounted leaves for testing purposes, have several specimens of each kind that you use on different occasions. Otherwise, students will come to recognize given leaves by the torn corner of the mounting page, for example, rather than by the characteristics of the leaves. Also, test students using fresh examples of leaves you bring in or you see on outings.

To Parents and Others Providing Support:

Your support and role modeling in noting various living things, using field guides to identify them, and learning along with your children will be invaluable. Every outing, even going shopping, provides an opportunity to sight and later identify trees, flowers, birds, etc.

When/if your child shows a particular interest in pursuing a given group of plants or animals, support them so far as possible by providing necessary materials, equipment, work space, etc. (Requirements to keep things organized and tidy can still be enforced.)

Moving on to observe and discuss with your children the habits and needs of particular plants and animals should follow naturally.

Connections to Other Topics And Follow-Up to Higher Levels:

Ecology
Animal and plant behavior
Plant and animal anatomy and physiology
Wildlife management and conservation

Re: National Science Education (NSE) Standards

This lesson is a steppingstone toward developing students' understanding and abilities aligned with NSE, K-4:

Content Standard A, Science as Inquiry
• Abilities necessary to do scientific inquiry
• Understanding about scientific inquiry

Content Standard C, Life Science
• Characteristics of organisms

Content Standard G, History and the Nature of Science
• Science as a human endeavor

Books for Correlated Reading:

Numerous titles in the "Eye Wonder" Field Guide series, DK Publishing.

National Audubon Society "First Field Guide" series

National Geographic "My First Pocket Guide" series

Lesson B-4B

What is a Species?

Overview:

The concept of a species is one of the most basic ideas in all of biology. But the word SPECIES, is commonly misunderstood and misused, causing much confusion. This lesson is to clarify the picture from the start. It will give students an understanding of what we mean by a SPECIES and its key defining feature. It will develop their understanding regarding the taxonomic organization of living things. Finally, it will make the distinction between a species and subdivisions such as breeds, varieties, races, and subspecies. This will lay the foundation for further study and understanding in genetics, heredity, plant and animal breeding, and other aspects of life sciences.

Time Required:

Short discussions (not more than a minute or so) that will be virtually automatic in the course of using field guides to identify living things (Lesson B-4A).
Some longer discussions (10-20 minutes) will be in order at certain points.

Objectives: Through this exercise, students will be able to:

1. Recognize and use the following words in their proper context: SPECIES, GENUS, FAMILY, BREED, VARIETY, RACE, SUBSPECIES, VARIATION, TAXONOMY, HEREDITY.

2. Recognize and state how the term SPECIES refers to a particular kind of plant, animal, or other organism.

3. Describe how the definition of a species hinges on the ability of members to mate and reproduce.

4. Describe how species generally occur in groups of similar species that we call a GENUS; give examples that illustrate such groups.

5. Describe what is meant by the term HYBRID and give the main attribute of a hybrid.

6. Recognize and describe how there is always variation among the members of a

species, although the differences are not as distinctive as those between species.

7. Give the distinction between a species and subdivisions of a species such as different breeds, varieties, races, and subspecies; give examples that illustrate the distinction.

8. Tell what is meant by heredity; give examples of hereditary characteristics.

9. Tell how certain features of an individual may be shaped by environment, as well as by heredity; give examples.

10. Describe how naming species is a human endeavor and subject to change.

Required Background:

Lesson B-4, Life Cycles
Lesson B-4A, Identification of Living Things

Materials:

This lesson is an extension of Lesson B-4A. No additional materials are required. Pictures of various breeds of dogs (Pictures of breeds of other animals may add interest but are not essential. Google " breeds of _____ ")

Teachable Moments:

Any time in the course of using a field guide to identify a plant, animal, or other organism will present an opportunity to talk about species and other aspects of this lesson.

Methods and Procedures:

For the Teacher

Because the word species is so commonly misunderstood and misused, let me begin by giving the basic definition. Then, we will turn to the techniques of presenting the concept to children.

A species refers to a particular KIND of plant, animal, or other organism. The "particular kind" is commonly based on shape, coloration, size, and/or other features that distinguish one "kind" from another. For field identification, we rely entirely on such features. But, there is variation among the individuals of a species; take humans for example. The problem becomes: Where do we draw the line between variations within one species and differences that define separate species?

The answer lies in the ability to interbreed. If members of the group do or potentially can interbreed and produce healthy young that can breed in turn, they are defined as a single species. Thus, there are many BREEDS of dogs, but they are all considered a single species because they can and do interbreed if they are allowed to do so. The same is true for breeds of cats, horses, and other domestic animals. Pure breeds are developed and maintained only by carefully controlling which individuals are allowed to mate and reproduce. The same holds for domestic plant species as well except that the word VARIETY is used in place of breed. Thus, there are many varieties of tulips, roses, peas, etc.

On the other hand, if such interbreeding cannot occur between groups, i.e., viable young are not produced even in the event of mating/cross pollination, then the two groups are defined as separate species. Of course, testing for reproductive compatibility is impractical except in highly controlled laboratory situations. In the case of fossil species, it is totally impossible.

Therefore, in field identification we come back to distinguishing species on the basis of certain visible characteristics. But underlying this should be the recognition that each species represents a reproducing group. The members of a single species can and do mate/cross pollinate and reproduce their kind. Members of separate species cannot/do not produce viable young even if mating/crosspollination is made to occur. Thus, different species remain separate.[19, 20]

Conveying the concept of species and higher orders to students

Come back to the actual teaching, simply use the word, species, in the course of using a field guide to identify a given plant or animal (Lesson B-4A). For example: It is a bird. What species is it? I don't think it is a ___; I think it is a different species, and so on. Very soon, children will gain the idea that a species refers to a particular kind of animal, plant, or other organism, which is visibly distinguishable from other kinds/species.

At the same time, students will become aware of (perhaps frustrated by) the fact that the specimen in hand does not exacly match the picture in the guide. Explain that there is VARIATION, i.e., minor differences, among the individuals of any species just as there is variation among individual humans. It will be gradually and with experience that students will learn the key features that distinguish one species from another and the insignificant differences that only imply variation among the individuals of the species.

By the same token, students will become experientially familiar with the fact that

---

[19] There are situations in nature that reveal that the situation is not always as clear cut as this definition makes it sound. Still, this is a working definition that will serve as a solid foundation for later studies.

[20] Another but more technical way of looking at this is in terms of gene pools. A species represents a certain gene pool in which the mixing of genes is continuously occurring through reproduction. Mixing of genes between the pools of different species does not occur because reproduction between them is blocked in one way or another. Of course modern science has developed techniques of transferring genes from one species to another but this is for later lessons; it does not change the basic definition of a species.

species tend to fall into groups of similar species. For example, they will discover that there are several species of maple trees, several species of oak trees, several species of pine trees, several species of tree squirrels, several species of sparrows and so on. This will be true regardless of the flora and fauna of your region. Whatever species you may be looking at, more often than not, you will find from your field guide that it will be one of several similar species. In other words, observing nature reveals that species tend to occur in groups of similar species.

As students become familiar with this fact, explain that the group of very similar species is called a GENUS (plural, GENERA). Some students may enjoy and really get into the technical scientific names of species. Where this is the case, point out that the scientific name is always given by two words: the first gives the name of the genus; the second gives the species within that genus. For most students, however, Latin names will be superfluous, but the main point is not. The observation that species occur in groups (genera) of similar species is highly significant. By the same token, you may carry this to the FAMILY level. That is, groups of genera that share certain common features lead us to place them in families that are distinct from other families. (Different types of insects—beetles, flies, grasshoppers, etc,—are actually differentiations at the next higher lever, ORDERS. Within each order, there are families, genera, and species.)

This may well sound overwhelming and confusing. Back up and remember that this not something to be taught in a single or even a few lessons. Basically, you are training children in developing the important life skill of discerning similarities and differences and to use this skill to gain foundational understanding concerning the diversity of living things. Like all skills, this depends on practice, practice, practice. The ability to recognize variations within a species, distinguish one species from another, group them together into genera, and so on will come naturally with practice and experience.

How do species remain distinct?

As children gain experience and become able to recognize and identify common species of the flora and fauna of their region, pose questions such as: What keeps different species separate and distinct? Why, in some cases, do we refer to differences as variations among individuals of the same species, and in other cases, use differences to designate different species?

Review and reemphasize that living things only continue their existence through reproduction. No one individual lives forever (Lesson B-4). The crux is that a species should be visualized as a reproducing group. Males and females of the same species mate (cross pollinate in the case of plants) and produce offspring of their kind that can mate and reproduce in turn. Males and females of different species do not mate or reproduce. (There are certain exceptions that we will come to shortly.) This fact leads to the following definition of a species: A species consists of all those individuals that do or potentially can mate/cross pollinate and produce offspring that reproduce in turn. Therefore, saying that they belong to the same species implies a reproductive potential among them. Saying that individuals belong to different species implies that there is lack such potential.

Bernard J. Nebel, Ph.D.

Guide students to reason how this definition explains why we refer to the different sorts of dogs, not as different species, but as different BREEDS. On the basis of appearance alone, we might well think that the different kinds were different species. However, we know that males and females of different breeds can and do mate and produce crosses or mutts of various sorts. Therefore, by definition, they all belong to the same species.

Students will very likely ask how the different breeds/varieties came about. It is through a process called SELECTIVE BREEDING. It is a process in which humans (animal/plant BREEDERS) skillfully select which individuals of the species will be mated and allowed to reproduce. Doing this over many generations gradually accentuates variations and leads to the development of desired traits, i.e., the particular breed. (The details of this will be the subject of another lesson.)

Then, pure breeds are preserved only by humans controlling that the parents of each generation are the same breed. The same is true for different breeds of cats, horses, cattle, and all other domestic animals. It is also true for all sorts of garden flowers and vegetables except that the word VARIETIES is used in place of breeds.

Be sure that students are clear on the words used. To summarize, a SPECIES is the entire group that has the ability to mate and reproduce. BREEDS or VARIETIES are subgroups within the species that have distinctive features, but still have the ability to mate/cross-pollinate and reproduce with members of other breeds/varieties within the species. Further, the terms BREED and VARIETY say that they have been developed by humans controlling the breeding process.

When such subgroups of the same species are observed in nature, they are usually referred to as SUBSPECIES. The term RACE deserves special mention. It is used (almost exclusively) in connection with humans. Most importantly, we are all one and the same species despite being of different RACES. This is confirmed by the fact that intermarriages among the races lead to perfectly normal, healthy children.

Continue discussion to have students consider what would happen if members of different species did freely mate and have offspring. Bring them to reflect on their experience of crosses between different breeds of dogs. The offspring show certain features or characteristics of both parents or perhaps a blend of the two. They may even take themselves as an example, and cite how they have certain features that are like those of their mother, but other features that are like those of their father.

With this background, ask students, "What would happen if members of different species did freely mate and reproduce?" They should reason that features of the two species would combine and mix and the result, after a few generations, would be single group of similar, reproducing individuals, i.e., a single species. This should serve to reinforce the conclusion that each species should be visualized as a group of individuals that mate with each other and reproduce their kind, but matings and reproduction between members of different species is blocked in one way or another.

170

Students may have fun imagining and suggesting names for the "critters" that might be produced in crosses between different species. Stress again, however, that crosses between species simply do not work. Offspring are not produced even if mating/cross pollination is conducted. But there are some notable exceptions to this rule. Probably the most familiar is the mule, which results from a cross between a female horse and a male donkey, closely related (same genus) but different species. But the mule, while a healthy, vigorous animal, is infertile. Mules are incapable of producing more mules. The only way to produce more mules is to cross horses and donkeys.

Offspring that are produced from crosses between species are termed HYBRIDS, and like the mule, they are infertile. (There are a few rare exceptions that can be omitted for now.)

Heredity and Environment

Through the work described above, it will become amply evident that features of parents are passed to offspring in the course of reproduction. It goes far beyond overall features; we can see how specific traits such as eye-color, hair-color, and countless other detailed characteristics are passed from parents to offspring.

It is only necessary to use the word HEREDITY in talking about this phenomenon. All the traits, features, and characteristics, that are passed from parents to offspring are called HEREDITARY traits. They are INHERITED from one or the other or both parents. Typically, we say such things as: I got my _____ (any given trait) from my mother/father. Have students now substitute "inherited" for "got." I INHERITED my _____ (any given trait) from my mother/father. Therefore, _____ (the given trait) is a HEREDITARY trait. (The whole subject of genetics can wait for later lessons. At this level it is sufficient simply to convey the concept that traits are passed from generation to generation.)

But add that not all traits are hereditary (inherited from parents). Many traits may be the product of, or at least significantly modified by, the environment. Health and learning are two areas to focus on in this respect.

Pose questions such as: How are our bodies a product of both heredity and environment? What aspects did we inherit? What aspects are a matter of environment? Guide students in recognizing that countless basic features of their bodies are hereditary, as well as the detailed features commonly pointed at. However, many features, particularly as they grow and develop, are/will be a product of eating properly, exercising, and otherwise taking care of themselves. (You can obviously pursue this discussion to any extent you desire.)

Likewise, pose the question, "How are our brains and what is in them a product of heredity and environment?" Guide students to recognize that we did inherit basic features of our brains, such as having the ability to learn. But what we finally end up knowing and being able to do with our brains will be largely a product of environment, the opportunities to gain education, and our own volition—the time and effort we are willing to expend on learning. A simple but profound example is language. We inherit the capacity to speak, but the first language we learn to speak, including subtleties of accents, depends on our environment.

One additional aspect is exceedingly important. Traits we acquire during our lifetime are not and do not become hereditary. Particularly, any body deformity that a parent may have due to an injury does not become hereditary; it will not be passed on to children. Nor will any deformity that you may have due to injury be passed on to your children. By the same token, however, skills and knowledge gained by our parents is not passed on either. Therefore, it is up to each individual to develop his/her own skills and knowledge through practice and study.

Questions and Flux in Classification

Coming back to our ongoing project of identifying the living things around us, what does all this mean? Yes, each species should be understood as a reproducing group, and there are many cases where this is observed. Have students note examples of animal parents with their offspring that they have seen.

Still, as a practical matter, we identify species on the basis of appearance, and this is not always clear-cut. A frequent and often frustrating fact is that you cannot find an exact match between the specimen you are examining and the picture in the guide. There is always VARIATION (slight differences) among the individuals of a species; take us humans, for example. Gradually and with occasional help from others, you and your students will become familiar with the variations that occur within a given species and the distinctive features that distinguish it from other species. Intrinsic rewards of being able to recognize and identify the common plants and animals that you encounter in your region will accrue.

However, be aware that you may come across some cases where species lines are not clear. Does this specimen being looked at represent a simple variation of a given species? Does it represent a subspecies? Or, is it another species altogether?

Have students reflect that reproductive capability or incapability would be the sure test. But this is not easy to determine. At best, it requires long-term observation and testing. In the case of dinosaurs and other extinct animals and plants it is totally impossible. Therefore, there are many cases where classification remains based on appearance, and here, there are differences of opinion. One scientist may think that the differences between two groups is great enough to classify the two groups as different species; another scientist may think they should be classified as subspecies; a third may think the differences only amount to variations within the same species. You may find these differences of opinion reflected when using different field guides. One guide will show these two birds, for example, as different species; another may show them as the same species.

As new information becomes available—DNA testing is the most remarkable new tool—designations of subspecies, species, and even larger groups may be subject to change. This will not detract from the usefulness of field guides or the joy and reward of becoming able to identify plants and animals by common names. It is only to say that you should be aware that designations of species, subspecies, and so on are subject to change.

When/if your students come across such "disagreements," take it as an opportunity to point out and discuss the fact that the whole naming and classification system is a human endeavor. Do things in nature ever come with labels? How did names come about? Guide children to recognize that it must have been a matter of people observing what they found, giving it a name, and then passing the names on, parents to children. This was the origin of common names. Gradually people recognized that living things could be organized logically into a system of groups and subgroups according to similarities and differences.

A Swedish biologist by the name of Carl Linnaeus (1707 - 1778) developed the system of organizing living things into families, genera, and species and giving each a Latin name. As new species are discovered—new species do continue to be discovered—they are fit into the scheme. Thus, Linnaeus's general scheme continues to be used. In Linnaeus' time the organization was done entirely on the basis of similarities and differences in appearance. Designations according to reproductive compatibility and actual genetic (DNA) comparisons is much more recent and is ongoing.

In conclusion, when students come across changes in nomenclature, help them understand that is reflects our ongoing human endeavor to understand, categorize, and name living things according to actual genetic relationships and reproductive compatibility. Someday, they too may play a role in this process. There are many potential careers in this field, which is called TAXONOMY.

Questions/Discussion/Activities To Review, Reinforce, Expand and Assess Learning:

The ongoing nature of this lesson/activity provides its own constant review, reinforcement, expansion, and assessment.

In small groups, pose and discuss questions such as:

What is a species? ... genus? ... family? ... breed? ... variety?
... subspecies? ... hybrid?
Is each species entirely unique, or do species tend to fall into groups? Give examples.
What are the common features that cause different species of oaks, for example, to be placed in the same genus? (Use an example of a genus that has several representative species that children are familiar with.)
What are the key features in appearance that distinguish different species of the genus (above)?
Breeds of dogs are conspicuously different. On what basis do we consider them all as one species?
Are all the members of a given species identical? Why do we not consider every

173

variation as a different species?

If members of two different species did interbreed freely, what would be the eventual result?

There are many riddles posing the question: What you would get if you crossed a _____ with a _____. The humor notwithstanding, do such crosses actually work? Is it possible to cross animals (or plants) of markedly different species?

What do we mean when we say a certain trait is hereditary? Give examples of traits that are hereditary and traits that are not hereditary.

What hereditary traits do you have? Where/how did you get them?

What traits do you have that are not hereditary? (Examples will be any knowledge or skill that was developed by practice or learning.)

How do the names we give plants and animals stem from human endeavors?

To Parents and Others Providing Support:

Continue the activity of working with your children to identify the common kinds of living things that you come across, as described in Lesson B-4A. In the process, discuss the concept of a species and its ramifications as described in this lesson.

Coach children in recognizing and identifying traits they inherited from their mothers/fathers.

Connections to Other Topics and Follow-Up to Higher Levels:

Genetics
Heredity
Plant and animal breeding
Hybridization
Genetic engineering
Clones and cloning
Asexual Reproduction

Re: National Science Education (NSE) Standards

This lesson is a steppingstone toward developing students' understanding and abilities aligned with NSE, K-4:

Unifying Concepts and Processes
- Systems, order, and organization
- Constancy, change, and measurement

Content Standard C, Life Science
- Characteristics of organisms
- Life cycles of organisms
- Organisms and environments

Content Standard G, History and Nature of Science
• Science as a human endeavor

Books for Correlated Reading:

See list under B-4A above.

Kalman, Bobbie. *Many Kinds of Animals* (What Kind of Animal Is It?). Crabtree, 2005.

Ribke. Simone T. *Grouping at the Dog Show* (Rookie Read-About Math). Children's Press, 2006.

Roy, Jennifer Rozines and Gregory Roy. *Sorting at the Ocean.* Benchmark Books, 2005.

Selsam, Millicent. *Benny's Animals and How He Put Them Together.* Harper & Row, 1966.

Lesson B-5

Food Chains and Adaptations

Overview:

In this lesson, students will learn the concept of food chains and how animals are particularly adapted to feed on plants or on other animals, or sometimes on both. Students will begin to observe how animals and also plants are adapted to their environments and other species with which they interact in numerous other ways as well. This lays a foundation that is integral to all of ecology and is central to many other life sciences as well.

Time Required:

Part 1. Food Chains (discussion and activity, 45-60 minutes)
Part 2. Adaptations (interpretive discussion based on children's experience of animals common to their region, 30-45 minutes; examination of herbivore and carnivore teeth plus interpretive discussion, 30-40 minutes)

This lesson can very easily be integrated into and carried along with Lesson B-4A, Identification of Living Things, as an ongoing study always with new examples to discover and learn. Plan on using the preclass period to insert aspects of this lesson in addition to the identification of species in your region.

Objectives: Through this exercise, students will be able to:

1. Recognize and use the following words in their proper context: FOOD CHAIN, ADAPTATION, HERBIVORE, CARNIVORE.

2. Give examples of food chains among species that are common in their region.

3. Define what is meant by being at "the top of the food chain."

4. Describe what happens to the amount of available food as you go up the food chain.

5. Tell how and why all food chains must start with plants.

6. Explain what is meant by the word, ADAPTATION. Give examples.

7. Explain how the features and behaviors of a familiar animal, such as a squirrel, adapt it to its particular environment and way of life.

8. Point out features that adapt herbivores and carnivores to their particular ways of obtaining food.

9. Describe adaptations of different sorts of plants.

Required Background:

Lesson C-1, Concepts of Energy I: Energy Is What Makes Things Go
Lesson B-3, Distinction Between Plants and Animals
Lesson D-4, Land Forms and Biomes, may be conducted concurrently
This lesson grows out of the ongoing examination and identification of living things (Lesson B-4A). Therefore, it should be conducted concurrently and integrated with that study.

Materials:

Part 1. Beyond materials being used for the examination and identification of living things (Lesson B-4A), the only items required are:
Pasta pieces of three different shapes, amounts in the ratio of 100:10:1.

Part 2. Beyond the above:
Actual skulls or models of skulls (if possible) of an herbivore and a carnivore in which teeth may be examined. Otherwise, pictures showing the structure of teeth will suffice. (Google "herbivore carnivore teeth")

Teachable Moments:

Consideration and identification of any species at hand provides an opportune teachable moment.

Methods and Procedures:

In this lesson, please note that the word "animal" refers to any member of the animal kingdom (Lesson B-3). Likewise "plant" will refer to any member of the plant kingdom.)

Part 1. Food Chains

In observing and identifying various animal species of your region (ongoing Lesson B-4A), it will be easy to consider, in each case, what each animal eats and what may eat it. Happily, this is a subject that intrigues most children. Thus, make it a habit to interject such discussion in the course of considering any species at hand. Such information may be evident from observations, common knowledge, field guides, or elsewhere. Or, it may remain a question mark for later discovery.

In the course of such study, it will become evident to children that some animals, such as horses, cows, deer, rabbits, and mice, feed directly on plants or their products (fruits, seeds, sap, pollen, nectar). Other animals, such as foxes, weasels, wolves, and lions and other members of the cat family, feed on other animals. As children gain this familiarity, introduce them to the more scientific terminology.

Instruct students that those animals that feed on plant material are called HERBIVORES, and they are said to be herbivorous (plant eating). Animals that feed on other animals are called CARNIVORES, and they are said to be CARNIVOROUS (meat eating). Finally there are OMNIVORES (all-eating) or OMNIVOROUS animals, which eat both plant and animal materials. What are humans?

Point out that this is a basic feature throughout the animal kingdom. In the insect world, there are those species that feed on plants and those species, such as the preying mantis, that feed on other insects. Among birds, there are species that eat fruits and seeds and those, such as hawks and eagles, which feed on other small animals. The same is true among fish, reptiles, and most other large groups. Thus, in looking at each major type of animal (class), one is likely to find some species that are herbivorous, other species that are carnivorous, and some that are omnivorous. (To be sure, there are many technical terms that define the feeding relationships more precisely, but save these for later. The basic concept of herbivores, carnivores, and omnivores will suffice at this level.)

As students gain the idea that there are herbivores and carnivores, guide them in reasoning that there is a natural sequence of feeding that exists. There are plants, which make their food (Lesson B-3), animals that feed on plants, then animals that feed on other animals. A specific sequence, clover to rabbits to a fox, for example, is called a FOOD CHAIN. Another might be algae to tiny "bugs" to small fish to larger fish to humans. From their knowledge of who eats what, students should be able to describe any number of additional food chains.

As students demonstrate their ability to describe various food chains, use Q and A discussion to guide them in reasoning certain features that apply to all food chains.

First, pose the question, "What category of living things do all food chains start with?" On reflection, students should note that it is always plant material of one sort or another. The essence of every food chain is plants, herbivores, and then carnivores. There may be more than one level of carnivore, but the starting two steps are always plant material, then some "critter" (herbivore) feeding on that plant material.

Second, pose the question, "Why should this be the case?" (Here students be likely need coaching in getting to the answer.) Review as necessary what they learned in Lesson C-1, every organism needs a source of energy to run on, and in Lesson B-3, plants make food using the energy of sunlight; animals obtain energy from "burning" food. Then, ask students to put this in terms of a food chain. Guide them in reasoning that plants make an abundance of food. Animals gain energy by feeding on plants; other animals gain energy by feeding on the animals that fed on plants.

Still, some students will wonder: Why not have a food chain in which one carnivore feeds on another in an ongoing circular sequence? Similarly, why not have a food chain go on indefinitely, carnivore after carnivore? Assure students that such self-perpetuating food chains are neither found in nature, nor anywhere else. Still, this begs the question: Why not?

Remind students that food consumed for energy is totally "burned" so that there is nothing left except the chemical waste products: carbon dioxide (a component of the air), water, and some "ashes." Very little (usually in the order of only 10 percent) of the food consumed is actually converted into the body of the consumer. Therefore, very little of what is consumed becomes available for another consumer. Food chains do not go on for more than a few steps because available food is quickly depleted toward zero.

Students will gain a better visual picture of this effect from the following activity. In small groups, have them spread one hundred or more pasta pieces on a table. These represent plant material that may be eaten by herbivores. Instruct them that herbivores now eat the plant material. To represent this they will gather up the "plant pasta pieces" and replace them with differently-shaped pasta pieces to represent the herbivores. Most importantly, however, they must pick up 10 plant pasta pieces for every herbivore pasta piece they put down. This is to represent that 90 percent of what is eaten is burned to provide energy. Only 10 percent of what is consumed is converted into the body of the consumer. The result is that all the herbivores are represented by just the 10 remaining pasta pieces.

Now have students visualize these herbivores being eaten by carnivores. To represent this, they will replace the herbivore pasta pieces with a third shape of pasta standing for carnivores. Again, however, they will put down only one carnivore piece for every ten herbivore pieces they pick up to represent the fact that 90 percent of what is consumed is burned for energy. Thus, they end up with just one carnivore pasta piece from 100 plant pasta pieces. Have students reflect on what would happen if they started with more "plant" pasta pieces. They might try it if they wish. They soon discover that, regardless of the starting number, it is reduced to near zero after only a few steps

(We have described this activity using the terms herbivore and carnivore. It will be more meaningful to children if you put it in terms of a specific herbivore and carnivore that children are familiar with—ones that are typical of your region.)

Re-ask the question, "Why don't food chains go on and on without plants? From the exercise, students should reason that available food is depleted to nil in only a few steps. Therefore, all food chains are and remain dependent on plants producing more food through photosynthesis.

Add here that the last animal in the chain is said to be at the top of the food chain. Plants, on the other hand, are said to be at the bottom of the food chain. Said another way, however, plants support all the animals in the chain. Remove plants and everything else would collapse.

Part 2. Adaptations

In the course of observing and identifying animals (ongoing Lesson B-4A), make a habit of calling children's attention to how, in every case, the animal has all the attributes that make it suited to obtaining food in the way that is does. For example, a squirrel doesn't just randomly choose to climb trees and eat nuts. It has the feet, claws, body form, teeth, agility, and balance that make it particularly suited and able to climb trees, gather, gnaw through, and eat nuts. Similarly, horses, cows, and deer have the attributes that suit them for grazing or browsing. Members of the cat family have the attributes that suit them for bringing down, tearing apart, and consuming prey, and so on.

Don't try to present this all in one lesson. It should be one aspect of the ongoing process of finding, identifying, and learning about animals that are common in your region as described in Lesson B-4A. Any species under consideration can be looked at as the example to note how its physical and behavioral features are particularly suited for its way of life, and this should be done on numerous occasions. You can have students make it part of their show-and-tells regarding any animal they are speaking about.

Along the way, introduce the word ADAPTATION. We speak of the squirrel's attributes as ADAPTATIONS and say that the squirrel is particularly ADAPTED for climbing trees and gathering and eating nuts. Horses, cows, and deer are adapted for grazing and browsing. Carnivores are adapted to capturing and eating other animals; consider the claws and teeth of members of the cat family.

Likewise, animals have specific adaptations that help them escape or defend themselves from predators—a skunk's smell and a porcupine's quills are notable in this regard. Additional adaptations help them find mates, reproduce, and cope with adverse environmental conditions.

Then, adaptations go far beyond the outward structure of the body. They include internal anatomy and physiology as well. Gills are an adaptation for obtaining oxygen from the water; lungs are for obtaining oxygen from the air. Different animals have modified digestive systems that enable them to digest different foods.

It is conspicuous that such study can easily balloon to become overwhelming. It will be important keep it down to what children can manage. The advice is this. Keep discussion centered on the species at hand and guide children in observing how its particular physical and behavioral features adapt it to its given way of life. Add additional features of adaptations only as they may come up in considerations of given species. But also let children know that such study can go to any depth and breadth including life careers.

This said, there is one adaptation that is of particular interest and importance that should be brought to children's attention in the course studying herbivores and carnivores. This concerns teeth. Show students pictures, or actual skulls if possible, of an herbivore and focus on the jaw and teeth. Ask students to describe the structure of the teeth and consider how it adapts the animal for eating and digesting grass. With Q and A discussion, guide them to note that the wedge-shaped teeth in front (incisors) are adapted for clipping off grass or small twigs. Then the broad, ridged molars toward the back are adapted for grinding the coarse plant material to make it palatable for digesting.

Do the same for a carnivore. Children will note that the teeth all the way around are more spike-shaped, with particularly pronounced fangs. These are adapted for gripping and tearing flesh. Anyone who owns a dog will attest that they don't chew their food; they simply gulp it down. Fortunately for them, meat does not need the chewing that grass and other coarse vegetation require.

Students are invariably intrigued when they hear this study related to dinosaurs. No one has seen a dinosaur eat. How do scientists know that some were herbivores and others were carnivores? From the above discussion they will correctly reason that it is by their teeth. There is no way that the teeth of a carnivore would enable it to eat and grind up grass for digestion. Similarly, there is no way that the teeth of an herbivore would enable it to tear into flesh.

The beaks of birds give similar clues as to what they eat. Short stout bills are adapted to cracking seeds; longer, thinner beaks are adapted to catching insects. The sharp, hooked bills of hawks and eagles are adapted to tearing apart prey.

Plants are likewise adapted to thrive best in certain conditions of light, moisture, and soil conditions. The natural plants (all members of the plant kingdom) that inhabit your region are, and inevitably must be, adapted to the climatic conditions that prevail. Thus, regions with different climates have different types and arrays of vegetation. Contrast, for example, deserts, grasslands, and forests; conifer forests, deciduous forests, and tropical jungles. Again, don't make an involved lesson of this; bring it up as it will occur naturally in the course of observing that different biomes exist in different regions of the world (Lesson D-4).

Interdependencies Among Species

A final aspect of this lesson is to call children's attention to the interdependencies among species. The dependence of every animal on those species that are its food supply may seem self-evident. But question students regarding examples such as: What would happen to birds that feed on insects, if all the insects were killed off by pesticides? Many students will suggest that they would eat something else—seeds perhaps. Remind them of adaptations. Could a bird with a long pointed beak, which is adapted for catching insects, crack seeds? Would its behavior, adapted to catching insects, even allow it to try eating seeds? We naturally don't like sad stories, but the fact is: If one link in a food chain is removed, everything above it will inevitably die of starvation.

There are non-feeding dependencies among species as well. Point these out as they may come up in the course of discussing the life cycle of various animals (Lesson B-4).

Bernard J. Nebel, Ph.D.

For example, many species of birds depend on trees for nesting sites. Likewise, they and other sorts of animals depend on vegetation for shelter and nesting materials. Spiders depend on shrubbery in which to hang their webs, and so on.

Questions/Discussion/Activities To Review, Reinforce, Expand, and Assess Learning:

Make books illustrating actual food chains among the plants and animals in your region.

Make posters illustrating food chains among the common wild plants and animals of your region.

Make bar graphs illustrating the relative amount of food available as you progress up any food chain.

Consider, discuss, and illustrate food chains supporting humans.

Have students pick their favorite real animal and write or tell about the adaptations that make it suited for its way of life.

For any plant or animal that comes to children's attention, inject a brief discussion regarding how it is adapted to doing everything that it does.

Mosquitoes are generally maligned, but their adaptations for their way of life are awesome. Consider them from this point of view.

In small groups, pose and discuss questions such as:

What is meant by a food chain? Give examples involving familiar species.
When you look at nature, why is it that the most abundant living things you see are plants?
In nature, why are herbivores seen more commonly than carnivores?
Why are hawks and eagles more rare than songbirds, which are mostly seed and insect eaters?
Do humans belong to food chains? What position(s) do they occupy?
What is different about human food chains that distinguishes them from natural food chains?
Aside from risk factors, why are carnivores not raised and sold for meat?
Call students' attention to any familiar animal, then ask and discuss how is it adapted?
Showing students a jaw with teeth, ask: "Did it belong to an herbivore or carnivore? How can you tell?"
How are we humans adapted?

To Parents and Others Providing Support:

One can't go on an outing without confronting natural species. In addition to working on identification, discuss how it fits into one or more food chains. Discuss the features of

182

any given species in terms of adaptations for a particular way of life.

At meal times, discuss how we fit into food chains. Discuss how agriculture and animal husbandry are a system of making food chains to support us.

Connections to other topics and follow up to higher levels:

All of ecology
Plant and animal breeding
How adaptations arise
Evolution
Genetics

Re: National Science Education (NSE) Standards

This lesson is a steppingstone toward developing students' understanding and abilities aligned with NSE, K-4:

Unifying Concepts and Processes
  • Systems, order, and organization
  • Form and function

Content Standard C, Life Science
  • Characteristics of organisms
  • Life cycles of organisms
  • Organisms and environments

Content Standard G, History and Nature of Science
  • Science as a human endeavor

Books for Correlated Reading:

Bailey, Jacqui. *Staying Alive: A Story of a Food Chain* (Science Works). Picture Window Books, 2006.

Hickman, Pamela. *Hungry Animals: My First Look at a Food Chain*. Kids Can Press, 1996.

Lauber, Patricia. *Who Eats What? Food Chains and Food Webs* (Let's-Read-and-Find-Out Science, Stage 2). Scott Foresman, 1995.

Numerous titles describing food chains in various ecosystems. Bobbie Kalman Books.

Lesson B-6

How Animals Move I:
The Skeleton and Muscle System

Overview:

In this exercise, students will gain the concept of how their bodies move. In addition to becoming familiar with the structure of the human skeleton, they will learn how body movements are produced by muscles pulling at various points on the skeleton to cause bending at the joints. This becomes the foundation for learning additional aspects of anatomy, physiology, and nutrition as we will connect these topics to providing energy, nutrition, and control to the skeletal-muscular system.

Time Required:

Part 1. The Human Skeleton (activity plus discussion, 30-45 minutes)
Part 2. How Muscles Move the Skeleton (activity, 35-45 minutes, plus interpretive discussion, 30-40 minutes)

Objectives: Through this exercise, students will be able to:

1. Point out the relationship between bones and joints in a model human skeleton and respective bones and joints in their own bodies.

2. Demonstrate how pulling their forearm toward the shoulder is done by their biceps muscle pulling between the forearm and the upper arm.

3. Demonstrate how pushing the forearm away from the shoulder is done by their triceps muscle pulling between their elbow and upper arm.

4. Demonstrate how other back-and-forth body movements are performed by paired muscles pulling on opposite sides of the joint.

5. Express how all physical body movements are produced by muscles pulling between two parts of the skeleton; give examples.

6. Express the generalization that muscles can only contract and relax; they cannot push.

7. State how animal meat, including chicken and fish (apart from organs), is muscle

of the given animal.

Required Background:

No particular background knowledge is required beyond the usual experience of a six-year-old.

Materials:

Realistic scale model of the human skeleton with movable joints
Model of human body showing muscles; pictures may suffice

Teachable Moments:

Bring out a model of the human skeleton and invite children to "play" with it, examining how it moves.

Methods and Procedures:

Part 1. The Human Skeleton

Bring out a model human skeleton with movable joints and provide time for children to examine and "play" with it as they wish, only being careful not to damage it. Ask them to particularly focus on how it moves at the joints.

Then, with model in hand, ask them to point at and feel various bones in their own bodies that correspond to particular bones that you point to in the model. (Don't bother with technical names of bones or get into the tiny bones of hands and wrists; common names of major bones will be fully sufficient, i.e., lower arm bones, hip bone, head bone, back bone, etc.)

Next, focus attention on joints and how they move. Have students follow your demonstration of opening and closing your arm at the elbow, and then observe on the model how the extent and limits of this movement is given by the hinge joint at the elbow. For example, they can't bend their elbows backwards because the extension of the lower arm bone at the elbow runs into the upper arm bone. Similarly, the fact that they can move their arm in all directions at the shoulder and even rotate it to some extent is enabled by the design of the ball-and-socket joint at the shoulder. The ability to twist and bend the back and neck is enabled by the individual bones (vertebrae) being able to bend and twist to some degree at each of the "joints" between.

Provide time and encourage students to work with the skeleton contrasting their own bones, joints, and movements with those of the skeleton. The lesson may or may not get into more detailed movements such as rotating the wrist, the ankle, jaw, and so on. Of course, you may point out anywhere along the way that forcing a "bending" of the skeleton in a way that the joints don't allow results in breaking bones or pulling them apart at the joints (dislocations).

Bernard J. Nebel, Ph.D.

Part 2. How Muscles Move the Skeleton

When students have gained an appreciation and understanding of their own skeletons, pose the question, "How do we make it move?" For example, we open and close our arm at the elbow as allowed by our skeleton, but how do we make the skeleton do that—or any other movement? Let's figure this out.

Have students pair up and do as you do in the following demonstration. (You may have an assistant or select one of your students to be your partner.) Each pair should sit close together facing each other at the side of a table or counter. Clasp hands and adjust yourselves so that your forearms are nearly vertical and your elbows are resting on the table. (You may need to put some books or something under your partners elbow to compensate for a difference in height.)

Now, each person pushes against the other's hand while using their free hand to feel the muscles in their upper arm as they do so. Admonish students that this is not an arm-wrestling contest. They are not to try to push their partner off his/her chair. The objective is to feel how the muscles in your upper arm behave as you go from relaxed into pushing (attempting to straighten your arm). Repeat the same process, pulling against each other's hand.

Have students tell and discuss with each other what they feel the muscles in their upper arm are doing as they alternatively push and pull. They may repeat the exercise as necessary to draw a conclusion. The observation will be that in going from relaxed to pushing (attempting to open their forearm) they feel the muscle on the back side of their upper arm (the triceps) flex, while the muscle on the front of the forearm (the biceps) remains soft and relaxed. In pulling (attempting to close the arm), the opposite occurs; the triceps remain relaxed and the biceps flex.

Explain to students that the flexing of a muscle (its getting hard and bulging) is a consequence of its getting shorter (contracting), and thus pulling between two points. Thus, their observation is that both pushing and pulling (opening and closing their arm) result from a muscle contracting and pulling, but it is a different muscle in each case. This conclusion is contrary to what is logically assumed. It will bear repeating and take some time to sink in. How is it possible that both the pushing and pulling actions are performed by muscles contracting? Let's take a closer look at the skeleton and where the two muscles are attached. (You may use or draw an enlarged diagram of the forearm, elbow, and upper arm.)

Have students especially observe that the elbow joint is not just a simple hinge. The bone of the forearm extends back, behind the hinge point forming the "funny bone" of the elbow. Thus, the forearm is really a sort of teeter-totter or see-saw arrangement, although one end is much longer than the other. Now explain that the biceps muscle is connected between the forearm and the upper arm. The triceps muscle is connected between the tip of elbow and the upper arm. Ask students if they can now explain how both opening and closing the forearm are performed by muscles contracting. Yes, opening the forearm is performed by the triceps muscle pulling at the tip of elbow; closing the forearm is performed by the biceps muscle pulling on the hand-side of the "teeter-totter." (Some students may elect to make a model of this arrangement.)

Ask students, "Do you think this idea applies to other sorts of movements?" Let's test it. Another simple action to test is this. Have children sit in chairs close to and facing a wall so that their toes are actually against the wall. Pushing feet hard against the wall, they will feel muscles on the top of their thighs flexing. Alternatively, in pulling the feet firmly back against the legs of the chair, they will feel the muscles on the under side of their thighs flexing. Similarly, students can test other sorts of movements.

Is there any situation where we find the SAME muscle both pushing and pulling? If students cite a movement where they think this is so, help them analyze it more closely.

The conclusion that should become evident is that MUSCLES WORK ONLY BY CONTRACTING. In contracting they can exert a great force. But beyond contracting, they can only relax and allow themselves to be stretched out again. They cannot exert any force in pushing between two points. All body movements, then, are performed by the contraction of muscles pulling between two parts of the skeleton. Back and forth movements, as demonstrated, are performed by paired muscles pulling alternately on opposite sides of the joint.

A point that is likely to need further clarification is our hands. They have great grasping power but there is very little muscle in them. Here is a neat "trick" to do with your students. Rest one forearm on a table, palm up, and hand relaxed. With your other hand, press on your forearm and feel your fingers close in. What does this mean? The muscles for grasping are actually in your forearm. They are connected to your fingers by cords called tendons. Your students can feel the tendons as the cord-like things on the inside of their wrists.

Have students observe a model or pictures showing the full muscular structure of the body. They will be impressed by the fact that there are hundreds of different muscles. Have them ponder the fact that all the flesh on their bodies, apart from skin and fat, is muscle. And all muscles work by pulling between two specific points on the skeleton. We have considered very simple, basic movements. Most movements, even those as seemingly straightforward as walking, require the coordinated contraction/relaxation of hundreds of different muscles in the legs, back, neck, and shoulders.

Questions/Discussion/Activities To Review, Reinforce, Expand, and Assess Learning:

Make books illustrating how a given movement is performed by a pulling action of given muscles.

Set up an activity center where students can continue to examine a model skeleton.

Invite students to make a model that demonstrates how a back and forth movement is performed by paired muscles pulling on opposite sides of a joint.

In small groups, pose and discuss questions such as:

What gives the basic structure and shape of your body?

Why/how does your body only bend in the ways that it does? For example, why does your arm only bend at the elbow? Why does your elbow only bend back and forth, whereas at the shoulder, you can move your arm in any direction?

What do muscles do? What can they not do?

Perform a simple movement and ask students to analyze it in terms of where the muscle is located and the points it is pulling between.

Every back-and-forth movement requires at least two muscles. Why?

Have students demonstrate a simple movement and point to their bodies to show the location of the muscle(s) that is/are performing that movement and the points it is pulling between.

To Parents and Others Providing Support:

Find opportunities to examine real or life-sized model skeletons with your children. Coach them as necessary in relating bones and joints in the skeleton to actual bones and joints in their bodies.

Repeat the activities described and have children go through the rationale in reaching the conclusions regarding the action of muscles.

When eating chicken or fish that includes the bones, talk about how the meat is actually muscle. Similarly all red meat is muscle of the given animal. Only organ meats, liver, kidney, etc., are not muscle.

Connections to Other Topics and Follow-Up to Higher Levels:

Similarities among all vertebrates

How movements are controlled (Lesson B-8)

Different body designs (Lesson B-7)

How/where muscles obtain energy (Lesson B-9)

Further studies into anatomy, physiology, and nutrition (Note that, in this text, we use supplying energy and nutrients for muscle function as the central focus for these studies.)

Health and nutrition

Re: National Science Education (NSE) Standards

This lesson is a steppingstone toward developing students' understanding and abilities aligned with NSE, K-4:

Unifying Concepts and Processes
- Systems, order, and organization
- Evidence, models, explanation
- Form and function

Content Standard A, Science as Inquiry
  • Abilities necessary to do scientific inquiry
  • Understanding about scientific inquiry

Content Standard C, Life Science
  • Characteristics of organisms

Content Standard F, Personal and Social Perspectives
  • Personal health

Books for Correlated Reading:

Catala, Ellen. *How Do You Move?* Capstone, 2006.

Nettleton, Pamela Hill. *Bend and Stretch: Learning About Your Bones and Muscles* (Amazing Body). Picture Window Books, 2004.

Rau, Dana Meachan. *What's Inside Me? My Bones and Muscles.* Benchmark Books, 2005.

Sweeney, Joan. *Me and My Amazing Body.* Dragonfly Books, 2000.

Balestrino, Philip. *The Skeleton Inside You.* (Let's-Read-and-Find-Out Science, Stage 2) HarperTrophy, 1989.

Barner, Bob. *Dem Bones.* Chronicle Books, 1996.

Krensky, Stephen. *Bones* (Step-Into-Reading, Step 2). Random House Books For Young Readers, 1999.

Rau, Dana Meachan. *What's Inside Me? My Bones and Muscles.* Benchmark Books, 2005.

Sweeney, Joan. *Me and My Amazing Body.* Dragonfly, 2000.

Lesson B-7

How Animals Move II:
Different Body Designs; Major Animal Phyla

Overview:

Within the animal kingdom, we find three fundamentally different body designs: skeletons inside, skeletons outside, and no skeletons at all. Students will investigate these three designs, how each works to perform movements, and animals that are representative of each. In turn, they will learn that the animal kingdom is divided into major subgroups, or phyla, on this basis.

Time Required:

Part 1. Animals with Skeletons Inside: Vertebrates (observation, analysis and discussion 40-50 minutes)

Part 2. Animals with Skeletons Outside: Arthropods (observations plus interpretive discussion, 40-50 minutes, plus activities as desired)

Part 3. Animals with No Skeletons: Snails, Worms, and Others (observing a worm move, 20-30 minutes, plus interpretive activity and discussion 30-40 minutes)

Objectives: Through this exercise, students will be able to:

1. Recognize and use the following words in their proper context: VERTEBRATES, INVERTEBRATES, ARTHROPODS, EXOSKELETON, PHYLUM, CLASS.

2. Describe the basic "skeleton inside" (vertebrate) body structure and name the types of animals (classes) that share this structure.

3. Point out how the presence of a backbone distinguishes these animals (2 above) from other sorts of animals that do not have a backbone.

4. Explain how the presence of a segmented backbone is the feature used to make a major subdivision (phylum) of the animal kingdom.

5. Describe the basic "skeleton outside" body structure and name the kinds of animals (classes) that share this structure.

6. Explain how the presence of a jointed skeleton outside is used to make another major subdivision (phylum) of the animal kingdom, namely arthropods.

7. Recognize and name the kinds of animals that have no skeleton; explain how they move; state how they form several additional phyla of the animal kingdom.

Required Background:

Lesson B-3, Distinguishing Between Plants and Animals
Lesson B-4A, Identification of Living Things
Lesson B-6, How Animals Move I: The Skeleton and Muscle System

Materials:

Part 1.   Access to real or model skeletons of a mammal, bird, reptile, amphibian, and fish. (Photographs of such skeletons may suffice.)

Part 2.   One or more specimens of a crab, crayfish, lobster, or large insect that students can examine closely, feel, and dissect.
If you and your students elect to build a model:
Paper towel rolls
String
Stapler
Scissors

Part 3.   One or more large, living earthworms. (may be purchased or dug from a garden and maintained and reared as you desire)
Egg-sized ball of clay or dough

Teachable Moments:

In reviewing the human skeleton and muscle system and how it works (Lesson B-6), invite students to consider other sorts of animals that share this basic structure. Have model skeletons (or photographs) of other vertebrate animals available for examination.

Methods and Procedures:

Part 1. Animals With Skeletons Inside (Humans and Other Vertebrates)

Revisit and review Lesson B-6, "How Animals Move I: The Skeleton and Muscle System." As students are up to speed, invite them to ponder other sorts of animals that have this same basic structure. From life experience, and especially from the ongoing activity of identifying local plants and animals (Lesson B-4A), they will be familiar with various mammals, birds, reptiles, amphibians (frogs, toads, and salamanders), and fish. Facilitate their examination of representative model (or photographs of) skeletons of each of these kinds of animals.

With Q and A discussion, have students cite similarities and differences among the skeletons of different kinds of animals. Make two columns on the board listing similarities and differences. Children are likely to cite differences in size, appendages, skull, and so on but, bring them to observe that all share the same basic principle of having a bony, jointed skeleton on the inside. Their movements are performed in the same manner as was discovered for humans in Lesson B-6; muscles pull on the skeleton at various points to perform bending at the joints.

Further, all the differences notwithstanding, have students note that they all share the feature of a backbone, which is made up of individual pieces or segments, each called a vertebra (pl. vertebrae).

Do all members of animal kingdom have a backbone of this sort? What about crabs, insects, clams, or worms? Students' experience should lead them to say no. Explain, then, that this feature leads us to make a basic subdivision of the animal kingdom. All of these kinds of animals that have a backbone—a skeleton on the inside—are put in a category we call VERTEBRATES. Children will readily see where this name comes from. Finally, such a major subdivision of the animal kingdom is called a PHYLUM (pl. PHYLA).

Fish, amphibians, reptiles, birds, and mammals (including humans) are subdivisions or categories of the vertebrate phylum. Major subdivisions of phyla are called CLASSES. Thus, fish, amphibians, reptiles, birds, and mammals (including humans) are CLASSES of the vertebrate phylum. All other members of the animal kingdom are commonly referred to as INVERTEBRATES since they do not have a segmented backbone. But, as we shall see in the following, there are several phyla of invertebrates.

You can think of and conduct activities that will call on children to sort various animals into vertebrates or invertebrates and into the various the classes of vertebrates as you see fit.

(Turtles may cause some confusion and need special attention. It does appear that their shell is a skeleton on the outside. Have students note, however, that neck, legs, and tail are typical of other vertebrates; the bones are on the inside. Then point out how the shell is really a fusion of vertebrae on the top-side and a fusion of the sternum (breast bone) on the underside, and both are still covered with a skin. Thus, turtles are vertebrates; they are a subdivision of reptiles.)

Part 2. Animals With Skeletons Outside (Arthropods)

Invite students to examine a crab, crayfish, a large insect such as a grasshopper, or other such animal. Does it have a skeleton, particularly a backbone on the inside, as vertebrates do? Allow students to touch and feel the exterior and note that it is hard and inflexible. Likewise, have them examine the joints of the appendages; they will observe both ball-and-socket and hinge joints clearly visible. Finally, if there is not enough material to have students make their own investigations, at least demonstrate how the inside of the "critter" is all soft tissues.

What does all this say about the overall design and structure of such animals? Through Q and A discussion, guide students to the conclusion that such animals have their "skeleton" on the outside, and muscles on the inside. The skeleton being outside is called an EXOSKELETON. It includes many jointed appendages, which are basically tube-like structures with joints. Muscles are inside the tubes and body; they pull on the "tubes" at various points inside and cause bending at the joints. You may have students who have enjoyed eating crabs or lobsters who can attest to the skeleton being outside and meat (muscle) inside.

Have students note that, here again, muscles only work by contracting. Back and forth movements are performed by paired muscles pulling on opposite sides, in this case opposite points inside the tubes that form legs and other body parts.

Instruct students that this basic body design of skeleton outside, or exoskeleton, leads us to make another basic subdivision of the animal kingdom. Such "critters" are put into another phylum called ARTHROPODS. (Translated, arthropod actually means "jointed feet.") Have students reflect on and cite the different types of animals that share this basic structure. The list should include crabs, crayfish, lobsters, insects, and spiders. Again these are different categories, or CLASSES, within the ARTHROPOD phylum. (There are a number of other classes of arthropods, such as centipedes and millipedes, but keep the discussion to "critters" that students are familiar with.)

Some students may suggest putting clams, snails, and other such animals here because of their shell. Point out that such animals do not have jointed appendages as do all arthropods. Therefore, clams and snails belong in yet another phylum known as mollusks, but let's save them for another lesson.

One may wish to go into and discuss of the advantages and disadvantages of the vertebrate internal skeleton and the arthropod exoskeleton. With Q and A discussion, you can bring students to note that the exoskeleton does offer some degree of protection, although many birds, frogs, and other animals eat the whole thing regardless. The major disadvantage of an exoskeleton is that it impedes uniform growth. The only way an arthropod can grow is to split its old exoskeleton, climb out, expand to its larger size, and then form a new one. As it comes out of its old exoskeleton, it is soft, nearly helpless, and very vulnerable. Also, in large sizes, hollow tubes are subject to buckling under heavy weight. Therefore, science fiction movies notwithstanding, very large (human-sized) land-dwelling arthropods are physically impossible.

Part 3. Animals With No Skeletons

Bring students to ponder: Do "skeletons inside" (vertebrates) and "skeletons outside" (arthropods) now cover all members of the animal kingdom? What about worms? Jellyfish? Clams and snails? Some students will declare that snails and clams do have a skeleton—the shell. Point out that shells are not jointed as is a true skeleton. Therefore, shells do not count as skeletons. Students should conclude, then, that such animals have no skeleton at all, which brings up the question: How do they move? Lets see if we can figure it out.

Place a large, healthy earthworm on a pad of moist paper towel and have children observe its "crawling." It will twist and wiggle at first, especially if poked, but let it sit quietly for a minute or more and it will begin to stretch out straight and move forward. This is the motion you want students to examine. Most children will be fascinated.

Coach students as necessary in analyzing and describing the worm's movement. There are basically two phases. It pushes its forward end out in front; then it pulls up its hind end.

Have students model this movement. Give them an egg-sized lump of clay or dough. As they squeeze the diameter of one end–call it the back–it oozes out at the front. Then, if the front is held in place and the back is pulled up, and the whole process repeated, they will have a working model of the worm's movement.

When children have a visual understanding of the worm's movement, you may give a description of how this actually works. The worm has "belt" muscles. These are like a series of belts down its entire length. It pulls in the thickness of its body at any point by means of these muscles. Then, it also has muscles running the length of its body. Shortening the length of its body is performed by these muscles.

Review and emphasize again that muscles only contract and relax. They cannot push. Thus, the pushing forward of the front of the worm IS NOT a matter of muscles pushing. It is a matter of the back portion squeezing in by means of the belt muscles and forcing the front portion, to bulge forward, just as in their model. The front portion grips the surface. (Students may actually feel the rows of tiny barbs along either side of the front portion of its "belly" that it uses for this purpose.) Then it pulls up its hind portion using the muscles along its length.

Have students observe their worm some more, noting how/where it squeezes itself in with its belt muscles and how/where it pulls itself up with its length muscles. They will observe that these two processes may occur at the same time along different portions of the worm's body—quite a feat of coordination.

Have students reflect on this sort of movement. How efficient is it? It may seem very slow, cumbersome, and awkward. Actually it can be very quick and accurate, and it can be seen in many places in the animal kingdom. Children will be excited to learn that their tongues work by same principle. Have them note that they have no bones in their tongues. But there are muscles crisscrossing in various directions. Muscles contract in one location forcing the tongue to bulge out in another location and thus move accordingly.

In conclusion, have children cite again the many sorts of animals that have no skeletons. They all move by this principle we have just observed. But among these animals there is still such diversity—contrast a snail with a jellyfish, for example—that we don't put them into a single phylum of "no skeletons." They are separated into several categories, or phyla, according to additional distinctive features of their bodies. This will be the topic of later lessons.

194

Questions/Discussion/Activities to Review, Reinforce, Expand, and Assess Learning:

Make books illustrating the three basic body designs.

Set up an activity center where students can continue to observe, compare, and contrast the kinds of animals and their respective movements.

Make models that demonstrate the structures and movements studied.

Set up a large bulletin board with columns labeled: Vertebrates, Arthropods, and Animals With No Skeletons. As children continue Lesson B-4A, have them include with their identification and show-and-tell the basic body design and the name of the phylum to which their "critter" belongs. Have them make a tag with the name of the animal and post it on the bulletin board in its proper column. As this activity continues, further separations of animals with no skeletons into respective phyla and separations of phyla into classes and orders may be conducted as fitting.

If some students wish to independently pursue collecting, identification and classification of given groups of organisms, facilitate their doing so.

In small groups, pose and discuss questions such as:

What are three basic body designs found in the animal kingdom?
How does this affect the way we place animals in categories?
What classes of animals share the basic vertebrate design?
What do mammals, birds, reptiles, amphibians, and fish have in common?
What classes of animals share this basic arthropod (skeleton outside) design?
Cite kinds of animals that have no skeletons.
How do animals without skeletons move?
How do we move our tongues?
Cite any "critter" that students are familiar with and have students classify it
    according to is body design and assign it, so far as possible, to phylum and class.
Cite any "critter" that students are familiar with, and have students describe its basic
    body structure and it mode of moving.
How did species names and the classification scheme come about? (Were they always
    there, or is it a human-made system? Go into the history of classification as you
    wish.)

To Parents and Others Providing Support:

When you and your children come across an invertebrate critter, invite them to inspect it more closely, especially its movements, and consider its body structure and how it moves. Another opportunity comes when eating a whole crab or lobster.

Visit natural history exhibits and ponder displays of various vertebrate skeletons, including human. Note and discuss their similarities in structure and mode of moving. Review as necessary how movements are performed by muscles pulling between particular points of the skeleton. Back-and-forth movements are produced by paired muscles alternately pulling on opposite sides of a given joint.

When you come across a large earthworm on the sidewalk after a rain or when digging in the garden, take time to observe and discuss its body structure and how it moves. Does it have a skeleton? How does it move?

Facilitate your children setting up an earthworm farm, if they wish. Earthworms can be kept and raised in a fish bowl or aquarium with moist (not wet) garden soil. Cover leaving a small opening, to preserve moisture but still allow air exchange. Vegetable peelings put on the surface will provide sufficient food and water. Indeed, many people and some municipalities use earthworms as a means of disposing of plant waste and generating rich topsoil (humus) in the process, but that is another lesson.

Continue the ongoing activity of using field guides to identify common plants and animals found in your region. Add to this the classification of animals according to phylum, class, and orders where given.

Connections to Other Topics and Follow-Up to Higher Levels:

The nervous system and how it coordinates the muscular system to give purposeful movements (Lesson B-8)
Providing energy to the muscles (further considerations of anatomy and physiology and nutrition, Lesson B-9)
Further classification/taxonomy of the animal kingdom
More detailed classification (orders through species) within a given group
How muscles work on the cellular level

Re: National Science Education (NSE) Standards

This lesson is a steppingstone toward developing student's understanding and abilities aligned with NSE, K-4

Unifying Concepts and Processes
• Systems, order, and organization
• Evidence, models, and explanation
• Form and function

Content Standard A, Science as Inquiry
• Abilities necessary to do scientific inquiry
• Understanding about scientific inquiry

Content Standard C, Life Science
• Characteristics of organisms

Content Standard G, History and Nature of Science
• Science as a human endeavor

Books for Correlated Reading:

Glaser, Linda. *Wonderful Worms.* Millbrook Press, 1994.

Himmelman, John. *An Earthworm's Life* (Nature Upclose). Children's Press, 2001.

Pfeffer, Wendy. *Wiggling Worms at Work* (Let's-Read-and-Find-Out Science, Stage 2). HarperTrophy, 2004.

Smithyman, Kathryn, and Bobbie Kalman. What is an Arthropod? (Science of Living Things), Crabtree, 2002.

See also list under B-3.

Lesson B-8

How Animals Move III: Coordinating Body Movements; The Nervous System

Overview:

Students have learned that body movements are performed by muscles contracting and pulling between two points on a skeleton (Lesson B-6). But movements would be random and meaningless were it not for the central nervous system. Here, students will learn principal aspects of this system: how it gathers information concerning our surroundings, interprets that information, and directs appropriate responses. Through this lesson, children will begin to appreciate the danger of mind-altering drugs.

Time Required:

Introductory demonstration and discussion (30-40 minutes)
Play-acting functions of the nervous system (as desired)

Objectives: Through this exercise, students will be able to:

1. Understand and use the following words in their proper context: NERVOUS SYSTEM, SENSORY ORGANS, BRAIN, NERVES, MOTOR NERVES, NERVE IMPULSES.

2. Describe the three basic roles performed by the central nervous system and name the parts that play each role.

3. Tell how their body senses the external environment by listing their five basic sensory organs.

4. Playact the functions performed by the central nervous system in responding to any given stimulus.

5. List aspects of the internal environment that are also sensed.

Required Background:

Lesson B-6, How Animals Move I: The Skeleton and Muscle System

Materials:

No particular materials are required.

Teachable Moments:

In the course of reviewing Lesson B-6 or otherwise talking about how movements are performed by muscles, guide discussion into how those movements are controlled.

Methods and Procedures:

In the course of reviewing Lesson B-6, "How Animals Move I" pose the question, "How are movements coordinated and controlled?" It might go something like this. Pretend you are eating; demonstrate and have students perform a pretend eating motion bringing their hand back and forth between a plate on the table and their mouths. What muscles are active in this motion? They might point at various muscles in their arms and shoulder.

But, quickly move to: How are those muscles coordinated and controlled? That is, without coordination and control, you might put the food in your eye, throw it over your shoulder, or whatever. Kids will have fun thinking and demonstrating other uncoordinated movements. (Interject, however, that this is no laughing matter. People with certain diseases or injuries do have trouble coordinating and controlling movements, and what a unfortunate and serious disability it is.)

Next, pose the question, "How do you think movements are coordinated and controlled?" Let kids ponder this a few moments, but then in Q and A discussion, guide them to recognize that there are three main components.

1) seeing the food on their plate.
2) recognizing what it is and deciding to eat it (or not).
3) the action of the muscles performing the movements that bring the food from the plate to your mouth.

Repeat this process and discussion using examples that involve other senses, such as hearing a phone ring and reaching for it; having a kitty on your lap and gently stroking it; and so on. In each case, have children note that the process involves the same three elements: sensing what is there, which may be seeing, hearing, feeling, tasting, or smelling; deciding what they wish to do about it; and activating muscles to bring about the action.

As students master these three basic parts to their actions, go on to explain that these are functions performed by their NERVOUS SYSTEM. The NERVOUS SYSTEM has three main parts:

1. SENSORY ORGANS (eyes, ears, nose, tongue, and those that sense touch). Discuss briefly how each is sensitive to one and only one thing: eyes sense light,

199

ears sense sound, and so on. In each case, on receiving the given stimulus, the sense organ sends signals to the brain. The signals are called NERVE IMPULSES and they travel along NERVES to the brain. It is similar to signals being carried over a wire.[21]

2. The BRAIN. The brain, on receiving nerve impulses from a given sense organ, translates them accordingly. Nerve impulses from the eyes are translated into seeing; those from the ears are translated into hearing, and so on.

3. The brain decides what to do about what you see, hear, etc., and it sends nerve impulses along another set of nerves called MOTOR NERVES leading from the brain to the muscles. As the muscles receive nerve impulses, they are caused to contract. Note that there are nerves leading from the brain to every muscle in the body. Therefore, this last phase involves sending just the right number of nerve impulses to exactly the right muscles to make them contract with proper force and amount to bring about the desired movement.

An instructive game that will help solidify this lesson is to have students take turns pretending to be a robot and playacting what is occurring in the nervous system in given situations. For example, in response to a plate of cookies, it might go something like this:

Bzzz—eyes scanning the room. Sending nerve impulses to the brain.

Bzzz—brain making pictures from nerve impulses.

Bzzz—brain making picture of small, brown, round, flat objects on plate. Ding, ding.

Bzzz—brain interpreting picture as COOKIES. Ding, ding.

Bzzz—brain saying, WANT COOKIE. Ding, ding.

Bzzz—brain sending nerve impulses to muscles to walk over, pick up cookie, and put
    it in mouth. Ding, ding. Doing so

Bzzz—taste buds in tongue, sending nerve impulses to brain.

Bzzz—brain interpreting nerve impulses as, "Yum. Tastes Good."

Students can start with any sort of sight, sound, or other stimulus and decide on any action. As they master this concept, emphasize again that seeing, hearing, smelling, tasting, and feeling are all the brain's interpretation of nerve impulses received from the respective sensory organs. In combination, they tell us what is going on outside our bodies, that is our external environment.

---

[21] Comparing a nerve impulse to a signal carried over a wire is only to convey a general concept. A nerve impulse actually involves a complex chemical-physical change transmitted along the membrane of a nerve cell. This is obviously for more advanced biology and chemistry. At this level we do best by just providing a conceptual picture.

Have children ponder that all sensory organs are simultaneously sending nerve impulses to the brain, indeed millions of nerve impulses per second from each sensory organ. It may seem as if our brains are receiving nerve impulses from just one sensory organ at a time, our eyes for example. But, with your students, close your eyes and focus your attention on all the sounds you are hearing. In turn, focus your attention on all the sensations you are feeling, the pressure of your seat in the chair, and so on. Then focus on smell and the taste in your mouth.

The conclusion of this is that, yes, all the sensory organs are continuously sending nerve impulses to the brain. But the brain is continuously sorting them out, making you consciously aware of only those that you are deeming important at the moment, and guiding your actions accordingly. Nerve impulses from other organs go unnoticed, unless they are especially severe. A clap of thunder, for example, will call your attention away from whatever else you may be focused on.

The sensory organs we have been focusing on give us information regarding what is outside our bodies, that is, our external environment. You may or may not go on to explain that there are also numerous sensory organs that send nerve impulses to the brain regarding conditions inside our bodies, that is, our internal environment. Hunger, thirst, temperature (being too hot or cold), pain, position, and tiredness are among the main ones. Again, it is a matter of nerve impulses being sent to the brain, the brain interpreting them, and the brain sending nerve impulses to conduct appropriate actions.

Coming back to the brain sending nerve impulses to the muscles to cause desired movements, we should note it does not occur automatically all at once. There is much learning involved. Consider, for example, a child learning to walk, learning dance movements, learning to play a musical instrument, learning to catch and throw a ball, learning to swim, and so on. In every case there needs to be much practice. What is practice? Explain that through practice, we are "teaching" the brain and the brain is "learning" just how many nerve impulses to send to which muscles to produce the desired movements.

As this learning occurs, actions become automatic. Take walking for example. After learning, we no longer have to think about just how much tension to put on which muscles to maintain balance and move forward. We just get up and walk. In short, the brain has learned just how many nerve impulses to send to which muscles to give the desired action, and it does this automatically without our even noticing.

Therefore, you may need to emphasize on repeated occasions that, although actions have become automatic, the brain is still the functional organ in sending the proper number of nerve impulses to the proper muscles to coordinate these movements as desired. This is why the result of brain or nerve damage is often paralysis. The muscles no longer get the nerve impulses from the brain that they need to function properly.

But even all of this is but a small portion of what the brain does. Most significantly, it works to store information from learning and experience. Then it is capable of scanning this information and using it to answer questions and solve problems. Finally, it is able to make connections and extrapolate from what it knows to new ideas. This is the essence of creative thinking. It is witnessed in statements such as: This idea just came to me. Let's try it this way.

(Even with all this, you may note that we have omitted mention of another side of the nervous system. What we have portrayed here is referred to as the voluntary nervous system. That is, we are conscious, more or less, of the what our sensory organs are perceiving, and we are voluntarily deciding how or if to act on it. In addition to this, there is another part called the autonomic nervous system. This part of the nervous system monitors all sorts of internal parameters such as blood pH, blood sugar, temperature, water content and adjusts the functioning of internal organs automatically without our being aware of its operation. But, this is best saved for higher levels.)

If you and your students are not boggled by all the brain does, you should be. The brain's functions and capacities really are phenomenal and most of how it actually works is still unknown, providing great opportunities for future research.

Finally, as students begin to appreciate the phenomenal capacities and functions of their brains, they may also be impressed with the utter stupidity and risk of "playing" or "experimenting" with mind-altering substances including excessive amounts of alcohol. In doing so, they are messing in unknown ways with the most phenomenal "machine" known to man, and results may be anything but good.

In the course of this lesson, you may be inclined to delve into the anatomy of eyes and ears and how they work. You may even be dismayed that such anatomy has not been included here. But, the omission is quite intentional. Experience shows that it is more productive to first give children an overview of the whole system. The more detailed workings of eyes, ears, etc. will follow logically in due course.

Questions/Discussion/Activities to Review, Reinforce, Expand and Assess Learning:

Make books illustrating the three parts of the nervous system and what they do.

Set up an activity center where children can look at pictures (or models if available) showing the anatomy of the eye and ear, nerve cords going from the sense organ to the brain, and from the brain to the body. (Google "anatomy nervous system")

Play the "robot game" described above for various sorts of stimuli. Coach students as necessary in stating all the stages.

Compare and contrast the functions and capacity of the human nervous system with that of other members of the animal kingdom. In every sort of animal, one can identify various sensory organs, a brain (of some degree), and nerves leading to muscles to initiate movements. However, the degree of thinking that non-human animals can exercise is debatable, and subject to experimentation. In most invertebrate animals, there is little brain at all. Signals from sense organs go directly to muscles to initiate movements. Still there is an astounding degree of coordinated movement. Consider, for example, how a butterfly must move its wings just so in order to land on a particular flower.

In small groups, pose and discuss questions such as:

> You hear the sound of your mother's voice calling you to dinner.
> What sensory organ detects the sound?
> What does it do?
> What does the brain do? (two things)
> What is the final response?
> You may conduct this sequence of Q and A for different sorts of stimuli leading to all sorts of responses.
> What happens if one of your sense organs is nonfunctional?
> Why does a broken neck or back frequently result in paralysis?
> The responses of a person who has drunk too much alcohol are slow and uncoordinated. Why? What has alcohol done to each part of the nervous system?
> Suppose you want to learn a new physical skill. What do you need to do? As you practice, what are you teaching your brain to do?
> What does the brain do in addition to receiving impulses from sensory organs, interpreting what they are, and sending impulses to muscles to produce responses?

To Parents and Others Providing Support:

As children respond, in one way or another, to a given sight, sound, or smell, take the opportunity (after the emotion has subsided) to coach them in analyzing the situation and their response in terms of the three parts of their nervous system.

Play the "robot game" with them, coaching as necessary.

Discuss capacities of the brain in addition to simple stimulus-response actions described here.

Take opportunities to discuss the effect of drugs and alcohol on the nervous system.

In looking at another member of the animal kingdom, discuss its nervous system. Does it have one? How is it similar to and different from ours?

Connections to Other Topics and Follow-Up To Higher Levels:

Anatomy and physiology of each sensory organ
Anatomy and physiology of nerve cells
Anatomy and physiology of the nerve-muscle connection (How does a nerve impulse lead to muscle contraction?)
Anatomy and physiology of the brain
Comparative anatomy stressing the nervous system of other members of the animal kingdom

Bernard J. Nebel, Ph.D.

The major role of the other organs of the body is to provide the nervous and skeletal-muscle systems with energy to operate, and also material for growth, maintenance, and repair. Lesson B-9, "Energy to Run the Body" will provide this overview.

Re: National Science Education (NSE) Standards

This lesson is a steppingstone toward developing student's understanding and abilities aligned with NSE, K-4:

Unifying Concepts and Processes
- Systems, order, and organization
- Evidence, models, and explanation
- Constancy, change, and measurement
- Form and function

Content Standard A, Science as Inquiry
- Abilities necessary to do scientific inquiry

Content Standard C, Life Science
- Characteristics of organisms
- Organisms and environments

Content Standard F, Science in Personal and Social Perspectives
- Personal Health

Books for Correlated Reading:

Aliki. *My Five Senses* (Let's-Read-and-Find-Out Science, Stage 1). HarperTrophy, 1989.

Berger, Melvin. *Why I Sneeze, Shiver, Hiccup and Yawn* (Let's-Read-and-Find-Out Science, Stage 2). HarperTrophy, 2000.

Curry, Don L. *How Does Your Brain Work?* (Rookie Read-About Health). Children's Press,   2004.

Nettleton, Pam Hill. *Look, Listen, Taste, Touch, and Smell: Learning About Your Five Senses* (Amazing Body). Picture Window Books, 2004.
_____. *Think, Think, Think: Learning About Your Brain.* Picture Window Books, 2004.

Rau, Dana Meachen. *Look!: A Book About Sight* (Amazing Body). Picture Window Books, 2005.
_____. *Shhhh...: A Book About Hearing* (Amazing Body). Picture Window Books, 2005.
_____. *Sniff, Sniff: A Book About Smell* (Amazing Body). Picture Window Books, 2005.
_____. *Soft and Smooth, Rough and Bumpy: A Book About Touch* (Amazing

Body). Picture Window Books, 2005.

_____. *Yum!: A Book About Taste* (Amazing Body). Picture Window Books, 2005.

Sweeney, Joan. *Me and My Senses*. Crown Books for Young Readers, 2003.

Lesson B-9

How Animals Move IV:
Energy to Run the Body

Overview:

This lesson will cover the major internal organs of the body and the role that each plays in serving the body's needs. First-grade science lessons often focus on a single organ or system of the body, such as the heart or the digestive system, and cover it in some detail. This approach has two downfalls: it tells children that science is full of details with big words, which frequently turns them off, AND it fails to give them any appreciation of their bodies as an integrated whole.

The approach we use here is quite different. It will give students a broad understanding of how the major organs work together as a team while at the same time minimizing complex detail. This will provide a unified foundation that invites further studies to expand the picture and fill in more details as students progress. The unifying theme is energy, how the body derives the energy it needs to function. Additionally, basic concepts regarding the need for good nutrition are introduced.

Time Required:

The totality of this lesson should not be attempted in one sitting. It will be best broken into 15-20 minute segments addressing different points and gradually weaving them together.

Objectives: Through this exercise, students will be able to:

1. Recognize and state that a primary need of the body is a source of energy, as well as nutrients for growth and maintenance.

2. Recognize and state that the source of energy for the body is food.

3. Name the following organs and point to the general location of each in their bodies: lungs, heart, stomach, intestines, liver, and kidneys.

4. Describe the primary role of each organ or system in the context of providing the body with energy.

5. Observe and state how pulse rate, breathing, and frequently appetite are related to

level of activity; state how this relates to the energy needs of the body.

6. Describe how weight gain or loss results from an imbalance between energy (food) intake and energy expended (activity of the body).

7. Contrast and distinguish between food for energy and food for growth.

8. Describe the basic features of good nutrition.

Required Background:

Lesson A-9, Matter IV: Dissolving, Solutions and Crystallization (and earlier lessons in the A-thread)

Lesson B-8, How Animals Move III: Coordinating Body Movements: The Nervous System (and earlier lessons in the B-thread)

Lesson C-4, Concepts of Energy III: Distinguishing Between Matter and Energy (and earlier lessons in the C-thread)

Materials:

A model of the human body showing internal organs (In the absence of a model, pictures may suffice. Google "human anatomy illustrations")

Teachable Moments:

Teachable moments may occur in the course of playing "robot," the activity described in Lesson B-8, or when vigorous activity brings on heavy breathing and works up an appetite. In other words, teachable moments are times when children are experiencing a connection between breathing, appetite, and physical activity.

Methods and Procedures:

In the course of playing "robot" as described in Lesson B-8, pose the problem: If you were building an actual robot, how you would you go about it? What are the features you would need to consider? With Q and A discussion, guide students to come up with the basic ideas:

They would need to construct a skeleton of sorts.

They would need to have a mechanism (to act like muscles) for moving the skeleton.

They would need to have a system of sensors (to act like the nervous system) to enable it to detect things and respond appropriately. (Otherwise, it would be always running into and falling over things.)

What else would it need?

Have children think back to their lessons concerning energy (Lesson C-1, in particular). Does anything go or work without a source of energy to run it? Students should reflect that the answer is NO and conclude that they would also need a source of energy for powering the "muscles" and "nervous system" of their robot. (The complexity of doing all this should speak to why we don't have human-like robots in real life.)

Go on to emphasize that the human body is no different. We and all other living things need a source of energy to make us go. Where/from what do all members of the animal kingdom get energy? Have students recall from Lesson B-3 that the answer is FOOD. (They will recall that plants get their energy from light.) Food is potential energy; food is the fuel we use to run our bodies—make muscles contract, make the nervous system send impulses, and so on—and is also the source of nutrients for growth.

How does food provide energy? This is what we will investigate for the rest of this lesson (or several lessons depending on your schedule).

(As you read the following description, you may well be dismayed by the tremendous simplification and omission of terminology. Keep in mind that the objective is to present a conceptual picture to grade-two-level students. It is NOT to overwhelm them with details and big words. As they gain the conceptual picture, they will more than likely want to know more and you can expand to fill in details and terminology as desired. Of course, this is an unending process leading all the way to careers in ongoing research.)

Pose the question, "How does food provide that energy?" Students are likely to answer, and it is a common misconception, that food is simply turned into energy. Stress the fact from Lesson C-4 that matter is not turned into energy, nor is energy turned into matter. However, food is potential energy and it does release stored energy on being broken down and "burned."

Burned?

Well, it is not burned like wood in the fireplace. If it were, we would glow with flame in the dark and the heat would cook us. Rather, the burning of food in our bodies goes on at a relatively low temperature (body temperature), and the energy is released in a form that will make muscles contract, send nerve impulses, and run all the other processes that occur in the body, including putting chemical building blocks together for growth.

While the burning of food/fuel in the body occurs at a modest temperature, stress that the overall requirements are the same. It requires oxygen and releases carbon dioxide, just like a flame. (Have students recall, or demonstrate again, how a candle flame is snuffed out by lack of oxygen, Lesson A-7.) Further, there are generally some "ashes" left over.

Another difference is that you don't have a central engine or furnace in your body as you do in a car or home. Rather, the "burning" takes place in each and every tiny bit—we call the tiny bits CELLS—of muscle, nerve, eye, etc. Every living cell in your body has its own places where it "burns" food to release its potential energy. That energy goes on to do the cell's work—contract if it is a muscle cell, send an impulse if it is a nerve cell, and so on.

Allow some time, with reviews, for students to absorb the idea that food is effectively burned to release the energy needed to run our bodies, and that the burning occurs in each and every living cell of the body. Then pose the questions such as: How do you suppose food gets from your mouth to all the cells of your body where it is "burned"? What processes does it go through? What else is required for burning to take place? Finally, how are the "ashes" disposed of?

Instruct students as follows, but go slowly, providing time for them to ponder and ask questions at each point. Break the lesson into parts as necessary, but keep coming back to each part so that students see it in terms of the overall picture of providing energy to run the body. At each step, use a model, or at least pictures, showing the internal anatomy of the body, and help students point to their own bodies to show the location of each organ.

Step 1. Food must be broken into tiny particles

The first step is to break up the food into particles that are suitable for burning. An analogy with firewood may be helpful. It is impossible to bring a whole tree into the house and put it in a fireplace. It must first be sawed and split into pieces that will fit and burn easily. The same is true for food; it must first be ground and broken into tiny particles.

The initial grinding occurs in your mouth as you chew your food. When you swallow, the food goes down a pipe to your stomach. In your stomach, food is treated with acid and other chemicals that break it down further into the tiny particles suitable for burning.

From the stomach, food particles proceed down the long, folded-up tube called the intestines. This tube is continuous all the way to and out the back end. Have children note that food in the mouth, stomach, and intestines is not really in the working part of the body. The body has the basic structure of a donut. The mouth, stomach, and intestines are a hole through the middle. Have students consider: Does something that goes through the hole in a donut ever enter the donut? Thus, anything that goes in the front end of the body and out the back end has never entered the body proper. It has simply passed through the hole in the donut.

To be sure, a portion of the food you eat—despite being chewed and treated with acid and other chemicals—is not broken into tiny particles. It does end up going through and out the other end. (Use whatever word you do to refer to fecal matter.) This is necessary to keep the intestines open and functioning, but it has nothing to do with the food that provides energy or body building nutrients. Again, it has simply gone through the hole in the donut.

Bernard J. Nebel, Ph.D.

Step 2. Transportation from the intestine to the cells of the body and on

To get back to the energy/nutrient story, we must return to the food that IS broken down into the tiny particles that are suitable for "burning." As they move down the intestine, they are taken through the wall of the intestine and enter the blood. The blood system carries the food particles to all parts of the body.

Have students note that blood is a liquid; it is basically water, but it is thick with all sorts of things in solution or carried along. Most prominently, it carries along cells filled with a red material that serves to carry oxygen. These so called red blood cells give blood its red color. Wherever you may be scratched or cut, blood starts coming out. But your blood is not just there, it is continually moving; it constantly flows to and from all parts of the body. Your heart (ka-bump, ka-bump) is the pump that keeps the blood flowing. The fact that blood is constantly being moved to and from every single part of the body can hardly be overstressed.

Your blood does not actually enter the cells of muscle, nerves, bone, and other tissue of the body. Rather, it flows by something like water flowing over the fingers of your hand. As blood flows by, food particles to be burned, and other nutrients, move from the blood into the cells. Just as importantly, emphasize that wastes from the burning—the ashes, so to speak—move from the cells into the blood.

Students may be puzzled by the idea of things going in and out of cells at the same time. Explain that it is like a revolving door that permits some people to go in at the same time others are going out. We will find many instances of this sort of two-way movement.

Step 3. The need for oxygen and elimination of carbon dioxide

Stress again that burning requires oxygen from the air as well as the material to be burned. Perhaps repeat the candle snuffing demonstration. Without oxygen, there can be no burning and no release of energy. This is what your breathing is about. Take a deep breath. With that and every other breath, you bring fresh air, which contains oxygen, into your lungs. The oxygen moves across the walls of your lungs into your blood. Thus, blood is not just carrying food from your intestines to the cells of your body. It is also picking up oxygen from your lungs and carrying it to all the cells of your body as well.

Thus, all the muscle, nerve, and other cells of your body are constantly getting food particles to burn and oxygen to burn them from the blood.

But, the story does not end here. There are waste products ("ashes") from burning. (Emphasize again that matter is not destroyed by burning or any other chemical reaction; it is only changed into different forms.) As oxygen and food particles move from the blood into cells, wastes move from the cells into the blood.

What happens to the waste products? Carbon dioxide is the main one. Carbon dioxide is a gas but it does dissolve in water and the blood. As the blood flows past the compartments of the lungs, carbon dioxide moves from the blood into the air spaces in the lungs and out with exhaling. Again, students can visualize a revolving-door process. Carbon dioxide is moving from the blood into the lungs at the same time oxygen is moving from the lungs into the blood. Breathing constantly refreshes the air in your lungs, bringing in more oxygen and expelling carbon dioxide.

Have children stand up and do jumping jacks for a full minute or so until all are breathing hard as a result. Proceed with Q and A discussion to guide them in reasoning why vigorous activity should result in faster, deeper breathing. They should conclude that their activity requires additional energy. The extra energy comes from additional fuel being burned. The additional burning requires additional oxygen and the disposal of additional carbon dioxide. Hence, you breath harder. Fortunately, you don't need to think about doing this; your nervous system (the autonomic system) detects the need and makes you do it automatically.

Step 4. Kidneys and urination

Additional wastes (ashes) from burning food are not gases; hence they cannot be eliminated through the lungs. But they can be carried in water solution in the blood. How is the body to rid itself of them? Here is where kidneys and urination come into the picture. (Use whatever word you do for urination.)

As blood circulates it goes through the kidneys as well as all other parts of the body. As it passes through the kidneys, it is filtered; some water with the "ashes" in solution move from the blood into tubes within the kidneys. This water with "ashes" in solution is urine. The tubes within the kidneys lead to the bladder where urine may be held for a time then released with urination. The kidney's filtering process goes on continuously; without the bladder, we would be dripping constantly.

Step 5. The role of the liver

Emphasize that all cells have a constant need for food to burn for energy and oxygen to burn it. Hence, we must breathe all the time, even when we are asleep. Going without breathing for more than a minute or so, as occurs in suffocation or drowning, is fatal. Use Q and A discussion to bring children to reason out the steps by which drowning causes death. It is not the water; it is the body not getting oxygen. Without oxygen, the burning of food immediately stops, energy release stops, and cells cease to function and die for lack of energy. All cells need a constant supply of energy to maintain the living state.

We do breathe all the time, but it is conspicuous that we don't eat all the time. How do cells gain a constant supply of food to burn? This is where the liver comes in. Have students note in models or pictures the relative size of the liver. In terms of mass, it is the largest organ in the body. (The lungs may be larger, but most of that space is air.)

Our livers are what enable us to go between meals without eating. The liver acts as a storage place for food particles to burn between meals. Shortly after a meal, there is more than enough food particles moving from the intestine into the blood to supply the energy needs of cells. The extra amount moves from the blood into the liver and is put into storage as the blood flows through. Between meals, food particles move from the liver back into the blood and then to the cells as necessary.

In addition, our livers are chemical factories that process many other nutrients and wastes as well, but this is for future lessons.

Step 6. Energy needs, food intake, and weight gain or loss

Excessive weight has been identified as a national epidemic and one with serious health consequences. Further, it is becoming increasingly common among children. Therefore, it is well to add this topic to the lesson, especially since it ties in so well with what we have been discussing.

Review again, that our bodies need a constant supply of energy to maintain the living state and to perform all the activities of the body. In eating we are taking on fuel to supply that energy. (Again we are leaving other nutritional aspects of food aside for the time being.)

Now, have students note that our cars have fuel tanks of a given size. We can't put in more fuel than the tank will hold, and the amount of fuel we put in over the long term depends entirely on how much we use for the miles we drive. It might be nice if our bodies functioned the same way, but they don't. It is easy and all too common to eat an amount of food with an energy content greater than what is required by the body. We have talked about how the liver stores food after meals and releases it to provide for energy needs between meals.

However, the liver is limited in its storage capacity; we are not so limited in our eating capacity. When we persistently eat more than enough to supply our energy needs, the body turns the extra food particles into fat and stores it in a layer on the outer body; one gains excess weight accordingly. (The conversion of non-fatty foods to fat occurs in the liver and the fat particles are transported by the blood.)

Conversely, if one persistently eats less than what is required to meet the body's energy needs, the reverse occurs. Fat is taken out of storage and burned to make up the deficit, and weight loss occurs. This can only go on as long as there is adequate fat stores, however. If an energy deficit continues after fat reserves are gone, the body will commence breaking down muscle and other vital tissues, burning itself so to speak. This leads to death by starvation. There is no way that the body can live without a constant supply of energy.

Use Q and A discussion to bring your child to recognize what a person must do to lose the extra pounds. The answer is two-fold; the person must increase physical activity to bring his/her need for energy up to a level above the energy value of food being eaten. This forces the body to use its stored fat to make up the difference.

The alternative is to eat less, especially less fats, carbohydrates, and sweets, which have a particularly high energy equivalent. Again, the result is to make the body's need for energy greater than the energy equivalent of the food being consumed, forcing the body to draw on body fat to make up the difference. Of course, doing both or any combination of the two works even better. The one and only issue is making the body's expenditure of energy greater than the energy content of the food being consumed so that the body is required to draw on fat to make up the difference.

Conversely, weight gain is achieved by eating more.

Step 7. Nutrients—vitamins, minerals, and protein

Many children, by the time they reach school age, have been given the belief that they must eat to grow and be healthy and strong. Therefore, it is more than likely that this topic will come up along the way in the above discussions. The potential confusion is that many children gain the notion that providing for growth is the only role of food; further, they are likely to think that energy comes from vitamins.

Therefore, whenever this topic comes up, it will be important to make clear to students that there are two distinct roles that food serves in your body. First is: A significant portion of what you eat is burned for energy (or converted into fat if there is an excess) as we have discussed above. second is: Food (at least a good diet) also provides the essential building blocks that go into making the actual muscles, bone, skin, blood, and other tissues of the body, and into the chemical "machinery" that maintains them in working order.

You may go on to explain that there are three major categories of the essential building blocks: vitamins, minerals, and proteins. VITAMINS are chemical compounds essential for the running of the bodies chemical machinery. They are essential for our health and well being and that includes feeling more energetic, but they do not provide energy as such. It is analogous to the oil and grease in your car. They make it run better, but they do not provide the energy; it is only the gasoline that releases energy as it is burned. With respect to food, it is mainly carbohydrates, fats, and sugars that are burned for energy. Calories are the units most commonly used in measuring the energy value of foods.

MINERALS, commonly lumped with vitamins, often play the same sorts of roles as vitamins, i.e., they are essential for many of the chemical reactions that occur within the body. Certain minerals serve as actual building blocks; calcium is a major constituent of bone, for example. Iron is an essential part of red blood cells, which carry oxygen from lungs to all parts of the body.

Students may relate the minerals here to the minerals of rocks discussed in Lesson A-10. Indeed, we are talking about the same thing. A number of the same elements that occur in "rock minerals" are also required for body functions, and rocks/soil is the original source. They are absorbed (via solution) into plants, and we get them through food chains (or vitamin/mineral supplements).

PROTEIN provides major building blocks for muscle, skin, internal organs, and all other tissues of the body. Indeed, protein may also be burned, for energy but its most significant role is a body building material.

Of course, the foods that supply essential vitamins and minerals are mainly fruits and vegetables. Meat and/or beans are the primary source of protein. Milk is the renowned source of necessary calcium.

Make clear to students that what we call a good BALANCED DIET means eating a variety of foods that will provide both the essential nutrients (vitamins, minerals, proteins) and material (but not too much) to burn for energy needs (fats, carbohydrates, and excess protein).

Conduct Q and A discussion to bring students to reason out what results from eating an overabundance of sugary-starchy-fatty things such as french-fries, potato chips, and donuts, but lacking fruits and vegetables. Such eating habits may well supply excess calories while at the same time lack essential nutrients. In short, excess weight is no indicator of having adequate nutrition. It is quite possible to be both overweight and malnourished.

An activity that will review and solidify this understanding is to have students write the names of various things they eat and drink on cards, one item per card. Make several copies of each. Lay the cards out on a long table and pretend it is a smorgasbord. Have children choose from the "smorgasbord" what they should eat to have a balanced diet. Have each student discuss his/her selections (or omissions) in terms of meeting that objective.

Finally, point out and emphasize to students that what we have been discussing and learning in this lesson applies to all members of the animal kingdom, invertebrates as well as vertebrates. All animals must consume a quantity and kind(s) of food adequate to meet their energy needs as well as to supply nutrients necessary for chemical functions of the body as well as growth. Hence, they all have a digestive system to break down the food to particles, a circulatory system that will transport the particles from the digestive system to the cells of its body, a means of attaining oxygen, and a system for excreting wastes. There is only a difference in the size and form these organs and systems take. The potential to extend this lesson to comparing the anatomy of different sorts of animals is obvious.

Questions/Discussion/Activities To Review, Reinforce, Expand and Assess Learning:

Our effort here has been to provide an overall framework of understanding concerning how the body functions. This framework will provide a much better foundation for further learning than going into more detail regarding given organs and systems. Getting into detail in the initial stages frequently confuses the picture as children "fail to see the forest for the trees." Therefore, conduct activities and reviews in a manner that continues to stress the connections and integration of the overall picture. Move into more detail of the various organs and systems only as students have gained a comprehension of the broad picture.

Write books illustrating various aspects of the lesson.

Set up an activity center where students can continue to examine a model or pictures of body showing its various organs.

Play a "Simon says" game asking children to place their hand(s) on the location of different organs.

Give students math problems to the effect: If they consistently eat ____ Calories worth of food, but only "burn" _____ Calories worth, what will be the consequence? What should/can they do to correct the situation?

Conduct the "smorgasbord activity" described in Step 7 above

In small groups pose and discuss questions such as:

Why do we need to eat? (Give two roles that food serves.)
How does food provide energy? (What do our bodies do with food to get energy from it?)
Where does the "burning" take place?
What must happen to food before it can be "burned"? What happens first? … second? …third?
What else is needed in order for food to be "burned?"
Where does that come from? How does it get to the cells?
Are there wastes from the "burning?" What are they?
How does the body get rid of them?
In answering and discussing these questions, name and point to the locations of the key organs.
Why do you have to breathe? How does the air you inhale differ from the air you exhale?
What is urine—in terms of wastes from energy metabolism?
What is the origin of fecal waste? (Emphasize as often as necessary that fecal waste is material that has simply gone through the "hole in the donut." It has not been in or come from the body proper.
What is the origin of urine? (It is a water solution of the actual "ashes" from metabolism that has been filtered from the blood by the kidneys.)
Why do your breathing and heart rate increase as you engage in vigorous activity? Why do you work up an appetite after a day of vigorous activity? (Relate this to the energy demands.)
How do exercise and eating relate to putting on or taking off pounds?
Why do we need fruits, vegetables, and meat or beans?

To Parents and Others Providing Support:

Meal or snack times, breathing hard after a vigorous activity, and bathroom functions all offer abundant teachable moments. Take such opportunities to discuss one or more aspects of this lesson as they apply. In each case, have children identify the major organs involved and point to where they are located in their bodies.

Connections to Other Topics and Follow-Up To Higher Levels:

Further detail concerning the structure and function of any or all organs and systems, including the reproductive system
Disease conditions involving impairment of particular organs
Comparative anatomy of various vertebrate and invertebrate organisms
Ecology and the functioning of ecosystems
Nutrition
Agricultural production

Re: National Science Education (NSE) Standards

This lesson is a steppingstone toward developing student's understanding and abilities aligned with NSE, K-4:

Unifying Concepts and Processes
   • Systems, order, and organization
   • Evidence, models, and explanation
   • Form and function

Content Standard C, Life Science
   • Characteristics of organisms

Content Standard F, Science in Personal and Social Perspectives
   • Personal health
   • Types of resources

Books for Correlated Reading:

Blevins, Wiley. *Where Does Your Food Go?* (Rookie Read-About Science). Children's Press, 2004.

Brandenberg, Aliki. *I'm Growing!* (Let's-Read-and-Find-Out Science, Stage 1). HarperCollins, 1992.

Fowler, Allan. *Energy from the Sun* (Rookie Read-About Science). Children's Press, 1998.

Nettleton, Pamela Hill. *Breathe In, Breathe Out: Learning About Your Lungs* (Amazing Body). Picture Window Books, 2004.
_____. *Gurgles and Growls: Learning About Your Stomach* (Amazing Body). Picture Window Books, 2004.

_____. *Thump-thump: Learning About Your Heart* (Amazing Body). Picture Window Books, 2004.

Showers, Paul. *What Happens to a Hamburger* (Let's-Read-and-Find-Out Science, Stage 2). HarperTrophy, 2001.

Lesson B-10

Plant Science I:
Basic Plant Structure

Overview:

In this lesson, students will learn the basic parts of plants, roots, stems, and leaves, and their basic functions. They will learn that flowers, and likewise cones and capsules of non-flowering plants, are reproductive structures and gain more insight into the life cycle of plants. In the process, students will observe that, while virtually every plant can be described in terms of roots, stems, leaves, and reproductive organs, these parts may have a wide variety of different forms or modifications. Thus, this lesson will lay the foundation for all further aspects of plant science, and furthermore, it will begin development of the concept of change through modification of existing structures.

Time Required:

       Part 1. Roots, Stems, Leaves and Their Modifications (activity, 20-30 minutes, plus interpretive discussion, 30-40 minutes)
       Part 2. Functions of Roots, Stems, and Leaves (Q and A discussion, 25-35 minutes)
       Part 3. Reproductive Structures (activity plus interpretive discussion, 40-50 minutes; 10-15 minutes on additional occasions as flowers, cones, or spore capsules are found or brought in)

Short discussions (5-10 minutes each) regarding modifications should be injected at various points in the three parts. The preclass time described in Chapter 1, page 19 may be utilized.

Objectives: Through this exercise, students will be able to:

1.   Recognize and use the following words in their proper context: ROOTS, STEMS, LEAVES, PHOTOSYNTHESIS, REPRODUCTIVE STRUCTURES, POLLINATION, SPORES, SPORE CAPSULES, MODIFICATIONS.

2.   Identify roots, stems, and leaves on any given plant.

3.   Point out how all plants (with few exceptions) have a basic structure consisting or roots, stems, and leaves.

4.   Describe in general terms the basic function(s) of roots, stems, and leaves.

5. Recognize and describe how flowers (cones and spore capsules on non-flowering plants) are reproductive structures.

6. Point out male and female parts of a typical flower.

7. Tell the basic story of pollination, fertilization, and development of fruits and seeds that occurs in flowering plants.

8. Recognize roots, stems, and leaves in their various modified forms.

9. Give examples illustrating the various forms or modifications that roots, stems, and leaves may take.

Required Background:

Lesson B-3, Distinguishing Between Plants and Animals
Lesson B-4, Life Cycles
Lesson B-4A, Identification of Living Things

Materials:

Part 1. Roots, Stems, Leaves and Their Modifications
Starter plants (peppers work particularly well), at least one for each student or group (Or, you may grow seedlings in plastic cups for this purpose, or dig weeds.)
Newspaper to contain soil mess
A variety of different leaves
Sprigs from pines, fir, and/or other evergreens
Tuft of grass with roots
Weed, such as a dandelion, with a tap root (a carrot with the top attached)
A sprig of a twining vine or ivy (BE CAREFUL NOT TO COLLECT POISON IVY FOR THIS PURPOSE. "Leaves of three, let it be.")

Part 2. Functions of Roots, Stems, and Leaves
No additional materials will be needed.

Part 3. Reproductive Structrues
An assortment of different flowers
Cones from pines or other cone-bearing trees
Moss with spore capsules (as you may come across them on nature walks)
Fern fronds with spore capsules (as you may come across them on nature walks)

Teachable Moments:

This lesson will be best conducted in the spring when gardening is in the air and starter plants are readily available. A tray of starter plants from which to pass out one to

Bernard J. Nebel, Ph.D.

each student/group will catch their attention.

Methods and Procedures:

Part 1. Roots, Stems, and Leaves and Their Modifications

Pass out a healthy, living, starter plant (seedling) in its "pot," one to each student or work group. On newspaper to contain the mess, have students take a plant from its pot and shake the soil away from the roots. Then, ask, "If you were to think of your plant in terms of parts, what parts would you name?" Provide time for students to observe, reflect, and discuss among themselves. Guide observations and discussion as necessary to bring students to conclude that there are three basic parts: ROOTS, STEMS, and LEAVES.

Giving some time for this to sink in, go on to ask, "Can all plants be described in terms of these same three parts?" Students are likely to say no and cite examples such as tree trunks and pine needles. Have students recall that leaves they have been identifying (Lesson B-4A) come in a wide variety of shapes and sizes. That is how they are able to identify them as different species. Explain that we refer to different shapes and sizes as MODIFICATIONS. The needles of various evergreens, blades of grass, lily pads, and other such things are really highly modified leaves.

(In this and following discussions, recognize that there is no way that you are going to cover all the modifications of roots, stems, and leaves that occur in nature. Therefore, keep discussion to examples that you may bring in to show students and/or examples that they will be familiar with or bring up themselves. Additional modifications may be discussed as you and your students come across them in the course of ongoing nature studies. Utilize preclass show-and-tell/Q and A discussion sessions described in Chapter 1, page 19.)

Stems and roots, point out, may also come in a wide variety of shapes and sizes or modifications. Tree trunks are giant stems. Vines (in most cases) have stems that twine and climb. Grass and other plants, such as dandelions, have an extremely shortened stem so that the leaves seem to all come from a central point. That central point is the stem, nevertheless.

Turning to roots, there are two major kinds or modifications. One is illustrated by the roots of their seedlings and, also, grass. Have students describe how such roots are like a network of strings. This is called a FIBEROUS root system. Contrast this with a carrot. Children may be surprised to learn that a carrot is a root, but it is; this kind of root is called a TAP root.

Pursue this discussion according to students' attention and interest, but conclude with the summary and conclusion that all plants have the same three basic parts: roots, stems, and leaves. But, these structures may come in different forms or modifications.

220

Part 2. Functions of Roots, Stems, and Leaves

Following a brief review of the above, use Q and A discussion to draw students into a consideration of the roles or functions of each major part. Pose questions such as: Why do plants need roots? What job(s) do they perform? Would a plant be able to stay in place without roots? Would it be able to obtain water and nutrients from the soil without roots?

With Q and A discussion pertaining to each part, bring students to reason their way to the following conclusions, some of which may be recalled from previous lessons:

Roots. The major job or function of roots is to absorb water and certain chemical nutrients from the soil. Roots also serve to anchor the plant in the soil and hold it upright.

Leaves. Their major function is absorbing sunlight and using the light energy to make sugar from carbon dioxide (a gas absorbed from the air) and water absorbed from the soil. The process is called PHOTOSYNTHESIS. In turn, the plant uses the sugar and certain nutrients from the soil as the raw materials for making all the other tissues of its body.

Stems. The functions of the stem are support and transport. Without stems, how could the plant get its leaves up to the sun? How could it transport water from the soil to the leaves? How could it transport food made by photosynthesis down to the growing roots? It will be important to emphasize that stems engage in a two-way transport; water and a few raw chemical nutrients are transported up the plant, and food made in the leaves is transported down to nourish the roots and/or other growing plant parts.

It is conspicuous that one might go much more deeply into the anatomy and physiology of root, stems, and leaves, and you may do so at higher levels. At the first-grade level, however, it will be necessary to keep it simple and very basic unless students' questions draw you to go more deeply.

Part 3. Reproductive Structures

We have already introduced children to the concept that species only maintain themselves through reproduction (Lesson B-4). Hence, every species has its particular reproductive parts or structures. Review this as necessary, noting that in humans and other mammals, babies are born with their special boy (male) or girl (female) parts. In young plants, however, reproductive parts are simply not present. Hence, we observed that roots, stems, and leaves are the only structures. In plants, reproductive structures only develop as the plant matures and at certain times of the year.

Emphasize that reproductive structures all accomplish the same basic function of producing eggs (female) and sperm (male), accomplishing fertilization, and commencing growth of the next generation. But, as with roots, stems, and leaves, there is a wide variety of different forms or modifications that reproductive structures may take. In fact, the plant kingdom is divided into major subgroups, or DIVISIONS, on this basis. (Note that in the plant kingdom, the term "division" is used in place of "phylum.")

The largest division of the plant kingdom, by far, is flowering plants, those plants that form flowers, and subsequently, fruits with seeds inside. In turn, different shapes or structures of flowers are used to separate this division into different classes, orders, families, and genera. Each different kind or structure of flower and, in turn, different sorts of fruit can be considered as variations or modifications on the theme.

Then, there are several divisions of non-flowing plants, plants that have reproductive structures that are nothing like flowers. Most prominent are:

Pines and related species: Cones of various types are the reproductive structures.

Moss and related species: Especially in the springtime one will find moss bearing small capsules on the ends of straight thin stalks. These are the reproductive structures of the moss. When they are ripe the capsules shed spores, which are the equivalent of seeds. Spores will germinate and start new moss plants.

Ferns: The brown or black spots on the underside of some fern fronds (leaves) or sometimes ridges along the margins are the reproductive structures of the fern. They are actually clusters of microscopic capsules that shed spores, as in moss. (A view of these capsules under the microscope reveals quite an elaborate and artistic structure.)

If children's interest warrants, you may go more deeply regarding plant reproduction. For flowering plants, the points to emphasize, without getting too detailed and confusing, are the following:

Just like animals, plants come in male and female, and fertilization is required in order to produce "babies." In most flowering plants, male and female parts are in the same flower. You may point these out in any simple flower that is large enough to observe easily. Lilly flowers are great. (A good diagram may be found by Googling: flower parts function/ Specifically see:
http://www.enchantedlearning.com/subjects/plants/printouts/floweranatomy.shtml/

The male parts are the anthers, the tiny slipper-shaped things on the ends of stalks in a ring around the center. The female parts are the ovaries, the club-shaped thing(s) in the center of the flower. Pollen produced in the anthers is transferred to the end of the ovary, usually by insects. This is POLLINATION. From there, the pollen grains sprout a microscopic tube that grows its way down into the ovary where the egg cells are located, and fertilization occurs. After fertilization has occurred, the ovary and fertilized eggs grow into the fruit. Each fertilized egg, with some additional material, develops into a seed, the seed being a baby plant or EMBRYO surrounded by stored food material and a resistant coat. Thus, a seed may be seen as a baby done up in a basket (the seed coat) with its lunch. The rest of the ovary develops into the rest of the fruit surrounding the seed(s).

It will take some emphasis and additional observations to have children recognize that all fruits start out as the ovary(s) of a flower. Yes, a watermelon started as the ovary in the flower of a watermelon plant. If you grow garden vegetables such as tomatoes, squash, peas, or beans (these are really fruits of various sorts) it will be good to show your students that tiny "fruit," the ovary in the center of the flower. Then have children keep an eye on this every few days and observe how it really does grow into the respective fruit with seeds inside.

A second avenue is to find a plant that bears multiple flowers along the end of the flowing stalk. In such plants, the lower flowers bloom first, while new buds and flowers are still forming at the tip. As such plants approach the end of their blooming cycle, one will commonly find mature "fruits" (pods of one sort or another with seeds inside) at the lower portion of the stalk while there are still flowers at the tip. All the stages of maturation are in between. Thus, in one flower stalk you can show your students all the stages of development of the ovary of the flower into the fruit.

Note that, botanically speaking, the word "fruit" refers to the ripened ovary(ies) of the flower, which contain seeds. There may be just one seed per fruit, cherries for example, but often there are hundreds as in the case of pumpkins. Again, any ripened ovary containing seeds is properly called a fruit from the botanical point of view. It is those plants with edible fruits and/or seeds that we grow agriculturally. The majority of wild plants do not have edible fruits or seeds. In every case, however, the structure of the fruit does play a role in dispersal of the seed; this can be another lesson.

In the course of seed development, the fertilized egg develops becoming an embryo and then goes into a "sleep" state called DORMANCY (and it may remain in this state for a matter of years). But, when temperature and moisture are sufficient, the "baby" wakes up, starts feeding on its "lunch," and grows (germinates), becomes a seedling, and on.

Turning to pines and related species, the clusters of small (1/2 – 3/4 inch long) segmented, egg-like things at the tips of many branches in the spring are the male cones. The amount of pollen they shed is often conspicuous if we park our car under a pine tree as it is shedding pollen in the spring. The pine cones one picks off the ground and collects are the female cones. The seeds, which are formed between the scales of the cone, have been shed. In searching, you will find the small immature female cones still on the tree. It is here that pollination, fertilization, and seed development occur.

Mosses, ferns, and other such plants do not form seeds. Spores, which are single cells with a highly resistant cell wall, play the role of seeds.

Of course, the steps in the reproductive cycles of all these plants is full of additional details and complexities that students may pursue at higher levels as their interest dictates. What we have given here, however, is generally sufficient for the K-2 level and provides a foundation that they can easily build on.

In summary, keep the focus of this lesson on the major plant parts (roots, stems, and leaves), their functions, and the various forms or modifications they may take. But don't try to cover it all at once. Every plant specimen that you or your students may bring in offers an opportunity to discuss its particular modifications. Utilize the preclass show-

and-tell/Q and A discussion period described in Chapter 1, page 19. There will be many modifications in addition to those mentioned above that may gradually be added to the list.

Likewise, bringing in flowers, fruits, cones, fern fronds or moss with spore capsules provide opportunities to inject the generalities of plant reproduction, modifications on the theme, and add details as children's questions may dictate.

Questions/Discussion/Activities to Review, Reinforce, Expand and Assess Learning:

Make books illustrating the main parts of plants; modifications of roots, stems, and/or leaves; modifications of flowers and fruits.

Facilitate students making collections or posters illustrating modifications of roots, stems, leaves, and/or features. For example, the natural function of fruits is to aid in seed dispersal. A collection of seed/fruits illustrating different mechanisms of dispersal is a splendid activity.

Engage students in activities that call them to match functions with plant parts.

Continue the practice of having children bring in samples (or sightings) of living things and doing a show-and-tell at the beginning of class. Add the concept of MODIFICATIONS to these discussions. That is, talk about how particular plants show modifications of roots, stems, and/or leaves. Note that different kinds of animals can be discussed in terms of modifications on the "animal theme" as well.

Name various plant items that children commonly eat and have them identify each item as (or from) roots, stems, leaves, flowers, fruits, or seeds. Alternatively, cite a plant part and have children name something they eat that is that part. You can think of a number of ways that this can be made into activities or games.

In small groups, pose and discuss questions such as:

What are the basic parts of all land plants?
What are the functions of each part?
Give examples illustrating how roots, stems, and leaves may come in a wide range of shapes and sizes, i.e., modifications.
What is the function of flowers?
Do all plants have flowers? Cite examples of non-flowering plants. How do they reproduce?
Name a familiar plant (or show students a picture of an unfamiliar plant) and have them describe it in terms of roots, stems, and leaves and how the given structure(s) are modified in the particular case.

To Parents and Others Providing Support:

In the course of gardening, weeding, transplanting, growing seedlings, or on an outing, have children identify roots, stems, and leaves and tell their functions coaching as necessary. In discussion, draw them to make the generalization that all plants have the same basic structure of roots, stems, and leaves. (There are some exceptions, but these may be left aside for the time being.)

A fun activity is "Grocery Store Botany." While in the produce area of a supermarket, have children identify the parts of a plant we eat. In the case of lettuce and spinach, it is leaves. Carrots and beets are roots. Celery and asparagus are stems. Broccoli and cauliflower are clusters flower buds at the tips of stems. Artichokes are single, large flower buds. Cucumbers, squash, melons, and string beans are fruits, as are apples and oranges. You may also have children identify seeds: peas, corn, beans, rice, wheat, etc.

Have children understand and use the term "modifications" in describing the multitude of shapes and sizes of roots, stems, leaves, flowers, etc.

On walks and outings, have a game of pointing out roots, stems, leaves, and reproductive structures as they are seen in various plants. Ask: What is the purpose of roots? stems? leaves? flowers?

You will encounter many plant items, such as onions, cacti with their needles, runners on various plants, etc. With your children, reflect on such things as modifications of roots, stems, or leaves. Check your conclusions in additional sources.

Facilitate children making collections where they have an interest in doing so.

Connections to Other Topics And Follow-Up to Higher Levels:

Further investigation of modifications of roots, stems, leaves, and reproductive structures
Further detail regarding plant reproduction
Mechanisms of seed dispersal
Photosynthesis
Further studies regarding the anatomy and physiology of the different plant of roots, stems, leaves, and flowers

Re: National Science Education (NSE) Standards

This lesson is a steppingstone toward developing students' understanding and abilities aligned with NSE, K-4:

Unifying Concepts and Processes
  • Systems, order, and organization
  • Evidence, models, explanation
  • Form and function
Content Standard A, Science as Inquiry
  • Abilities necessary to do scientific inquiry

Bernard J. Nebel, Ph.D.

Content Standard C, Life Science
  • Characteristics of organisms
  • Life cycles of organisms

Books for Correlated Reading:

Bergen, Lara Rice. *Looking at Trees and Leaves* (My First Field Guide). Grosset, 2002.

Numerous titles by Vijaya Bodach in the Plant Parts series.

Bulla, Clyde Robert. *A Tree Is a Plant* (Let's-Read-and-Find-Out Science, Stage 2). HarperTrophy, 2001.

Dorros, Arthur. *A Tree Is Growing.* Scholastic, 1997.

Gibbons, Gail. *From Seed to Plant.* Holiday House, 1993.
_____. *Tell Me, Tree: All About Trees for Kids.* Little Brown & Co., 2002.

Heller, Ruth. *The Reason for a Flower,* Putnam, 1999.
Jaspersohn, William. *How the Forest Grew.* HarperTrophy, 1992.

Lauber, Patricia. *Be a Friend to Trees.* (Let's-Read-and-Find-Out Science, Stage 2). HarperTrophy, 1994.

Maestro, Betsy. *How Do Apples Grow?* (Let's-Read-and-Find-Out Science). HarperTrophy, 1993.

Lesson B-11

Plant Science II:
Germination, Seedling Growth And Responses

Overview:

In this lesson, students will observe the germination of seeds and the early stages of seedling growth. They will contrast growth of seedlings in the dark with that in the light, and they will observe the plant's responses to gravity and direction of light. Through these activities, students will gain an understanding and appreciation for a plant's ability to perceive and adapt to its environment. Learning to record results in an orderly fashion will be an integral part the lesson.

Time Required:

Part 1. Seeds and Seed Germination (set up, 30 minutes, plus observations, recording, and discussion of results,15-20 minutes on each of the next 5-10 days)

Part 2. Seedling Growth in the Light and the Dark (set up, 30 minutes, plus observations, recording, and discussion of results, 40-60 minutes, roughly two weeks later)

Part 3. Responses of Plants to Gravity and Direction of Light (set up, 30-40 minutes, plus observations, recording, and discussion of results, 30-40 minutes, at later times)

Objectives: Through this exercise, students will be able to:

1. Recognize and use the following words in their proper context: EMBRYO, DORMANT, GERMINATION, TROPISMS, GEOTROPISM, and PHOTOTROPISM.

2. Describe a seed in terms of the state of its embryo, food supply, and protective coat.

3. State what is required for germination to occur.

4. Describe the process of germination in terms of what emerges from the seed first, and what emerges second.

5. Tell how the sequence of events (4 above) make sense in terms the plant's needs.

6. Describe how seedlings grown in the dark differ from those grown in the light.

7. Tell how the differences (6 above) make sense in terms of the plant's needs.

8. State the source of energy for seed germination and early seedling growth, especially for seedlings grown in the dark; state why seedlings will not grow indefinitely in the dark.

9. Tell how roots and shoots respond to gravity; give the term that describes this response.

10. Tell how shoots respond to where light is coming from; give the term that describes this response.

11. Describe how these responses (9 and 10 above) adapt the plant to living in natural environments.

Required Background:

Lesson B-4, Life Cycles (particularly the plant portion)
Lesson B-10, Plant Science I: Basic Plant Structure   —
Lesson C-3, Energy II: Kinetic and Potential Energy
Lesson A-8, Evaporation and Condensation

Materials:

Seeds of various kinds: radish, sunflower, beans and peas work especially well, but other species will also serve and may add to the lesson.
Saucers or dinner plates
Paper towels
Potting soil
Clear plastic cups with strongly tapered sides
Wax pencil
Onion bag netting material
Clear or semi-clear plastic "baggies"
Cardboard box roughly a foot on each side or larger

Teachable Moments:

This lesson will be best conducted in the spring of the year when gardening is in the air and seeds of all sorts are available. Passing out a few seeds to each student will catch their attention.

Methods and Procedures:

Part 1. Seeds and Germination

Pass out a few seeds to each student. Seeds of radish, sunflower, and peas work especially well in that they are large enough to handle and germinate fairly rapidly, but other kinds of seed will serve and may add to the lesson. Allow children to compare and reflect on their seeds a few moments, then pose the question, "Are seeds living or dead?"

In all likelihood, many students will declare that they are dead, and, to be sure, they show none of the attributes of living things. Therefore, you will have to stress that DEAD means more than just not showing signs of active life. It means that all potential of reviving or resuming a living state is also gone. Review as necessary the information from Lesson B-10 that a seed is a "baby plant," (embryo) in a basket (seed coat) with its lunch (stored food). Therefore, what does it mean if that embryo will revive and start growing again? Draw students to conclude that it means that the embryo is not dead; it only "asleep," or DORMANT. On the other hand, if the embryo will not start growing again, it means that it is really dead.

Thus, the only way we can determine if the embryo is really dead or just DORMANT is to test if it will start growing again. Instruct students that we speak of the embryo within the seed "waking up" and resuming growth as SEED GERMINATION. So, let's determine if the embryos within our seeds are alive or dead by testing if these seeds will GERMINATE.

Have students place several layers of paper towels on a saucer or dinner plate; thoroughly wet the towels and drain off excess water. (Seeds want plenty of water but they don't want to be drowned.) Place several seeds on the paper spaced about an inch apart and cover with a similar plate inverted. Be sure that edges of towel are folded in so that they are not sticking out. (You might ask students why covering is important. Help them recall from Lesson A-8 that it is to prevent evaporation.) Set the dishes aside in a location where they will be at regular room temperature.

On the following days, briefly remove the cover, observe, and record the results. (Coach children as necessary in setting up a table with dates down the left hand column and additional columns labeled for each seed. Coach them as necessary in recording results in the table. They may use small diagrams rather than words.) After observing and recording results, replace the cover and set it aside for the next day.

The first day after starting, students will observe that their seeds have swelled up considerably. This is NOT germination. It is similar to a dry sponge swelling as it soaks up water. You may wish to demonstrate this for students. Occurring as it does in an inanimate sponge, it has no bearing on the embryos being alive.

After about three days to a week, depending on the seeds and temperature, students will see a root tip protruding from the seed. This is GERMINATION. The embryo has indeed broken dormancy (woken up) and resumed growth. Have children reflect back to the beginning and recognize how the result proves that their seeds were actually alive; the embryos were only dormant.

Continue to have students make daily observations of their germinating seeds, calling their attention to the sequence of events. They should note that regardless of species, it is the root portion that comes out first. The root usually reaches considerable size before the shoot portion emerges. Challenge them with the question: Why should this be?

Use Q and A discussion to guide students to deduce that the most immediate need for a germinating seedling is to establish a system for obtaining water and nutrients for the rest of the plant. The shoot, much less leaves, could hardly survive and grow without a supply of water. Thus, the growth of the embryo is programmed to have the root system develop first. Of course, the roots will also play an essential role in anchoring the plant in the soil.

Germinating seeds on a moist paper towel facilitates students being able to make close-up observations of germination, and it may take as much as three weeks for some seeds to germinate, that is, before declaring them dead. But, it is evident that seedlings cannot be sustained in this manner.

For longer-term observations, especially those of the shoot portion, have students fill one or more clear plastic cups, the sort with strongly tapered sides, with potting soil. Thoroughly wet the soil; drain away excess water, and insert three or four seeds around the perimeter between the soil and the side of the cup so that the seeds remain visible. Water as needed to keep the soil moist but not saturated, and observe and record the progress of germination as before, As roots grow down some will stay more or less pressed against the side of the cup and be clearly visible. The shoots will begin to grow as they do.

Part 2. Seedling Growth in the Light and the Dark

Invite children to ponder the mystery: What is the effect of light on germination and early seedling growth? Will seeds germinate in the dark? Will seedlings grow in the dark? If so, how much? Have them make and record their guesses (hypotheses) and even give their reasoning, but there is nothing like determining the answer through experimentation. What should the experiment involve? Students should reason that it is simply a matter of keeping some in the dark and having others in the light. Okay, let's do it!

Have each student/group fill two plastic cups or larger pots with potting soil; wet and drain the soil. Plant three or four seeds in each pot (beans give a particularly dramatic result in this activity). Then, put one pot in a completely dark location, or cover it with a box, being sure to block light from the cracks; place the other where it will get plenty of light, preferably sunlight. The light and dark locations should be where the temperature is the same.

Students may recognize a problem: Observation requires light. If we regularly observe those in the dark, we no longer have the experiment of keeping them in the dark. Therefore, explain that they should watch only those in the light. When they have sprouted and spread their first leaves, they can reason that those in the dark have had time to grow also if they are going to do so at all. (One may take an occasional peek in dim light at those in the dark to be sure the soil is remaining moist.)

When seeds in the light have germinated and spread their first leaves, usually in about two weeks, have students remove their seedlings from the dark and compare results. They are usually quite amazed at what has occurred.

Contrary to the usual expectation, it is the dark-grown seedlings that will have grown the tallest. But guide students to note and record other significant differences as well. The dark-grown seedlings, while tall, are very weak and spindly; leaves are very reduced in size, folded, and yellowish. In the light, the seedlings, while shorter, are more sturdy; the leaves are more expanded and definitely green. Why should these differences be?

Use Q and A discussion to bring students to consider what the seedling must accomplish to achieve longer-term life. First, review what they learned from previous lessons, namely that plants get their energy for growth from light. How does the seedling grow at all in the dark? They should recall the seed contains "lunch," as well as the "baby." Thus, students should reason that during germination and early growth, the embryo is feeding on its "lunch," stored food in the seed. It is only later that they require light energy for photosynthesis.

(In many species, including peas, beans, and sunflowers, the stored food is transferred to and stored in the embryo itself in the course of seed development. The two fleshy structures on either side of the sprout are the stored food. In sunflowers and beans, these storage "leaves" are carried upwards with the sprout; in peas, they remain at the position of where the seed was planted. There are technical terms for these and other portions of the seedlings, but they become burdensome and are not necessary to convey the conceptual understanding.)

So, where should a plant put its efforts when there is no light? Expanding leaves? Useless! Making chlorophyll, which imparts the green color, for photosynthesis? Useless! Growing tall to where light is usually to be found? Ah ha!

Thus, guide students to conclude how plants are remarkably adapted to cope with conditions of their natural environment. In the absence of light, germinating seedlings are programmed to put all their energy (from their stored "lunch") into growing tall, as this usually brings them into the light. But this can only go on as long as the energy from lunch lasts. Follow those seedlings kept in the dark for another week or so. Students will find that they collapse and die as their food supply is exhausted.

On the other hand, when seedlings have light there is no immediate need to grow taller, but there is a need to put out leaves for photosynthesis. Therefore, seedlings grown in the light are not as tall, but leaves are expanded and green to take advantage of the available light energy.

Bernard J. Nebel, Ph.D.

In this discussion, it is tempting to make it sound as if the plant is actually seeing and thinking, "Ah, there light, I should …," or, "It is still dark. I need to …." It will be important to emphasize that the plant is not thinking; indeed it has no nervous system to think with. However, it does have a mechanism that perceives light versus dark and a genetic program that guides its growth and development accordingly.

Just in passing, if children note a similarity between the dark-grown bean seedlings and the sprouts seen in Chinese cooking that is exactly right. Dark-grown bean spouts and alfalfa sprouts are commonly used in salads and cooking.

Part 3. Response of Plants to Gravity and Direction of Light

Another fascinating aspect that may be observed easily in the course of this work is the seedling's response to gravity and to the direction of light.

Have students start another set of seeds in tapered plastic cups with potting soil as before. As seeds germinate, the roots will grow down and remain pressed against the side of the cup because of its taper. When roots have grown 3/4 of an inch or so, trace the position of the root on the side of cup with a wax pencil. Carefully tip the cup on its side so that the root is positioned at the side of the cup. Leave the cup in that position for a day and observe the position of the root again. Students will note that it has grown from where it was, and made a sharp bend downward according to its new position. Set the cup upright again and go another day. Students will find another bend. Have students reflect that the root is able to ascertain which direction is down and grow accordingly.

As sprouts emerge above the soil, have children tip the cups as before and note what the shoots do. Even within an hour or so, students will note the shoots beginning to curve upward.

Have students ponder the question: Do these responses indicate that the plant is sensing the soil and growing into or away from it? Or, is it a response to gravity alone? See if they will come up with the following idea to test this question.

Fill a cup or small pot to the brim with potting soil. Insert seeds and fasten a piece of onion bag mesh over the top of the pot to hold the soil in place. In addition, fasten a clear plastic bag over the top of the pot to conserve moisture, but still leave space for whatever emerges from the soil. Finally, suspend the arrangement such that the pot with its seeds is hanging upside down. The onion bag mesh will hold the soil in place. Students can guess (hypothesize) what will happen.

Students will discover that roots grow down, even though in the inverted pot this means coming out of the soil into the air, and the shoots will grow up into the soil. In other words, they grow upside-down relative to the pot and soil. Students should reason from this that the seedlings are responding to gravity alone.

Use Q and A discussion to bring children to note that plants do not have conspicuous sense organs or nervous systems, much less brains. Nevertheless, they are able to sense gravity, and they direct the growth of their stems up and their roots down accordingly. When their environment is upside down, this leads them to do exactly the wrong thing. But, what would be the state of plants in the natural world if they did not have this remarkable adaptation?

Another intriguing observation is the response of plants to the direction of light. Put some pots with growing seedlings in a covered cardboard box with a slit at one side to admit light. After a few hours, a day at most, observe the seedlings. Students will find that they have all bent toward the light. Again, use Q and A discussion to bring students to reflect on the capacity of plants to sense the direction of light and orient themselves accordingly. Likewise, have children reflect on the advantage of this adaptation for the plant.

A final example of plants' ability to sense their environment may be seen in tendrils of grape vines curled about wires or twigs, or climbing beans and other vines twined around poles or the trunks of small trees. (Keep your eyes open for examples to bring in and show students.) The action is harder to show or demonstrate in a classroom, unless you have a setup for time-lapse photography, but you can explain the action. The tendril or stem constantly sweeps around in a broad circle as the shoot grows upwards. When it touches an object, the sweeping motion stops and it proceeds to curl around the object. Scientists have shown that it is a response to touch. The touch stimulates the tendril/stem to start bending around toward the side of the touch.

Explain that the capacity of plants to grow in ways that orient them to gravity, light, or touch are called TROPISMS. The response to gravity is called GEOTROPISM; the response to the direction of light is called PHOTOTROPISM; the response to touch is called THIGMOTROPISM. Of course, these responses may be observed in many situations beyond the growth of seedlings. Many houseplants, for example, persistently turn their leaves away from the dimly lighted room toward the greater light from a window.

Questions/Discussion/Activities to Review, Reinforce, Expand, and Assess Learning:

Write a book illustrating seed germination. ... contrasting seedling growth in the light and dark. ... tropisms.

Some students may wish to test the germination and growth of additional kinds of garden seeds and/or seeds collected from wild plants. Facilitate their doing so.

Facilitate students' further experimentation and testing regarding light and dark growth of seedlings and tropisms.

In small groups, pose and discuss questions, such as:

What is the state of the baby plant or embryo in the seed?
What is required for germination to occur?
A plant does not have conspicuous sense organs. Can it nevertheless sense certain

aspects of its environment. What aspects? What is the evidence that it can sense light versus dark? … gravity? … where light is coming from?

How do seedlings grown in the dark differ from those grown in the light? Will growth in the dark continue indefinitely? Why not?

Describe ways in which a plant can sense its environment and adapt its growth accordingly.

To Parents and Others Providing Support:

Facilitate children repeating any of the activities described and discuss results with them. Help them test additional kinds of seeds and make and test variations as they may desire. Growing large seeds such as acorns, chestnuts and avocado pits is especially intriguing to children. Be patient however; they will take a long time (sometimes months) before they germinate. Help them keep records along the way.

Draw children's attention to cases such as a fallen-over plant bending upward and houseplants orienting their leaves toward the window. Ask them to explain what is going on and what is it called.

On outings, be on the lookout for examples that illustrate geo- or phototropism. Guide children in analyzing them. For example, in woods it is not uncommon to find a tree with an older part of its truck more or less horizontal and newer "trunk(s)" growing up from its side. This is a form of geotropism. The tree was knocked down but not killed. Side branches took over in growing up.

Connections to Other Topics and Follow-Up to Higher Levels:

Many seeds have chemical or physical features that prevent immediate germination, adapting the plant to its particular environment. Seeds of some plants actually require some light to germinate. Thus, how seeds break dormancy, what they require to do so, and how it adapts the plant to a particular environment may be pursued to any depth desired.

Plant physiology: How plants perceive gravity and the direction of light and respond to them can be studied to any depth desired, including research careers.
Additional aspects of plant physiology

Re: National Science Education (NSE) Standards

This lesson is a steppingstone toward developing student's understanding and abilities aligned with NSE, K-4:

Unifying Concepts and Processes
  • Systems, order, and organization
  • Evidence, models, and explanation
  • Form and function

234

Content Standard A, Science as Inquiry
  • Abilities necessary to do scientific inquiry
  • Understanding about scientific inquiry

Content Standard C, Life Science
  • Characteristics of organisms
  • Life cycles of organisms
  • Organisms and environments

Content Standard G, History and Nature of Science
  • Science as a human endeavor

Books for Correlated Reading:

Bergen, Lara Rice. *Looking at Trees and Leaves* (My First Field Guide). Grosset, 2002.

Bodach, Vijaya. Numerous titles in "Plant Parts" series. Capstone, 2007.

Bulla, Clyde Robert. *A Tree Is a Plant* (Let's-Read-and-Find-Out Science, Stage 2). HarperTrophy, 2001.

Dorros, Arthur. *A Tree Is Growing.* Scholastic, 1997.

Gibbons, Gail. *From Seed to Plant.* Holiday House, 1993.
_____. *Tell Me, Tree: All About Trees for Kids.* Little Brown & Co., 2002.

Heller, Ruth. *The Reason for a Flower,* Putnam, 1999.

Jaspersohn, William. *How the Forest Grew.* HarperTrophy, 1992.

Lauber, Patricia. *Be a Friend to Trees.* (Let's-Read-and-Find-Out Science, Stage 2). HarperTrophy, 1994.

Maestro, Betsy. *How Do Apples Grow?* (Let's-Read-and-Find-Out Science). HarperTrophy, 1993.

Lesson B-12

Plants, Soil, and Water

Overview:

It is no stretch to say that soil, in harmony with water, is the most important of all resources because it is the base for nearly all food production. Without maintaining productive soils, agriculture would crash, taking civilization with it. Yet, worldwide, one sees too many cases of soils being lost and degraded for lack of proper management. There are many facets to understanding soils and appropriate conservation measures. This lesson will reveal the needs for drainage, water holding capacity, and prevention of erosion. This is far from the whole story, but it will provide a foundation for further investigations into all aspects of soil science.

Time Required:

Part 1. Dirt Versus Soil, Good and Poor (discussion, 15-20 minutes)

Part 2. Soil, Water, and Seedling Growth (set up, 40-50 minutes, observation and recording of results, 5-10 minutes every other day over the next 3 weeks or so; interpretive discussion of results, 30-40 minutes)

Part 3. Erosion and the Soil Resource (activity/demonstration, 20-30 minutes, plus interpretive discussion, 20-30 minutes)

Objectives: Through this exercise, students will be able to:

1. Distinguish between the terms SOIL and DIRT.

2. Explain what is meant by a WATERLOGGED soil; tell why seeds do not germinate nor seedlings grow on a waterlogged soil.

3. Tell why it is important for excess water to be able to drain from soil.

4. Tell what is meant by soil AERATION and why it is important.

5. Explain what is meant by the WATER-HOLDNG CAPACITY of a soil.

6. Tell why seeds may not germinate nor seedlings prosper on a sandy soil, especially when there is several days between being watered.

7. Define what is meant by the term DROUGHT.

8. Describe what happens to bare soil exposed to the elements; name the effect.

9. Tell how a soil is changed by erosion and how this change diminishes a soil's ability to support plant growth.

10. Describe techniques for reducing erosion.

11. Discuss the importance of soil as a resource, defining terms SOIL MANAGEMENT and SOIL CONSERVATION.

Required Background:

Lesson A-10, Rocks, Minerals, Crystals, Dirt and Soil
Lesson B-5, Food Chains and Adaptations
Lesson B-9, How Animals Move IV: Energy to Run the Body
Lesson B-11, Plant Science II: Seed Germination and Seedling Growth
Lesson D-8, Rocks and Fossils

Materials:

Part 1. Dirt Versus Soil, Good and Poor
No materials required

Part 2. Soil, Water, and Seedling Growth
Seedling starter pots (2-3 inches)
Coarse sand/fine gravel (grains 1/16 – 3/16 inch)
Rich garden topsoil or potting soil
Bowls or pans in which pots with soil can be immersed in water
Packets of seeds (peas, beans, sunflowers, squash and/or other sizable, fast-germinating seeds)

Part 3. Erosion and the Soil Resource
Bucket of garden top soil
Dish drainer pad
Sprinking can and water

Teachable Moments:

In the early spring of the year, set out materials and a variety of seeds and invite children to participate in growing starter plants of a number of different vegetables and flowers. In the course of doing this, or any other gardening activity, bring up and discuss the attributes of soil and move into activities as described.

Alternatively, engage students in a Q and A discussion of where food comes from, bring them to the conclusion that nearly all food production depends on soil (in harmony with water), and move into activities from there.

Methods and Procedures:

Part 1. Dirt Versus Soil, Good and Poor

In the course of starting seeds, or any other sort of gardening activity, pose the question, "What do we call this "stuff" that we are planting seeds in?" Students will invariably reply, "dirt," because that is the commonly used word they undoubtedly know. Explain: Yes, but dirt refers to any sort of "earth stuff" underfoot. When it comes to growing plants, we use the word SOIL. Dirt is any sort of earth material that you might dig out of a hole or fill a hole with. When one says the word "soil," one should immediately think in terms of how well suited that earth material is for growing plants, including trees, shrubs, etc. Farmers and gardeners never speak of DIRT; they always speak of the SOIL of their fields or plots.

In turn, one may qualify a soil with any adjective ranging from excellent to terrible depending on how well suited it is for supporting plant growth. One may talk about soils in terms of specific attributes, such as fertile or infertile, compacted or loose, heavy or light. One also assigns names to soils depending on physical properties, sandy versus clayey, topsoil versus subsoil, for example. All of these attributes have an impact on plant growth. Conversely, dirt is just dirt.

(Direct this discussion in a general manner noting that the purpose here is not to have children learn specific facts. It is only to get them thinking in terms of soil having many characteristics and attributes, all of which have bearing on supporting plant growth. You may add that there is a whole branch of study called soil science and career opportunities as soil scientists.)

Review and emphasize that plants get their energy from sunlight (Lesson B-11 and earlier lessons). They use the energy from sunlight to construct their tissues from the ingredients, carbon dioxide, which they take from the air, and water. They DO NOT eat soil in any manner like we and other animals eat food. Soil has NO energy value for green plants.

Nevertheless, plants do depend on soil in various ways beyond simply having something that anchors them in place. What are some of the attributes of soil that bear on supporting plant growth? Let's do an activity that will shed light on a couple of critical aspects.

Part 2. Soil, Water, and Seedling Growth

Each student or group should set up three seedling starter pots of the same size: one with coarse sand and two with rich garden or potting soil. Let each student/group select what kind of seed they wish to use, but they should plant two seeds of the same kind in each of their pots. Water all pots and let any excess water drain out.

Then, each student/group should place one of their pots with potting/garden soil in a bowl or pan of water such that the water level comes even with the surface of the soil. Be careful to immerse the pot in the water slowly; otherwise one is likely to float the soil out of the pot, especially if the soil has not been thoroughly wetted ahead of time.

Label and set all the pots where they will get equal light (either sunlight or bright artificial light) and have the same temperature. In summary, for each kind of seed used by the class, there will be one pot of coarse sand (watered and drained), one pot of potting soil (watered and drained), and one pot of potting soil immersed in a bowl of water. Guide students in making a table for recording results. (The table should have four columns, one for the date and one for each of the three pots.)

Water every other day (Mon., Wed., Fri.) and skip watering on Sat. and Sun.) For each of the pots in the air, water such that the soil is thoroughly wetted and water drains from the bottom, but then empty the excess water that drains through. Don't leave these pots sitting in a saucer of water. For the pot in the bowl/pan of water, keep the water level at the top of the soil. Plan to keep this activity going for three to four weeks. (If the plants in the coarse sand seem to grow as well as those in potting soil, reduce the frequency of waterings to just once per week.)

Results

The results, of course, will be what they are. However, you may anticipate the following: After a week or so, students will see the seeds in the potting/garden soil (air) sprouting. Seeds in coarse sand may or may not germinate; those immersed in water, with rare exceptions, will not have germinated.

Keep going for another two weeks or so, watering as before. Seedlings in the potting soil (air) will be seen to grow and prosper, but seedlings in the coarse sand, if they have germinated at all will likely have wilted and died over a weekend. If they have not died, their periodic wilting will have slowed their growth so that they are behind those in potting soil/air. If the seedlings in coarse sand are on a par with those in potting soil/air, continue with the activity, but reduce the frequency of waterings to once per week. The wilting/death of those in coarse sand will be observed given more time.

Those immersed in water will not have germinated. Have students remove the pot from water, drain it, and carefully dig down to find and determine the status of the seeds. They will likely find a "yucky mess"; the seeds have rotted.

Interpretation

Assuming that students have obtained results that are parallel to what is described above, bring them into Q and A discussion for interpretation.

What does wilting of a plant indicate? (Lack of water)
Why were seedlings in coarse sand subject to wilting and dying whereas those in
   potting soil (air) survived? (Potting soil holds enough water to carry seedlings

over several days without watering. Coarse sand doesn't.)

You or your students can readily demonstrate this effect. With two additional pots, fill one with the coarse sand (DRY) and another with the potting/garden soil, also DRY. Take a graduated cylinder with 100 ml of water and carefully pour the whole 100 ml onto each pot while holding it over a cup with a pour lip to catch the water that drains through. Return the water that drains through to the graduated cylinder. Subtracting the amount returned to the cylinder from 100 ml will give the amount held in the soil. Students will note that the potting/garden soil holds considerably more water than the coarse sand.

Explain that this is a characteristic of soil called WATER HOLDING CAPACITY. Potting or rich garden soils have good water holding capacity; coarse sand has very poor water-holding capacity. Hence, plants on sandy soils wilt and die unless they are watered or irrigated at frequent intervals. Of course, even on soil with good water holding capacity, plants will run out of water, wilt, and die if the time between rainfalls or irrigation is extended too long. We speak of such extended dry periods as DROUGHTS.

(Depending on students' interest and familiarity, you may or may not get into the topic of desert plants. In deserts, extended periods without rain are the rule, not the exception; this is what makes them deserts. Explain that, plants that thrive in desert environments have special adaptations for such conditions. Cacti, for example are able to soak up water and store it in their own bodies. This is the function of their thick, fleshy stems. Other desert plants are able to go into dormancy during prolonged dry periods. It is evident that this can become an extended lesson/discussion in itself. How far you wish to go with it here will be your option.)

Coming back to the lesson at hand: What about the pot immersed in water? Seeds were in the same soil as proved supportive, and they certainly had plenty of water. What is another factor? If students draw a blank, give them a hint. We need water, but what happens if we are held under water too long? So, what is needed in addition to water? AIR!

Guide students in reasoning that the experiment shows that the same is true for land plants as well. Roots must obtain water from the soil but they also need air, specifically oxygen, just as we and other animals do.

This may well strike students as a contradiction. They have learned that green plants take in carbon dioxide and give off oxygen in the course of photosynthesis. Now roots require air/oxygen. (???) Yes! Roots, being below ground, do not have light. Nor, do they have chlorophyll (green color) to provide capacity for photosynthesis even if light were present. But, they do need energy for growth and other functions. Roots get that energy from respiration (breaking down food using oxygen) the same as we and other animals do (Lesson B-9). Roots can get sugar, made by photosynthesis from the top the plant; it is transported down through veins (phloem). But land plants, in general, cannot transport oxygen through their veins. Therefore, the roots of most land plants depend on oxygen being able to move through open spaces in the soil to the roots. This property of soil is spoken of as AERATION. A good soil must allow AERATION.

When a soil is saturated with water, it is said to be WATERLOGGED. In a waterlogged soil, seeds and roots cannot get sufficient oxygen. They suffocate and die as result, and proceed to rot.

(Students may be familiar with the fact that there are many plants such as cyprus trees, water lilies, and reeds that do grow in swamps, bogs, and other situations where soil is perpetually waterlogged. Rice is a crop plant that requires such conditions. Explain that such plants do have special adaptations that enable them to cope with this condition. A common feature is a large open space or hole through the center of the stem that enables passage of oxygen from the top of the plant down to the roots. Again, how far you wish to take this, will be your option. The fact remains that most land plants do not have such adaptations and, consequently, cannot survive in such conditions.)

Help students review and summarize the lesson to this point. From their results they should reason their way to three attributes of a good soil:

1) it must have drainage that will prevent WATERLOGGING
2) it must have a loose structure that allows AERATION
3) it must have good WATER HOLDING CAPACITY.

Of course, let children know that this is not the end of the story. There is also the matter of particular chemicals necessary for growth: nitrogen, phosphorous, potassium, calcium, and iron, as well as small amounts of a number of others. Importantly, these are the same elements required for our bodies and those of other animals. They are passed up food chains from soil to plants to animals (Lesson B-5). Note that this assumes we eat a diet that includes good helpings of fruits, vegetables, dairy products, as well as meat. (You may include that chemicals are returned to the soil through animal wastes, at least in natural ecosystems. These chemical aspects of soil and the utility of fertilizers will be examined further in future lessons.)

Then there is the whole topic of SOIL MANAGEMENT. What practices do/should we use to maintain soils in optimal condition? In short, this is only the beginning of a whole area, SOIL SCIENCE. There is one aspect of this subject that can be addressed here, however. That is erosion.

Part 3. Erosion and the Soil Resource

In Lesson D-8, we described an activity of sprinkling water on soil to demonstrate EROSION. The focus there was to observe soil particles being carried away and eventually settling and burying organisms in the process. It will be worth repeating that activity here using garden topsoil. This time, however, focus children's attention on the soil that is left behind as erosion occurs. You may have them record what they see.

Children will see, as before, that fine soil particles are carried away as "rainfall" impacts bare soil and runs off the surface. Now, what about the soil that is left? They will note that coarse sand, pebbles, and any larger stones stay more or less in place on the soil.

Pose the question, "How does this change the quality of the soil and its ability to support plants?" With Q and A discussion, guide students in reasoning that, as erosion occurs, the remaining soil becomes more sandy and stony. At the very least, this diminishes the soil's water holding capacity—its ability to absorb and hold water—and, hence, its ability to support plants. (A large portion of the chemical nutrients are also lost because they are largely connected to the small soil particles. Add this at your discretion.)

Point out that erosion is not just caused by water; in dry regions or during droughts, wind erosion may be equally severe. Take some dry topsoil, crunch it into a loose pile, and let students blow on it. (Be sure that no one is in the way of the blowing. Soil particles can be blown into someone's eyes.) Students will observe a cloud of the fine soil particles being blown away. Again coarse material will remain.

Stress that this exercise is far more than just a classroom activity. In the coarse of agriculture, particularly plowing, cultivation, and after harvesting, soil is bare, exposed, and subject to erosion. Erosion is a huge worldwide problem effecting and diminishing the quality of agricultural soils. What is the danger of this? Where does all our food come from?

Bring students to focus on the fact that soil, in harmony with water, is the most fundamental and important of all resources. We tend to think that all our food simply comes from the supermarket. But how does it get to the supermarket? Guide them in recognizing/reviewing that all food, directly or through food chains, comes from plants, and those plants depend on soil. Thus, erosion is a potential threat to all of civilization.

Hasten to add that this is not meant to frighten them; it is meant to stress the importance of conducting farming and gardening in ways that protect the soil from erosion. It all comes under the heading of SOIL MANAGEMENT and SOIL CONSERVATION. Chief among soil conservation/management practices is maintaining a vegetative cover. Use Q and A discussion to guide students in reasoning why this is so. Bring to light that erosion occurs mostly when rain, running water, or wind impacts bare soil. A vegetative cover helps in two ways: it shields the soil from direct impact, and roots hold soil particles in place.

There are numerous other methods as well, but it is not practical to pursue them here. Emphasize again, however, that there are numerous career opportunities in the field of soil conservation. Invite students to pursue this independently if they may wish.

Questions/Discussion/Activities to Review, Reinforce, Expand, and Assess Learning:

Make books illustrating any one or more of the effects observed here.

Invite a person from the Agriculture Extension Service to address your class (usually listed under "Extension Service" in the county government sections of telephone directorys)

On playgrounds, the soil gets packed down (compacted) by many feet. A common management practice is to go over the ground with a machine that makes numerous small holes in the soil. What is the problem with compacted soil? How does making holes help the grass? What is the operation called?

Test and compare the water holding capacity of various soils in your locality. Contrast them with peat moss or well-rotted compost.

Observe and compare different soils in terms of sand, silt, clay, and organic content.

What is meant by soil erosion? What are the causes? What is the best way to reduce erosion?

Invite a person from your local Agriculture Extension Service/Soil Conservation office to speak to your class regarding soil management/erosion control practices in your region.

Conduct a field trip to point out instances of erosion and erosion prevention measures in your area.

In small groups, pose and discuss questions, such as:

What were the results of germinating and growing seedlings on coarse sand, on drained potting/garden soil, and on waterlogged potting/garden soil?
What do these results tell us about the needs of seed/roots?
What do these results tell us about characteristics of soil that are required to support plant growth?
Why does sand make a poor soil?
Why does clay, which waterlogs readily, make a poor soil?

In what way does erosion change the character of the soil? (What is removed? What remains in place?)
What is the single best way to prevent/minimize erosion?
Why is soil conservation important for all peoples/nations of the world?

Tremendous practical experiential knowledge combining soil science, plant science, pest management, weather, and more can be gained through having students maintain a small garden. Form a committee with your colleagues to negotiate with school authorities to set aside and fence a portion of the schoolyard for garden plots. It works best if students can keep the same plot throughout their school years. In this way, they can reap the benefits, year to year, of their inputs regarding soil management. If your schoolyard is entirely paved, tub gardens are an option (see addendum at the end of this lesson).

To Parents and Others Providing Support:

If you do any gardening at all, even keeping house plants, engage your children in the activity and in discussion regarding having and maintaining good soil.

The practical learning that comes from growing a few vegetables is so great that the experience should not be overlooked. A few square feet of ground is sufficient for a start. In the absence of this, the tub gardens described at the end of this lesson are a practical option.

On outings, among other things, observe the soil. What are its characteristics? Are these good or bad for supporting plant growth?

On trips and outings, note visible instances of erosion. Discuss the damaging effects of that erosion and how it might be prevented. Also note soil conservation practices such as strip cropping, terracing, maintaining grass waterways, and so on.

After heavy rains, the water of streams and rivers is frequently muddy brown with soil particles being carried along. Discuss: Where did that soil come from? (Erosion!) What does that say is happening to the quality of soil that is eroding?

Connections to Other Topics and Follow-Up to Higher Levels:

Soils and ecosystems
The role of organic matter and soil organisms in soil formation and maintenance
Soil conservation/soil management
All aspects of soil science

Re: National Science Education (NSE) Standards

This lesson is a steppingstone toward developing students' understanding and abilities aligned with NSE, K-4:

Unifying Concepts and Processes:
  • Evidence, models, and explanation

Content Standard A, Science as Inquiry
  • Abilities necessary to do scientific inquiry

Content Standard D, Earth and Space Science
  • Properties of earth materials

Content Standard F, Science in Personal and Social Perspectives
  • Types of resources

Books for Correlated Reading

Bailey, Jacqui. *Cracking Up: A Story About Erosion* (Science Works). Picture Window Books, 2006.

Riley, Joelle. *Erosion* (Early Bird Earth Science). Lerner, 2007.

Rosinsky, Natalie M. *Dirt: The Scoop on Soil* (Amazing Science). Picture Window Books, 2003.

Stewart, Melissa. *Down to Earth* (Investigate Science). Compass Point Books, 2004.
_____. *Minerals*. Heinemann, 2002.
_____.*Soil*. Heinemann, 2002.

Tomecek, Steve. *Dirt* (Jump Into Science). National Geographic Children's Books, 2002.

Walker, Sally M. *Soil* (Early Bird Earth Science). Lerner, 2006.

Addendum: Tub Gardens

Almost any tub will suffice for a tub garden, but the following has been found to be exceptionally serviceable, convenient, and low cost. Purchase a child's ridged, plastic wading pool, which is about five feet in diameter and one foot deep. Cut a ring of holes (about one inch in diameter and spaced about 16 to 18 inches apart) in the side of the tub, about one inch from the bottom. Place the tub in the desired (sunny) location and fill with equal parts of peat moss, well-rotted manure or compost (**compost** is well rotted plant wastes) and garden (top) soil, mixing them well as they are added. All these ingredients may be purchased at a garden store if necessary. An advantage of purchased ingredients is that they are virtually free of weed seeds, which otherwise germinate and may easily be confused with what you have planted.

After seeds are planted, gently sprinkle until water drains from the holes you have made in the side. The reason for having the holes up about an inch from the bottom is that good drainage will be allowed, which is a must, but a reservoir of water will be maintained in the bottom that will diminish the frequency of waterings required. Plant seeds desired. Only re-water at intervals as the soil begins to feel somewhat dry to the touch. One does not want to have the soil remaining soggy.

Thread C

Physical Science

Pursue Threads A, B, and D in Tandem

For a flowchart of the lessons and an overview of concepts presented see pages 8-13

Lesson C-1

Concepts of Energy I:
Making Things Go

Overview:

Energy is involved in everything we do or see happen. Yet, systematic instruction regarding energy is generally lacking at the elementary level, and most students receive little if any at higher levels. The result is that misconceptions and erroneous thinking regarding energy are rife, even among highly educated people. This and subsequent lessons in this thread will systematically build the foundation of clear thinking regarding energy.

In this lesson, children will gain the understanding and appreciation that energy is required to make anything go, move, work, or change. Further, they will learn the different forms of energy and appreciate how one form of energy may be converted to another form, but energy is not converted to matter nor is matter converted to energy (at least not in usual chemical, biological, or physical processes). Finally, they will apply their newly acquired concepts of energy to the functioning of their own bodies, other animals, and to plants, thus laying the foundation for understanding the function of living systems, as well as much of the physical world.

Time required:

Part 1. Forms of Energy and What They Do (game/activity plus discussion, 40-50 minutes)
Part 2. Storage and Release of Energy (continuation, 40-50 minutes)
Part 3. Energy Changes From One Form To Another (continuation, 40-50 minutes)

Objectives: Through this lesson, students will be able to:

1. Recognize and describe how things do not go, work, move, or change by themselves. Something additional is required.

2. In any given situation, identify the "something additional required" as heat, electricity, light, or movement.

3. State how heat, electricity, light, and movement are known as forms of energy.

4. Describe how many things, such as gasoline, firewood, batteries, and food, are not

energy by themselves, but contain stored energy that may be released.

5. Recognize and tell how various devices, such as light bulbs, motors, and engines, change one form of energy into another.

6. Recognize that ENERGY is not created from nothing. It must always come from another source.

7. Recognize and state how ENERGY (in everyday actions and processes) is never changed into solid, liquid, or gas, nor is solid, liquid, or gas changed into energy.

Required Background:

No particular background is required. However, it will be helpful to integrate this lesson closely with:

Lesson A-2, Solids, Liquids, and Gases

Lesson B-3, Distinguishing Between Plants and Animals

Lesson D-1, Gravity I: The Earth's Gravity, Horizontal and Vertical

Materials:

Energy cards. Prepare 3 x 5 cards, each with one of the following words and/or symbols depending on your students reading ability: HEAT ENERGY, ELECTRICAL ENERGY, LIGHT ENERGY, MOVEMENT ENERGY. You will need several cards of each.

Action cards. Prepare 3 x 5 cards, each with one simple familiar action, such as:

making a lamp light
making a bike go
making a toaster work
making a ball sail through the air
making water boil
making the toy go
making plants grow
turning a windmill
making a swing swing
making a solar powered calculator work
baking cookies
making your parent's car run
You can continue using any number of simple, familiar examples of making things go, work, move, or change.

Various items or devices that will illustrate the above. You may also rely largely on children's familiarity and experience with such things. Thus the following list is optional.

Magnets

Birthday candles and matches
Spring operated toys or devices
Battery operated toys or devices
Electrically operated appliances, such as a plug-in lamp, toaster, plug-in radio
Toys or devices, such as balls or toy cars, that require a push or pull to move
Calculator and/or other device(s) powered by solar cells

Teachable Moments:

Seat children in a semicircle to play a "thinking game." An array of items such as those listed above will help to attract and hold their attention.

Methods and Procedures:

Begin by dispelling any feeling or notion that energy is too abstract for young children to grasp, and be willing to learn along with your students. The various aspects of ENERGY addressed here are perfectly evident from everyday experience. It is simply a matter of becoming aware of that experience and using "energy words" as they apply.

Part 1. Forms of Energy and What They Do

Seat children in a semicircle with the "action cards" and "energy cards" you have prepared in the center. Begin by having children recall some of the magical things they have seen in "Harry Potter" or other fanciful movies or cartoons, things such as: flying through the air with or without a broomstick, causing explosions with a shake of wand, etc. Then pose the question, "Do or can any of these things happen in real life, or are they entirely make believe?" Some children will undoubtedly have tried to do such things and can share their experience that their attempts just didn't work. Thus, draw children to the conclusion that all such things are in the realm of make believe.

Next, pose the question, "What does it really take to make something go, work, move, or change?" Immediately, break this down into specific questions. Using one of the "action cards" or picking up one of the actual items that you have assembled, ask, "What does it take to make this ___ go/work/move/change?"

Have children ponder the item and the "energy cards" and choose the one that fits. (Cards for each form of energy should be in separate stacks so they can be seen clearly.) For example, the toaster requires electrical energy. To make water boil requires heat energy. To make a ball go sailing needs a "push" or movement energy. The solar calculator needs light energy, and so on. Set the paired cards, or items with their energy cards, where they can be seen by all.

Note that the number of examples may be infinite, but it does not get more complex. The forms of energy required for making them go/work/move/change remain the same.

Notes to teachers:

If you have some physics background, it will be conspicuous that we are taking some liberties and shortcuts with definitions in the above discussion in order to convey the basic concept of energy without getting bogged down in details. Let us be clear about the shortcuts we are taking. This is not to be part of the lesson; it is only to show that we are on the same page with more technical descriptions of physics. The shortcuts are:

1. What we are calling movement energy is, in the more technical language of physics, mechanical energy. However, since it is experienced as one moving thing pushing or pulling on something else, children grasp it more readily if it is called MOVEMENT energy.

2. Likewise, in the more technical language of physics, to get something moving requires a force. But force and movement are the two sides of a coin. A force against a movable object causes it to move. A moving object, flowing water for example, exerts a force against any object(s) in its path. Hence, to reduce the complexity of the concept, we are referring to any push or pull that puts something in motion as applying movement energy. Conversely, something moving—wind for example—contains movement energy that can turn windmills or do great damage, as in the case of hurricanes or tornadoes. Again, we will find that children grasp the energy idea if we refer to any push or pull as applying movement energy, as well as things in motion having movement energy.

3. There are two cases, however, where you will need to speak of FORCE apart from energy: gravity and magnetism. The force of gravity conspicuously imparts movement energy as things fall or start rolling/sliding downhill. But it can't be considered a source of energy because things invariably reach the bottom and stop. One must use another source of energy to get things back to the top to start over. You can use children's experience of playing on a slide as an example.

   Magnetic force should also be familiar to students from preschool play with magnets. (Allow them to play/experiment with magnets if this is not the case.) Like gravity, magnetic force will cause movement as it pulls or pushes on other magnets or iron containing materials. Still, by itself it is not a source or form of energy because it will only move things so far, unless there is another source of energy to move the magnet or reverse its poles.)

4. Light is but one small portion of the electromagnetic spectrum, which also includes microwaves, radio and TV waves, ultraviolet waves,   x-rays, and more. All together, this is spoken of as radiant energy. Thus, light is one form of radiant energy. If any of these other forms of radiant energy come up in discussion, you can instruct students that they are similar to light energy, but not visible to the human eye. The usefulness of leaving them out of the present discussion is self-evident.

5. Sound is sometimes included as a form of energy. We will treat it as a special

case of movement energy (vibrations), and address it in Lesson C-2.

6. Chemical energy. Burning, in all cases, is a chemical reaction; therefore, the stored energy in fuels, food, and other materials is often referred to as "chemical energy." But, this is meaningless to young children until they have knowledge of the actual chemistry involved. (Then, the release of energy from the breaking of certain chemical bonds can still be viewed as analogous to the release of energy from springs under tension as they are triggered. The distinction is that the energy comes out in the forms of heat and light rather than movement.) Therefore, we leave chemical energy out of the present discussion and refer to fuels, etc., simply as containing stored energy.

Returning to the lesson at hand, you will discover that children will have various misconceptions regarding energy that will need to be brought out and corrected. Common misconceptions that deserve further explanation and discussion, include:

1. Many children are likely to have the notion that simply flipping the switch or pushing the button is what makes electrical devices work. Point out how it must also be plugged in or have batteries. Instruct them electricity comes to our homes, schools, and other buildings and then to all the plugs through wires. You may be able to show children the overhead electrical line and/or the electric box that brings electricity to the building. With ceiling lights, there is no obvious plug; explain that are they are already connected into the system of wires hidden in the wall. Flipping the switch or pushing a button is like turning on a water tap. It simply allows the electricity to flow. If there is no electricity, the switch has no effect at all. Some children may be able to attest to this from their experience of power outages.

   SAFTY LESSON. Instruct children that since there is electrical power behind every outlet, they should take care never to stick their fingers or anything else into a plug or light socket. The electrical power can severely burn and even kill them. Likewise, they should keep hands away from electrical lines and fuse/circuit breaker boxes.

   Remote control devices present the same problem to a greater degree. Explain that the remote has a battery that enables it to send a signal. A device in the TV, car door, or whatever, receives the signal and turns on an electrical switch to power the TV, open the lock on the door, etc. Have students note that the TV must be plugged in, or the car's battery must be charged for the remote to have an effect.

2. For cooking anything, children will want to say that a stove/oven is required. Guide them to recognize that it is not the stove itself; it is the heat provided by the electricity or burning gas that does the cooking.
   Two common experiences regarding heat may need special attention:

   a) Ice melting. Explain that heat energy from the surrounding water or air is moving into the ice causing it to melt.

b) Water freezing. Heat energy is moving out of the water toward cooler surroundings. Point out that the refrigerator/freezer uses electricity to run a motor that effectively pumps heat from the inside to the outside of the box.

3. For many battery-powered devices, children will declare simply that it takes batteries to make it go. It will be important to make the point that batteries are devices that store electrical power. The concept is similar to storing water in a bottle. When it is hooked up properly electricity flows out of the battery into the device making it work. Guide students to reason why batteries go dead. Like the bottle, which only holds so much water, a battery will only hold so much power; when that is gone, it is empty, or "dead."

The story is similar for fuels. For example, children will declare that it takes gasoline to make a car run. Again, point out that gasoline and other fuels are forms of stored energy. That energy is released as the fuel is burned in the engine. Demonstrate with a birthday candle. Candle wax is a form of fuel; as it burns, it releases energy in the forms of heat and light. The same is true of gasoline. As it is burned in the engine, the energy released powers the engine. (Again, the energy is actually released in the forms of heat and light. In the cylinders, the heat causes an expansion of gases, which drive the pistons, but such details can be saved for later.) The topic of stored energy will be addressed further in Part 2 (below).

4. Some may desire an explicit definition of energy. The technical definition is that ENERGY IS THE ABILITY TO DO WORK. But what does this mean? It simply says that energy is what makes things go, work, move, or change. What makes things go/work/move/ change? Heat, light, electricity, and motion. In short, there is no way of defining or understanding energy beyond recognizing it as heat, light, electricity, or motion and one or another of these are needed to make things go/work/move/change.

Summarize and review that there are just four "types of stuff" that make things go/work/move/change. These are HEAT, LIGHT, ELECTRICITY, and MOVEMENT ITSELF. We group these together as active forms of ENERGY. In addition, energy may be stored in various systems or devices, but when the energy is released, it will be in the form(s) noted.

Part 2. Storage and Release of Energy

In a following lesson, after a short review of the forms of energy that are necessary to make given things "go," bring students to focus on ways energy may be stored and later released. For example:

Fresh batteries will RELEASE electrical energy for the toy or flashlight.
Gasoline will RELEASE energy to make our car run.

Wax, wood, paper, natural gas, and other fuels RELEASE heat and light energy as they are burned.

A cocked spring RELEASES movement energy when it is triggered.

A reservoir behind a dam contains stored movement energy that may be RELEASED by letting the water flow down through the dam where it may turn generators.

The stress on the word RELEASE in the above statements is highly significant. It is very important to AVOID "turns into." It is all too common for students to develop the notion that in burning something, the fuel, firewood, for example, is "turned into" the heat and light energy. This idea is just plain wrong and causes all kinds of confused thinking later on.

It will be later that students learn the chemistry behind burning, but when they do, they will learn that all the matter making up wood or other fuel is accounted for in the byproducts of the burning (largely carbon dioxide and water vapor that go off in the air). In the meantime, it will be important to give the proper concept that burning is a process of RELEASING stored energy, NOT matter turning into energy.

The situation is analogous to a cocked spring. In releasing its energy, the spring is unchanged except in shape. It is obvious that the spring does not turn into the MOVEMENT energy of what it caused to go flying. Similarly, it is obvious that a battery does not turn into electrical energy; it remains the same size, shape, and weight; it only releases stored electrical energy. Thus, emphasize that the same is true in burning. Like springs, the material being burned only changes shape, so to speak; the energy, heat and light, is released as this occurs.

Energy and Food

This concept of energy being stored in and released from various burnable materials has particular application to FOOD. Children have invariably been told and have learned that they must eat to grow and be healthy and strong—period.

Now it is critically important to explain that every movement and function of the body, from active running to simply keeping the heart pumping and the mind working (or dreaming), like everything else we have discussed, requires a source of energy to make it work and keep going. Ask, "What is the source of that energy?"

With Q and A discussion bring children to the conclusion that it is FOOD. Food is our fuel. It contains stored energy. In a manner that is like gasoline being burned in the engine of our car to make it go, food is broken down in our bodies, and its stored energy is released in a way that makes us "go." The same is true for all other members of the animal kingdom. Food is fuel, the source of energy that keeps us alive and "working."

You may demonstrate the high, stored energy content of food by demonstrating that any foodstuff, when dried, will burn.

To overcome the pre-existing notion that food is required only for growth, you will need to give repeated emphasis and review to the fact that food serves also as fuel. Indeed, the major portion of what we eat (90% or more) goes toward providing energy. Again, food is not turned into energy. All the particles making up the food remain and are eliminated from the body. Of course, the 10 percent of food that is essential for growth of actual body tissues is equally important.

Energy and Photosynthesis

Along with discussing the role of food as fuel, introduce the fact it takes energy to make those plant tissues that become food. Plants use light energy to make all the tissues of their bodies, which we and other animals, in turn, consume as food. Again, be careful to avoid "turned into." Light energy is NOT turned into sugar or any other products. Light is the energy that runs the "chemical machinery" (photosynthesis); it is the photosynthetic "machinery" that takes the ingredients, carbon dioxide from the air and water, and makes them into sugar. Subsequently, the plant makes the sugar and certain chemicals from the soil into all the tissues of its body.

Instruct students that it is something like an electric mixer. In making cake batter: Is electricity for the mixer turned into the batter? NO! The electricity only powers the mixer; the mixer churns the ingredients into batter. Again, in plants, light energy just powers the chemical "machinery" of photosynthesis; that machinery makes the ingredients into plant tissues, hence, food.

Part 3. Energy Changes From One Form To Another

Through the preceding portions of this lesson, it should be conspicuous that much of what we have considered may be viewed in terms of changing energy from one form to another. Have students reflect on examples such as the following. Then cite additional familiar things and have students name the energy conversion that is occurring.

Sunlight shining on the pavement makes it hot. LIGHT ENERGY is being converted to HEAT ENERGY.

A motor converts ELECTRICAL ENERGY into MOVEMENT ENERGY.

A light bulb converts ELECTRICAL ENERGY into LIGHT ENERGY (and also heat energy).

A toaster converts ELECTRICAL ENERGY into HEAT ENERGY.

A photovoltaic cell (solar cell) converts LIGHT ENERGY into ELECTRICAL ENERGY.

HEAT ENERGY will boil water and create steam pressure that drives a turbine, MOVEMENT ENERGY.

Turning a generator converts MOVEMENT ENERGY into ELECTRICAL ENERGY.

STORED ENERGY IN FOOD is released and is transferred into MOVEMENT ENERGY, all the internal as well as external movements that your body performs.

STORED ENERGY IN FUELS is released as HEAT AND LIGHT ENERGY as the fuel is burned.

Hitting a ball with a bat transfers MOVEMENT ENERGY of one object into the MOVEMENT ENERGY of another.

In photosynthesis, LIGHT ENERGY is stored in the making of food.

You may make this into a game/activity by having children use "energy cards" to fill in the blanks in the following statement.

A _____(item or device) changes _____ (what form of) energy into _____ (what form of) energy

In this activity/discussion two additional facts become experientially evident, but will need special emphasis.

First, energy always remains a form of energy. It may be converted from one form to another; it may be stored in various ways and subsequently released. But, energy is never converted into matter (gas, liquid, solid), nor is matter ever converted into energy.[22]

Second is the fact that energy must always come from somewhere. It is not and cannot be created out of nothing or pulled out of nowhere, cartoons and fanciful stories notwithstanding. To store energy (to cock a spring, for example), have students note that an equivalent amount of energy must be put into the system. A spring will not cock by itself. Likewise, if they want to experience the fun of zipping down a slide (movement energy), what must they do first? The stored energy in food and firewood is no more than (in fact it is much less than) the amount of light energy that went into its photosynthesis.

This lesson will only become imbedded by continued use of the "energy vocabulary" as described above. As you go about the activities of the day, you will doubtlessly come across any number of things in addition to what is mentioned here. The rule to keep in mind is that anything that goes, works, moves, or changes has a source of energy behind it. Thus, each presents an opportunity for reviewing and amplifying the lesson. Take advantage of these opportunities.

Of course, many children will be likely to ask how and why (do such changes take place)? For example: How is one form of energy turned into another? Why can't we create energy out of nothing? Be prepared to say in a quizzical way, "That is a good question and I really don't know. Perhaps you can figure it out later." Indeed, answers regarding how and why are elusive to scientists as well. We can only say that this is what we observe.

---

[22] In the realm of high-energy, nuclear physics, there is a conversion of matter into energy and this is put to use in nuclear power plants and nuclear weapons; it is also the source of energy from the sun. However, this can well be left aside until physics classes in high school or college. The fact remains that mass-energy conversions do not occur in the physics, biology, or chemistry of everyday life.

Bernard J. Nebel, Ph.D.

Questions/Discussion/Activities To Review, Reinforce, Expand, and Assess Learning:

Make a book depicting: the four forms of energy; the form of energy being used to make a certain thing go/work/move/change; the storage and release of energy; the change of energy from one form to another.

Set up an activity center with various "wind-up" toys or devices and stress that students focus on their experience that energy outputs (running of the toy) is only equivalent to the amount of energy input (their winding or pushing).

In the course of usual classroom activities and in using pieces of equipment, inject the question: What form of energy does this _____ require to make it go/work? Further ask: What form of energy is that energy changed into? Take time for review and discussion as necessary.

Discuss with students how every advance in technology has hinged on designing and building devices that use energy to accomplish a given purpose. Consider cars, airplanes, computers, etc. Consider how this is likely to remain true in the future as well. (Will we ever be able to ride around on magic broomsticks like Harry Potter?)

Many children are inclined to invent great things. Draw their attention to the need for energy to make their invention go/work. Where is that energy going to come from?

In small groups, pose and discuss questions such as:

Holding any given device/appliance, ask questions such as: What does it take to make it work? What form of energy is required? What is that energy changed into? Again be particularly watchful for children skipping from energy to object. For example, children are likely to declare that a toaster changes electricity into toast. Be sure they back up to include: The electrical energy is changed into heat energy. The heat energy "cooks" the bread making it into toast.

What do such things as natural gas, gasoline, firewood, reservoirs behind hydroelectric dams, and food represent in addition to the material itself? How is the energy released from _____?

What does food provide for your body (in addition to nutrients for growth)?

Watch a short cartoon video and probe if that could that really happen in real life? If that situation occurred in real life, What would actually happen?

To Parents and Others Providing Support:

The routines of every day life (driving, refueling the car, cooking, using any appliance, etc.) afford innumerable teachable moments for conducting short Q and A discussion to review and reinforce the lesson. Take advantage of such opportunities. But this is unlikely to happen automatically. Give it some forethought. Think of things you may do in the day, how they involve energy, and what you will say to your children as they participate in or watch these activities. Then go on to make "energy conversation" a part of everyday talking. By doing so, children will gain comprehension. You may check their comprehension by periodically asking them what is required to make the ____ go/work/move/change. It will be energy in every case. Then discuss with them: What is the form and source of that energy? What are the changes that take place?

Watch and analyze cartoons with your children, asking if that could happen in real life. What would actually happen in real life? Why? Review and discuss energy aspects as necessary.

Connections to Other Topics and Follow-Up to Higher Levels:

Considerations of energy are an integral part of all chemistry (chemical reactions), most biology, and ecology, to say nothing of physics. Therefore, this and following lessons in this thread should be considered as building a basic foundation for these other subjects.

Much of the world's economy and important social/political issues revolve around energy resources. Energy lessons will be a foundation for these as well.

Many students will become intrigued with: How does it work? How does a motor turn electricity into movement? How does a light bulb turn electricity into light? Depending on the interest and ability of your child, you may use other sources to delve into these in as much detail as you wish. An excellent resource is www.howthingswork.com/.

Re: National Science Education (NSE) Standards

This lesson is a steppingstone toward developing students' understanding and abilities aligned with NSE, K-4:

Unifying Concepts and Processes
- Systems, order, and organization
- Evidence, models, and explanation
- Constancy, change, and measurement

Content Standard A, Science as Inquiry
- Abilities necessary to do scientific inquiry

Content Standard B, Physical Science
- Properties of objects and materials
- Position and motion of objects
- Light, heat, electricity, and magnetism

Content Standard C, Life Science
- Characteristics of organisms

Content Standard E, Science and Technology

Bernard J. Nebel, Ph.D.

- Abilities of technological design
- Abilities to distinguish between natural objects and objects made by humans
- Understanding about science and technology

Content Standard F, Science in Personal and Social Perspectives
- Personal health
- Types of resources

Books for Correlated Reading:

Bailey, Jacqui and Matthew Lilly. *Charged Up: The Story of Electricity* (Science Works series). Picture Window Books, 2004.

Berger, Melvin. *All About Electricity* (Do-It-Yourself Science). Scholastic, 1995.
_____. *Switch On, Switch Off ?* (Let's-Read-and-Find-Out Science, Stage 2). HarperCollins, 1989.

Bradley, Kimberly Brubaker. *Energy Makes Things Happen* (Let's-Read-and-Find-Out Science, Stage 2). HarperCollins, 2003.

Murphy, Patricia J. *Around and Around* (Rookie Read-About Science). Children's Press, 2002.
_____. *Back and Forth* (Rookie Read-About Science). Children's Press, 2002.
_____. *Up and Down* (Rookie Read-About Science). Children's Press, 2002.

Olien, Becky. *Electricity* (Our Physical World). Capstone, 2006.

Royston, Angela. *Using Electricity* (My World of Science). Heinemann, 2002.

Stille, Darlene. *Electricity: Bulbs, Batteries, and Sparks* (Amazing Science). Picture Window Books, 2004.
_____. *Energy: Heat, Light, and Fuel* (Amazing Science). Picture Window Books, 2004.

Storad, Conrad J. *Fossil Fuels* (Early Bird Earth Science). Lerner, 2007.

Trumbauer, Lisa. *What is Electricity?* (Roobkie Read-About Science). Children's Press, 2004.

Twist, Clint. *Electricity* (Check It Out!). Bearport, 2005.

Webster, Christine. *Energy* (Our Physical World). Capstone, 2005.

Young, June. *Energy is Everywhere* (Rookie Read-About Science). Children's Press, 2006.

Zemlicka Shannon. *From Oil to Gas*. Lerner, 2003.

Lesson C-2

Sound, Vibrations, and Energy

Overview:

In this lesson, students will observe that sound comes from the rapid back and forth movement, or VIBRATIONS, of strings, wires, and other things. They will discover that pitch depends on the rapidity or frequency of the vibrations. They will analyze various musical instruments in terms of their producing vibrations. Finally, they will model how vibrations are carried through air or other media, picked up by the ear, and translated into hearing.

Time Required:

Part 1. Sound and Vibrations (activity plus interpretive discussion, 40-60 minutes, plus analysis of musical instruments as desired)

Part 2. Transmission of Sound (activity plus interpretive discussion, 40-50 minutes)

Objectives: Through this exercise, students will be able to:

1. Understand and use the following words in their proper context: VIBRATIONS, TONE, PITCH, FREQUENCY, and AMPLITUDE.

2. Demonstrate and tell how sound and vibrations are related.

3. Analyze various musical instruments and voice in terms of producing vibrations.

4. Describe how we make sounds with our voices.

5. Demonstrate and tell how pitch depends on the frequency of vibrations.

6. Model how sound is transmitted.

7. Describe how the ear picks up vibrations and translates them into hearing.

Required Background:

Lesson A-3, Air is a Substance

Lesson A-4, Matter I: Its Particulate Nature

Bernard J. Nebel, Ph.D.

Lesson C-1, Energy I: Making Things Go

Materials:

Part 1. Sound and Vibrations
   Rubber bands of various sizes
   Stringed instruments of any sort, bells, tuning forks, and/or any other assorted things that produce a tone when plucked or struck
   Illustration of the human larynx

Part 2. Transmission of Sound
   A three to four foot strip of molding with a groove such that marbles can be rolled down the groove
   Marbles
   Diagram showing the anatomy of the human ear

Teachable Moments:

Pass out rubber bands and ask children to experiment as described below.

Methods and Procedures:

Part 1. Sound and Vibrations

Pass out rubber bands so that each student or small group has a variety of sizes. Demonstrate how a band may be stretched between middle fingers, held close to the ear, plucked with an index finger, and a resulting TONE heard. (Experiment with this beforehand so that you can guide children accordingly.) Have children do this, experimenting with different rubber bands and stretching them to different degrees. You may have them make tables listing rubber bands from small to large and record the relative pitch of each. Likewise, you may have them make another table listing various degrees of stretch for a given rubber band and recording relative pitch produced.

In the course of doing this, you will have to clarify the meanings of tone and pitch. TONE refers to any prolonged note. PITCH refers to exactly what the note is: A, B-flat, C, etc. You can demonstrate the idea with your voice even if you can't hit particular notes.

Have students discuss their experience with rubber bands in small groups and report their conclusions. The consensus should be that smaller bands produce higher pitches than larger bands, and, for any given band, the pitch goes up the more the band is stretched, although the heightening of pitch may only become noticeable as the band is stretched very tightly. If any student disagrees, have him/her retest the situation.

Next, pose the question, "What produces the tone?" Instruct children to watch a rubber band closely as it is plucked. Observation is most clear when the band is in strong direct light and viewed against a dark background. (SAFTEY! Caution children to NOT

262

hold a tightly stretched band less than a foot or so from their eyes. It is possible for rubber bands to snap and whip out.) Students will observe that the band appears as a blur for an instant after it is plucked but then it calms down to stillness. Have students focus on the relationship between this appearance and the tone produced. They will note that the two go hand and hand.

What causes the blurred appearance of the rubber band as it is plucked? With Q and A discussion, guide students to recognize that the blur is caused by the band going back and forth so fast that it can't be seen except as a blur. Explain that such rapid back and forth movement is called VIBRATION. The plucked band VIBRATES. Our ears are able to hear the vibrations.

(Some children have trouble accepting the idea that we hear vibrations. It appears that the vibrations are in one location while the sound seems to be in our ears. How can this be? If children are having this difficulty, it may be prudent to skip to Part 2 of this lesson and then return this part. Otherwise, continue.)

Students have observed that sound is associated with the vibrations in the case of rubber bands. Invite them to consider: Is this true for other sounds as well? So far as possible, invite children to examine and test various musical instruments: stringed instruments, bells, a xylophone, a tuning fork, and/or other things that produce a tone when plucked or struck. With stringed instruments, they can actually see the strings vibrate. With bells, they probably cannot see the vibrations, but they can feel them. As they lightly touch a toning bell, they will feel a "zip" in their fingertips as the vibrations and the tone are damped.

Call children's attention to their own voices. Are they exceptions? Show children a picture of the human larynx (voice box). Explain how, by passing air through the larynx and at the same time putting tension on the vocal cords, we cause them to vibrate, thus, giving off sound. We change the pitch of the sound by tightening or loosening the tension on the cords much like stretching the rubber band or tuning a stringed instrument. We do this automatically, however, so that we are not conscious of doing so. We only have to think of raising or lowering the pitch of our voice. The sound from our vocal cords is further modified by altering the shape of the cavity of our mouth, position of the tongue, and so on.

Children may be able to feel the vibrations of their vocal cords by tilting their heads far back, placing a finger very lightly on their larynx (voice box or Adam's apple) and giving a prolonged "Ahhhhh." (Practice this on yourself beforehand so that you can help students accordingly.)

What about the different pitches produced by the rubber bands, musical instruments, or the human voice? This is not easy to discern from simple observations, although some children may do so. In any case, explain that higher notes are given by faster vibrations, that is, going back and forth more times per second. The faster vibrations are termed HIGHER FREQUENCY. Higher frequency vibrations, hence, higher notes, are given by shorter, lighter strings, smaller bells, shorter tuning forks, etc. Likewise, frequency is increased and notes are raised by putting more tension on any given string(s), including vocal cords.

Guide students to reexamine stringed instruments. Especially, have them note and compare length of a given string and the note produced. Likewise, contrast the tension on a given string and note produced. Can they predict which of two strings will give the higher note before actually testing them? Why does pressing a string against the neck of an instrument like a guitar yield a higher note?

A final point to have students observe, and for you to emphasize, is that for any given string, bell, or tuning fork, the note given off (pitch) remains constant. It is louder when first being struck or plucked, but as the tone fades, the PITCH REMAINS CONSTANT. Use Q and A discussion to bring students to reason what this means in terms of vibrations. It means that the frequency (the number of back and forth movements per second) remains constant. Only the AMPLITUDE (the distance traveled from one side to the other) becomes less and less.

Students can observe the relationship between loudness and amplitude in their rubber bands. After plucking, the width of the blur decreases hand in hand with the fading of the sound.

Children enjoy playacting these terms. Have them move their hand and forearm up and down to represent a vibration. Increasing the speed of up and down movement corresponds to increasing the frequency, hence represents raising the pitch. Increasing the distance that the hand is moved up and down corresponds to increasing the amplitude, hence, the loudness.

The generalization to derive is that in every case, VIBRATIONS ARE THE SOURCE OF SOUND. But how does this work? How do the vibrations of the string, bell, or whatever get to your ear where they are heard as sound? The following activity will be revealing.

Part 2. Transmission of Sound

Have students in small groups or as a demonstration do the following. Carefully level a 3 to 4 foot strip of grooved molding so that gravity will not cause marbles to roll down the grove. Place three or four marbles in the groove at roughly the center such that they are touching or nearly touching one another. Give another marble a speedy roll down the groove toward the marbles in the center and observe what happens.

Children and adults too get a kick out of the result. As the "shooter" marble strikes those in the center, it stops dead (almost) and it is the marble at the other end of the group that commences rolling on down the groove. Invite children to play/experiment with this device as they wish with different numbers of marbles in the center, rolling the "shooter" at different speeds, and so on. You may choose to formalize this activity by having students make tables, record results, and so on. In any case, the crux of the activity will be the interpretive discussion.

Pose questions such as: Why do the marbles behave as they do? Can we interpret it in terms of energy? What sort of energy is involved? How is that energy passed from one marble to the next?

From Lesson C-1, students should recall that movement is one of the forms of energy. (Review as necessary.) Then guide students in reasoning that the shooter marble has a certain amount of movement energy. As the shooter marble hits the first marble in the row, its energy is transferred to that marble. With that transfer, the shooter stops and the hit marble would roll on, but it hits the next marble, transferring the energy again. Thus, the movement energy of the shooter marble is passed from marble to marble as each one bumps the next. At the end of the chain, the last marble receives the energy, and being free to move, rolls on with nearly the same energy (speed of roll) as originally contained in the shooter marble—"nearly the same energy" because some energy is lost in friction along the way. (If the same activity is done with rubber balls, it will barely work at all because there is much more friction in rubber balls hitting one another.)

Now, what does all this have to do with how vibrations get from the vibrating object to your ear? In Q and A discussion, coach students in recalling that air is a substance and it has a particulate nature (Lessons A-3 and A-4). Do you think that the particles of air might behave much like the marbles? From here, students will generally catch on. Yes, the vibrations of the string or whatever bump the air particles next to it. Each air particle bumps the next, all the way to your ears. Explain that, unlike the marble situation, the chain of air particles is continuous all the way from the vibrating object to your ears. Thus, the vibrating object causes a corresponding back and forth movement (vibration) of air particles all the way from the object to your ear.

At your ear (show students a cross-sectional diagram of the ear), the vibrations of the air particles cause a corresponding push-pull (vibration) of the eardrum resulting in our hearing. (How much detail you wish to give regarding the connections between the eardrum, the inner ear, and triggering nerve impulses will be up to you. At this stage, it is probably best skipped except to answer questions.)

(You may also include that sound is transmitted through liquids and solids in the same manner and even more efficiently since the particles are closer together, enabling the wave of "bumping" to be transmitted faster. Have children put their ear against one end of a table while you tap gently with your fingernail on the other end. Children with their ears against the table will hear the tapping clearly while other children standing at the same distance may not hear it at all.)

Summarizing the entire lesson, the key points are: a) there is the rapid back and forth movement or vibration of an object; b) the vibrating object bumps air particles and air particles bump each other, carrying vibrations from the object to our ears and causing our eardrums to vibrate in turn; c) the vibrations of our eardrum are perceived as hearing; d) pitch depends on the frequency of vibrations, higher frequency being perceived as higher pitch, lower frequency as lower pitch; d) loudness depends on the amplitude of the vibrations. In short, sound is the result of vibrations of one sort or another affecting our eardrums.

A health consideration should be inserted somewhere along the way. The habit of listening to very loud music (or other loud noise) makes the eardrums and parts of the inner ear vibrate with such amplitude that they are actually damaged over time. This reduces hearing ability.

Also, make students aware that there are all sorts of careers in the production, transmission, recording, and replaying of sound. Additionally, there are careers in eliminating sounds, i.e., noise, we don't want to hear. It is a science of discovering where the undesirable vibrations are coming from and damping them out in one manner or another.

This is the essence of the lesson, but there are a number of related points that may come up in the course of activities or discussion. We enumerate them in the following so that you may address them as is fitting.

1. Bangs versus pure tones. Striking a bell or plucking a string starts it vibrating. It vibrates at a given frequency depending on its physical nature (length/size, weight, and other factors). The result is a more or less pure note prolonged over a second or more only fading away as vibrations cease.

   Hitting something with a hammer or a firecracker, on the other hand, produces a sudden, single, big wave of air particles hitting one another as seen in the marble activity. When this sudden, sharp wave of movement hits and pushes the eardrum, it is perceived as simply a "bang." There is no continuing tone.

2. Whistles and flutes. Whistles and musical instruments such as organs and flutes emit great sounds in the absence of anything noticeable vibrating. Are they an exception to the rule? No, the air blowing over the opening causes a compression of air within the tube. That air pops out again, is recompressed, pops out, over and over. In other words, there is a vibration of the air column itself within the tube of the instrument as it "bounces" between compression and release. Note that the frequency of the vibrations and hence the pitch of the sound varies with the size of the instrument, lower frequency, deeper notes being produced by larger tubes. Pitch may be varied by opening or closing certain holes along the tube, which change its effective size.

3. Sound waves versus light waves. Both sound and light are spoken of as traveling as waves. However, it is important not to confuse the two. Sound waves traveling as the successive bumping of particles is totally different from light waves, which travel as electromagnetic radiation. Radio and TV signals and other forms of radiation also travel as electromagnetic waves. It follows that sound will only travel through gas, liquid, or solid where there is capacity for one particle to hit another as described. Sound will not and cannot travel through a vacuum. Light and other forms of electromagnetic radiation, on the other hand, travel very will through a vacuum.

4. Speed of sound versus light. Sound "waves" traveling by one air particle hitting

the next is markedly slower than light and other forms of radiation, which travel as electromagnetic waves. (The speed of sound is about 1,000 feet or about one fifth of a mile per second; that of light is 186 thousand miles per second.) Thus, in watching fireworks, for example, you commonly see the explosion, the burst of light, but only hear the resulting boom some moments later. The separation between lightning and thunder may be even more pronounced. Indeed, you may estimate how far away a given lightning strike is by counting the seconds between seeing the flash and hearing the resulting boom of thunder. A five-second separation translates as one mile.

5. Echoes. When the bumping of air particles comes against a solid surface, they bounce back and cause the chain of bumping to proceed in another direction. The echo comes after the initial sound because of the additional distance traveled.

6. Telephones. It is not sound (vibration) itself that is carried from the speaker to the listener. At the speaker's end, voice vibrations are changed into electrical signals, which are carried over wires; at the listeners end, the electrical signals are used to make a diaphragm vibrate, converting them back to sound. The same principle holds for cell phones, radio and TV; only the vibrations at the speaker's end are changed into and carried as respective electromagnetic waves, and these are changed back to vibrations at the listeners end.

7. Recording and playing sound. In modern equipment, vibrations at the recording end are changed into magnetic or light pulses and these make "footprints" on the disc or tape. In the player, a sensor "feels" the "footprints" and changes them back to vibrations and, hence, sound.

8. Inaudible "dog whistles." The human ear is sensitive to vibrations with a frequency between about 20 and 20,000 per second. (This varies in different people and the range generally decreases with age.) Higher or lower frequencies simply do not trigger the ear to send impulses to the brain. Hence, such vibrations go unheard. However, the ear systems of many animals, most notably dogs, are triggered by vibrations beyond the range of human hearing. Hence they can hear high-pitched sounds that humans can't.

Explain that, with more technical instrumentation, we have determined the frequency of oscillations (back and forth movements) per second for various notes. The oscillations per second for some notes are: low C = 261.63; D = 293.66; E = 329.63; F = 349.23; G = 392.00; Middle A = 440.00 B = 493.88.

Questions/Discussion/Activities to Review, Reinforce, Expand, and Assess Learning:

Make a book illustrating vibrations, the transmission of sound through air, and/or other aspects of the lesson.

Set up an activity center with items that will vibrate such that students can continue to review and investigate interactions between vibrations and sound.

Analyze musical instruments in terms of: What is the origin of the sound? How is the sound amplified? … transmitted?

Have students do show-and-tells analyzing any particular instrument they have examined or other sounds they may have experienced.

Facilitate students in designing and making musical instruments.

Study radio or TV speakers and how they produce sound.

In small groups, pose and discuss questions such as:

What is the connection between vibrations and sound?
Why will a "bell" made from soft clay not give a tone when struck?
Will non-elastic objects, such as a bell made from soft clay, vibrate?
What is the reason for two tones having different pitch? (Explain in terms of vibrations.)
What is the reason behind a tone being loud or soft? (Explain in terms of vibrations.)
How do we make sounds with our voice?
Why/how do different strings on a guitar give different pitch?
Why/how does pushing on different frets of a guitar produce different tones?
How does sound get from a vibrating object to your ears?
How does an ear change vibrations into sound?
Why is there often a separation between a flash of lightning and thunder?
How do tones differ from bangs?

To Parents and Others Providing Support:

For any tone or sound that may catch your child's attention, coach them in analyzing the source. (What is vibrating?)

Coach them in analyzing why/how different instruments give sounds.

Facilitate them in repeating any/all of the activities described.

Facilitate children in designing, making, and testing different things that will vibrate and produce sounds.

Coach them in analyzing how frequency of vibrations relates to pitch and how loudness relates to amplitude.

Make a tin-can telephone (For instructions, google "tin can telephone")

Connections to Other Topics and Follow-Up to Higher Levels:

Anatomy and physiology of the ear
Anatomy and physiology of the larynx

Causes of hearing loss

Comparative anatomy and physiology of the sound emitting and hearing abilities of different animals, including invertebrates

Harmonics

The operation of recording, amplifying, and transmitting systems

Audio engineering

Acoustics and acoustical engineering

Re: National Science Education (NSE) Standards

This lesson is a steppingstone toward developing students' understanding and abilities aligned with NSE, K-4:

Unifying Concepts and Processes
- Evidence, models, and explanation
- Constancy, change, and measurement

Content Standard A, Science as Inquiry
- Abilities necessary to do scientific inquiry
- Understanding about scientific inquiry

Content Standard B, Physical Science
- Properties of objects and materials
- Position and motion of objects

Content Standard C, Life Science
- The characteristics of organisms

Content Standard E, Science and Technology
- Abilities of technological design
- Understanding about science and technology

Content Standard G, History and Nature of Science
- Science as a human endeavor

Books for Correlated Reading:

Olien, Becky. *Sound* (Our Physical World). Capstone, 2003.

Pfeffer, Wendy. *Sounds All Around* (Let's-Read-and-Find-Out Science, Stage 1). HarperCollins, 1999.

Rosinsky, Natalie M. *Sound: Loud, Soft, High, and Low* (Amazing Science). Picture Window Books, 2003.

Royston, Angela. *Sound and Hearing* (My World of Science). Heinemann, 2001.

Trumbauer, Lisa. *All About Sound* (Rookie Read-About Science). Children's Press, 2004.

Twist, Clint. *Light & Sound* (Check It Out!). Bearport, 2005.

Walker, Sally M. *Sound* (Early Bird Energy). Lerner, 2005.

Whitehouse, Patty. *Loud Sounds, Soft Sounds* (Construction Forces). Rourke, 2007.

Lesson C-3

Concepts of Energy II:
Kinetic and Potential Energy

Overview:

Through "play" with simple things, such as swings, slides, and ramps, this lesson will bring students to a fuller understanding and appreciation regarding the interplay between active and stored energy that was introduced in Lesson C-1. They will learn two new words defining active and stored energy: kinetic energy and potential energy. Further, they will gain experiential understanding of the equivalence of energy inputs and outputs. Finally, they will apply these concepts to plants and animals. This builds a foundation for future understanding concerning all of chemistry, all of physics, most of technology, and numerous aspects of biology, ecology, and other subjects as well.

Time Required:

Activity, 20-30 minutes, plus interpretive discussion, 30-40 minutes

Objectives: Through this exercise, students will be able to:

1. Understand and use the terms KINETIC ENERGY and POTENTIAL as they apply to active energy and stored energy.

2. Identify examples of kinetic energy and potential energy and distinguish between them.

3. Recognize and give examples illustrating how potential energy may be converted into kinetic enegy and how kinetic energy may be changed into potential energy.

4. Recognize and give examples illustrating that, in conversions between kinetic energy and potential energy, there is no creation of new energy; there is an equality between energy inputs and energy outputs.

5. Recognize and give examples illustrating that, in conversions between kinetic energy and potential energy, there is no conversion of matter to energy nor energy to matter. (Mass-to-energy conversions that occur in nuclear power plants and nuclear weapons are for studies in advanced physics. The fact remains that in everyday biology, chemistry, physics, etc., such conversions do not occur.)

6. Relate kinetic energy and potential energy to photosynthesis, food, and animal activity.

Required Background:

Lesson A-2, Solids, Liquids, and Gases
Lesson B-2, Living, Natural Nonliving, and Human-made Things
Lesson C-1, Concepts of Energy I: Making Things Go
Lesson D-1, Gravity 1. The Earth's Gravity, Horizontal and Vertical
Lesson B-3, Distinguishing Between Plants and Animals (may be given concurrently)

Materials:

Ramp (any broad flat surface on an angle so that things can be rolled or slid down)
Balls and/or other toys to roll/slide down the ramp
Pendulum (any weight on the end of a string attached to the edge of a table where it will swing freely)
Rubber party balloons
Spring powered toy or other device
Rubber bands
Hand crank generator (if one is available)

Teachable Moments:

Assemble the items listed above. Invite children to join you in playing with them.

Methods and Procedures:

Invite children to play/experiment with:

Shooting rubber bands off the end of their thumb (Admonish SAFETY)
Blowing up balloons, releasing, and allowing them to zoom around the room
Rolling and sliding things down a ramp
A pendulum
Spring powered toys (Admonish SAFETY)
A hand crank generator (optional)

After children have played for 20 minutes or so, settle them down and pose the question, "How can we explain each of these things/actions in terms of energy?" Give them a quick review of Lesson C-1, as necessary. Namely, active forms of energy are heat, light, electricity, and motion. Then there is stored energy in various things.

Instruct students that we are now going to refer to the active forms of energy (heat, light, electricity, and movement) as KINETIC ENERGY, and we are going to refer to any form of stored energy as POTENTIAL ENERGY. We call stored energy potential energy because it has the ability, or POTENTIAL, to come out in one or another form of kinetic energy.

We are going to look at each of the things that they have played with in terms of changes between kinetic and potential energy. Explain one item, then guide students with Q and A discussion to explain each of the others in turn.

In the case of the ramp, for example, we use kinetic energy (lifting motion) to raise a ball to the top of the ramp. Some of that energy—a portion is lost in friction—becomes the potential energy of the ball being at a height where it will roll down. Then, with the pull of gravity, that potential energy becomes kinetic energy, the active energy of the ball's movement as it ball rolls down the ramp.

In blowing up a balloon, you are using energy to force air in. The air under pressure in the balloon then has potential energy. As that pressure is released, it becomes the kinetic energy (movement) of the air coming out and the balloon zooming about the room.

A pendulum or swing illustrates a continuing exchange between kinetic and potential energy. At the end of a swing's arc, there is an instant of no movement; but the weight on the swing has the potential energy of height. That potential energy of height becomes kinetic energy (movement) as the swing descends. At the low point of the arc, the potential energy of height is gone; it has been converted into kinetic energy of motion. That kinetic energy of motion then carries the swing up the other side and, in doing so, it is converted back to potential energy of height.

In this lesson, it may be effective to have children say or sing in chorus the energy change that is occurring as they watch or perform various actions. For example, in the case blowing up the balloon, they might recite, "Kinetic energy going in, becoming potential energy of pressure." As the balloon is released and jets itself about the room, they might sing, "Potential energy of pressure is becoming kinetic energy of movement."

Finally, draw students to recognize and make the generalization: FOR ENERGY TO COME OUT OF A SYSTEM, WE MUST FIRST PUT ENERGY IN. This conclusion is the crux of what we want students to observe and appreciate. Energy will not magically come out of a wand or broomstick. For energy to come out, we must, in one way or another, put that energy in.

Have children ponder this in relation to their own experience. To make any of the items they have just played with go: What did they have to do first? Have them consider the items one at a time. In each case, they should note that they first had to put energy in to raise the item, to stretch the rubber band, cock the spring, and so on. The concept, point out, is not really different than getting water out of a bottle. To get water out of a bottle, water must first be put in.

Furthermore, guide students to recognize and appreciate that the amount of energy that comes out is only equivalent to the energy that they put in. They might get a greater snap from a rubber band by pulling it back further, but this means putting more energy in. They might get more speed from a higher ramp, but this means using more energy to raise the ball, or climb to the top. Again, the analogy of the water bottle will help. You can't get more water out of a bottle than what you put in.

A pendulum also illustrates this principle elegantly. Hold the bob of a pendulum at a given position, and ask students, "How far up the other side will it swing when it is released (without any additional push)?" Students can make their guesses (hypotheses) as they will and then test the result. They will discover that the bob never goes or returns to a point higher than where it started. Why not? Guide students in reasoning that it starts with just so much potential energy. On release, that potential energy is converted to kinetic energy and back to potential energy as discussed above, but there is no additional energy added along the way. Thus, it cannot achieve potential energy (height) greater than what it started with. (In fact, it never returns to quite as high, because there is an inevitable loss of energy from friction and air resistance, but this is for another lesson.)

Some children may bring up the fact that they can make themselves go higher and higher on a swing. Ask, "But what do you have to do to make yourself go higher and higher?" Pump! Okay, what is pumping? Bring them to recognize that their pumping is a way of adding energy to the swing. What happens without pumping?

Go on to relate these ideas to additional things children experience. Roller coasters and bouncing balls may be analyzed in the same manner as the pendulum or swing. They involve a repeated exchange between kinetic and potential energy. Also, observe that the height they achieve on the second hill/bounce is never quite as high as the starting height (assuming no extra push). Again, without additional energy being added, they gradually run down.

Finally, ask children, "In any of these activities, did you observe a change in the item itself?" For example: Did the ball change size as it was raised or rolled down the ramp? Did the spring change in any way other than shape as it was cocked and released? … and so on. They should conclude that the answer is always NO. Again, emphasize that energy may easily be changed from kinetic to potential and from potential back to kinetic and from one form of kinetic energy to another. But energy is not turned into matter (solid, liquid, or gas) nor is matter turned into energy.

Students may cite instances where they seem to get energy out with no apparent input—batteries, for example. Explain that the manufacturer put the energy in. The manufacturer actually expended more energy making the battery than the battery contains. Likewise, wind and waterfalls may seem to be exceptions. Explain that the kinetic energy of wind comes from sunlight heating the atmosphere. The sun's heating also raises water up to the top of the falls via evaporation and subsequent condensation and rainfall. Similarly, major energy resources are coal and crude oil obtained from the earth. Explain that these are derived from ancient plant and animal material; thus, the energy input was the kinetic energy of sunlight through photosynthesis as noted below.

Relating Energy to Plants and Animals

Invite students to consider the exchange between kinetic and potential energy in relation to plants and animals. Can we think of what plants do in terms of kinetic and potential energy? Start with: Is light energy kinetic or potential energy? (Kinetic!) Is the sugar, made in photosynthesis, kinetic or potential energy? (Potential!) Thus, the plant can be viewed as capturing kinetic energy, light, and storing some of that energy as the potential energy in sugar, wood, and other tissues.

Is food (plant tissues) kinetic or potential energy? (Potential!) So what do humans and other animals do in terms of energy? What happens to the potential energy in food? By breaking the food down, that potential energy becomes the kinetic energy of all body movements.

Of course, some of the food eaten is remade into the body of the animal as well. This portion of food, now the animal body, still contains potential energy. Thus, some animals, including humans, feed on other animals. But the concept regarding energy remains the same. Guide student in tracing a food chain backwards. They will discover that food chains always lead back to animals that fed on plants. Thus, the original and continuing source of energy for all life on Earth is the kinetic energy from sunlight. (Some isolated, deep-ocean ecosystems derive energy from certain inorganic chemicals, but these can be saved for more advanced classes.)

Questions/Discussion/Activities to Review, Reinforce, Expand Assess Learning:

Make a book illustrating one or more of the items worked with indicating the input and output of energy in making it go.

Set up an activity center where students can continue to experiment with some of the items described. Their focus should be the observation that energy outputs are never greater than energy inputs.

In the course of usual classroom activities, pause and have children analyze given actions in terms of kinetic and potential energy. Take time for review and discussion as necessary.

As various devices/machines come up in discussion, have students consider their operation in terms of potential and kinetic energy.

Have students reflect on and discuss their own eating and activities in terms of energy inputs and outputs and conversions between kinetic and potential energy.

Discuss energy resources.

In small groups, pose and discuss questions such as:

Holding/demonstrating/citing any one or more of the things discussed, ask students to analyze its action in terms of kinetic energy input, potential energy, and kinetic energy output.

How does kinetic energy output, e.g., how high and far a ball goes, relate to energy input?

Can you get more energy out of a system than you originally put in? Cite examples that illustrate the equivalence of energy inputs and outputs.

In such activities, is matter changed to energy or energy to matter?

Watch short cartoon video clips focusing on energy inputs and outputs. Have students analyze: Would/could that happen in real life? Why not? For example, cartoons will commonly depict a huge response from a trivial energy input. (Of course, in setting off explosives there is a huge energy output from a trivial energy input, just pulling the trigger for example. Have students note that the energy input was in making the explosive materials. Pulling the trigger is like letting go of the ball at the top of a ramp. It just starts the release of that potential energy. Ultimately, the fact remains that one doesn't get something for nothing.)

To Parents and Others Providing Support:

In the course of recreational time with your children, you may well play on swings, slides, ramps, etc. Take advantage of such teachable moments to have them review how kinetic energy is changing to potential energy and back.

While eating, discuss the energy aspects involved.

Draw their attention to how getting more energy out of a system, a longer, faster ride on a slide, for example, depends on a greater energy input— climbing the higher slide.

Watch and analyze cartoons with your children by asking questions such as: Could that happen in real life? What would actually happen in real life? Why?" Review and discuss energy aspects as necessary.

Connections to Other Topics and Follow-Up to Higher Levels:

As well as the concept of energy itself, these changes between kinetic and potential energy provide a foundation for understanding all of chemistry, all of physics, most of technology, and numerous aspects of biology, ecology and other subjects as well.

Re: National Science Education (NSE) Standards

This lesson is a steppingstone toward developing students' understanding and abilities aligned with NSE, K-4:

Unifying Concepts and Processes
- Systems, order, and organization
- Evidence, models, and explanation
- Constancy, change, and measurement

Content Standard A, Science as Inquiry
- Abilities necessary to do scientific inquiry

Content Standard B, Physical Science

• Position and motion of objects
• Light, heat, electricity, and magnetism

Content Standard C, Life Science
• Characteristics of organisms

Content Standard E, Science and Technology
• Abilities of technological design
• Understanding about science and technology

Books for Correlated Reading:

Olien, Becky. *Motion* (Our Physical World). Capstone, 2005.

Royston, Angela. *Forces and Motion* (My World of Science). Heinemann, 2002.

Stewart, Melissa. *Energy in Motion* (Rookie Read-About Science). Children's Press, 2006.

Young, June. *Energy is Everywhere* (Rookie Read-About Science). Children's Press, 2006.

Whitehouse, Patty. *Energy Everywhere* (Construction Forces). Rourke, 2007.

Lesson C-4

Concepts of Energy III:
Distinguishing Between Matter and Energy

Overview:

In Lessons C-1 and C-3, we have emphasized that matter is not turned into energy nor is energy turned into matter (at least not in usual chemical, biological, or physical processes). Energy and matter remain separate. Yet, a misunderstanding of this separateness often remains and becomes the source of much confused and erroneous thinking. This lesson will bring students to understand that the attributes of matter are very different from those of energy, and these distinctions will strengthen students' understanding of their separateness. This will be an important stone in the foundation for critical thinking.

Time Required:

Part 1. Attributes of Matter (interpretive discussion, 35-45 minutes)
Part 2. Attributes of Energy (interpretive discussion, 35-45 minutes)

Objectives: Through this exercise, students will be able to:

1. Recognize and state the three attributes common to all forms of matter (gases, liquids, solids).

2. Recognize and state the attributes common to all forms of energy.

3. Recognize and discuss how the attributes of energy are very different and distinct from those of matter.

4. Use examples to tell how energy may affect matter.

Required Background:

Lesson A-2, Solids, Liquids, and Gases
Lesson A-3, Air Is a Substance
Lesson A-4, Matter I: Its Particulate Nature
Lesson C-1, Concepts of Energy I: Making Things Go
Lesson C-3, Concepts of Energy II: Kinetic and Potential Energy
Familiarity with weighing things

278

Materials:

Items illustrating solids, liquids, and gases. For example: wood blocks, container filled with water, balloon filled with air

Bathroom or more accurate scale or balance

Items that depict forms of kinetic energy such as a: lamp, heater, spring or rubber bands

Battery

Teachable Moments:

With the listed items at hand, call children into a circle for a "think game."

Methods and Procedures:

Part 1. Attributes of Matter

Have children focus on the items illustrating solids, liquids, and gases, and, reviewing Lessons A-2, A-3, A-4, bring them to recall that solids, liquids, and gases are referred to as THREE STATES OF MATTER, and ALL MATTER IS MADE OF PARTICLES.

Then, pose the questions such as: What else can we say about matter that will pertain to all its forms? Are there additional attributes or characteristics that apply to all solids, liquids, and gases? It is unlikely that children will simply come out with correct answers. Therefore, proceed with Q and A using questions such as the following:

Is there a limit to how many/much ___(cite any solid material) you can put in a box?

Is there a limit to how much water you can put in a bottle?

Is there a limit to how much air you can put in a balloon without it bursting?

In each case allow students to reflect that the answer is, "Yes!" Okay, what can we say about all matter? Guide students to reason and conclude: ALL MATTER TAKES UP OR OCCUPIES SPACE. Explain that this relates to its particulate nature. One can push particles closer together, as in compressing a gas, but two particles cannot be in the same place at the same time. Each particle takes up a certain amount of space; therefore, all matter takes up space.

Pursue this same technique and sort of questioning with regard to weight and bring students to conclude that another attribute is that ALL MATTER HAS WEIGHT (mass).[23]

---

[23] Of course, weight only exists when gravity is present. The more technically correct term would be MASS, which is independent of gravity. However, all the practical experience of young children is with gravity present. Trying to introduce mass at this point simply causes unnecessary confusion. Therefore, it is most straightforward to use the term weight and save the distinction between weight and mass for another lesson (see Lesson D-7)

Bernard J. Nebel, Ph.D.

Summarize and have students rehearse that all forms of matter have these three attributes:

a) All forms of matter have a particulate nature.
b) All forms of matter occupy space.
c) All forms of matter have weight (mass).

In all probability, children will need a break here, but pursue Part 2 without too much lapse of time.

Part 2. Attributes of Energy

Begin with a review of the three attributes of matter (above). Then pose the question, "Can these attributes of matter be applied to the forms of kinetic energy: heat, light, electricity, or movement?"

Students are likely to say "yes" and point to a lamp as representing light, for example. Emphasize that the lamp is NOT light. It is simply a device that converts electrical energy into light energy. Go on to pose questions, such as:

Can you put light itself in a container, cover it, take it into a dark room and let it out?
Can you blow up a balloon with light?
Does a bucket sitting in sunlight fill up and overflow with light?
Is a ball sitting at the top of a ramp larger or heavier than when it is at the bottom?
Does a ball get heavier or larger as it is thrown?
Does a spring under tension take up more space and/or weigh more than when it is not under tension?
Does a battery get smaller or lighter as electricity is drained from it?

Some students may wish to check these things out; facilitate their doing so. The more they reflect and check, however, the more firmly they should be able to conclude that the answer in each case is "No." The concepts of weight (mass) and occupying space simply do not apply to heat, light, electricity, or motion.

This is puzzling to minds, young or old. It is difficult to conceive of something that has no mass and takes up no space. So what are heat, light, electricity, and movement? You can only say that they are what they are. You experience them as you do; you just have to give up the idea that everything has to have the attributes of taking up space and having mass. (Actually, this is much easier for children than adults to accept because they have not had the long history of thinking in terms of things occupying space and having weight.)

While kinetic energy does not have weight or take up space it makes its existence clear in other ways. Proceed with Q and A to bring out how kinetic energy of one form or another affects matter. For example, ask questions such as:

How will heat change raw eggs? … cookie batter? … meat?

What will heat do to ice? ... water?
What does electricity do as it goes through a light bulb? ... a motor? ... a heater?
What does flowing water do to a waterwheel?
What does wind do to a windmill?
What does light do for plants?
What does light do for a solar calculator?

Students will be able to draw on their experience to say, in their own words, what happens in each case. As the final step, ask, "How can we summarize all these answers into one simple statement?" Allow students to reflect and coach them with examples such as: Heat turns ice into water and water into steam. Electricity makes the solid mass of a motor turn. Moving water makes a waterwheel turn, and so on. In short, guide students to derive the conclusion: Energy will do something to matter. ENERGY WILL CAUSE MATTER TO CHANGE OR MOVE IN SOME WAY. (We are including a change in temperature as a change.)

GuideHelp students in summarizing the total lesson by listing statements such as the attributes of matter and energy. Their lists should includefollowing:

Matter is solids, liquids, and gases.
All matter has is made up of particles.
All matter occupies space.
All matter has weight (mass).
One state of matter may be changed to another state, solids to liquid for example.

Energy is heat, light, electricity, and movement.
Energy does NOT occupy space.
Energy does NOT have mass/weight.
Energy DOES affect matter causing it change and/or move in some way.
One form of energy is readily changed to another form of energy.

In wrapping up, we should not let students persist in thinking that solids, liquids, and gases cover everything that exists in the physical world. There is one more, but only one more thing: ENERGY. Matter and energy do cover everything in the universe. There is nothing (so far as we know) that cannot be described and explained in terms of matter and/or energy.

Matter and energy can be thought of separately, but they are commonly combined in the same thing. The energy inherent in movement, for example, can hardly be separated from the mass that is moving. Likewise, potential energy will always be tied up with matter in a certain state: springs under tension, air under pressure, a rock at a high elevation, and all forms of fuel are examples.

Stress again that in all biological organisms and systems, and in all chemical and physical processes, the matter and energy, while mixed together so to speak, remain separate things. One form of matter may be changed to another form or state of matter; one form of energy may be changed to another form of energy. But energy is not changed to matter, nor is matter changed to energy.

(Again, it will be best to save nuclear power plants and nuclear weaponry, which do utilize a mass to energy conversion, for later lessons. Further, you may be aware that light is sometimes described in terms of particles, photons. This comes from the fact that under certain observational techniques, light exhibits the attributes of particles. Likewise, it has been shown that light from distant stars is bent to a slight degree by the sun's gravity as it passes close, indicating that it has a slight mass component. Save this, along with nuclear reactions, for higher level physics classes. The fact remains that for all practical purposes, light and other forms of kinetic energy can still be considered as having neither mass nor occupying space, and mass-energy exchanges do not occur.)

Questions/Discussion/Activities to Review, Reinforce, Expand and Assess Learning:

Make books illustrating the attributes of matter. ... the attributes of energy. ... how energy may affect matter.

Make posters showing states and attributes of matter, forms of kinetic energy, and how energy may affect matter.

Teach students to play the game, "I am thinking of something," where the something may be a form of energy as well as a form of matter. Coach them in asking an initial question that will distinguish between matter and energy.

In small groups, pose and discuss questions, such as:

What is matter?
What are the forms of matter?
What are the attributes common to all matter?
What are the forms of kinetic energy?
How can you distinguish between matter and energy?
Contrast energy and matter.
How is energy different than matter?
What will energy do to matter?

To Parents and Others Providing Support:

In the course of daily routines, cooking for example, you and your children will see many instances of energy affecting matter. Take advantage of these teachable moments to review:

What is the form of matter?
What is the form of energy?
How is the energy affecting the matter?

Review and reinforce other aspects of the lesson as opportunities avail.

Connections to Other Topics and Follow-Up to Higher Levels:

Having the distinction between matter and energy clearly in mind is a central pillar for the understanding of all of chemistry, physics, technology, most of biology, ecology, and countless other subjects, as well as numerous aspects of everyday life.

Understanding living systems (both animal and plant) hinges on understanding interplays between matter and energy.

Re: National Science Education (NSE) Standards

This lesson is a steppingstone toward developing students' understanding and abilities aligned with NSE, K-4:

Unifying Concepts and Processes
- Systems, order, and organization
- Evidence, models, and explanation
- Constancy, change, and measurement

Content Standard A, Science as Inquiry
- Abilities necessary to do scientific inquiry

Content Standard B, Physical Science
- Properties of objects and materials
- Position and motion of objects
- Light, heat, electricity, and magnetism

Content Standard D, Earth and Space Science
- Properties of earth materials

Content Standard E, Science and Technology
- Abilities of technological design
- Understanding about science and technology

Books for Correlated Reading:

Curry, Don L. *What Is Matter?* (Rookie Read-About Science). Children's Press, 2005.

Stille, Darlene R. *Matter: See It, Touch It, Taste It, Smell It* (Amazing Science). Picture Window Books, 2004.

Lesson C-5

Inertia

Overview:

    Everyday experience shows us that stationary things stay put unless we or something else moves them. Then again, swings, spinning wheels, and thrown balls keep moving after the initial push only slowing down gradually. In this lesson, students will learn that this feature of motionless things to stay motionless and moving things to keep moving is another basic attribute of all matter, an attribute we call INERTIA. Furthermore, they will learn that inertia is the expression of movement energy. This understanding will provide a foundation for interpreting countless everyday events, predicting outcomes, explaining numerous aspects of technology, and supporting critical thinking in general.

Time Required:

    Part 1. The Attributes of Inertia (games/activities, 30-40 minutes, plus interpretive discussion, 20-30 minutes)
    Part 2. Relating Inertia to Energy (interpretive discussion, 25-35 minutes)

Objectives: Through this exercise, students will be able to:

1. Demonstrate and give examples of how motionless objects remain still unless something moves them.

2. Demonstrate and give examples of how things in motion tend to keep moving. Explain how their slowing down/stopping is because of friction or running into something.

3. Describe how this characteristic of things to stay put or to keep moving is another basic attribute of all matter, an attribute we call INERTIA.

4. Enumerate various sorts of everyday actions/events that demonstrate or utilize the phenomenon of inertia.

5. Describe how inertia is an expression of energy being required to make anything go/work/move/change.

Required Background:

Lesson C-1, Concepts of Energy I: Making Things Go
Lesson C-3, Concepts of Energy II: Kinetic and Potential Energy
Lesson C-4, Concepts of Energy III: Distinguishing Between Matter and Energy

Materials:

Small ball (1-2 inches in diameter)
Tray with a rim
Ice cube
Cookie sheet
Beanbag or lump of clay
Small bowl

Teachable Moments:

Engage children in one or more of the games described.

Methods and Procedures:

Part 1. The Attributes of Inertia

The formal definition of INERTIA is: The property of a stationary object to remain stationary and a moving object to remain moving unless acted on by another force. However, don't try to give a formal lesson of inertia. Rather, engage children in one or more of the following "games" and follow up with interpretive discussion.

1. Place a small ball in the center of a tray on the floor. Give the tray a sudden jerk of a few inches in any direction. Have students note and describe what happens to the ball as you jerk the tray. As the tray moves in one direction, the ball moves in the other relative to the tray. But, in terms of room, the ball tends to remain stationary. (It will roll around some after the jerk so it is critical to focus attention on what happens immediately with the jerk.)

   Then place the ball against the rim of the tray and move the tray at an even rate (not jerking), such that the rim pushes the ball along with it. Stop the tray suddenly and have students describe what happens. They will observe that the ball continues moving in the direction of the motion.

   Essentially the same activity can be performed with an ice cube on a cookie tray. A challenge is marking a path on the tray and seeing if one can make the ice cube follow the path by jerking the tray in various directions—not lifting or tipping it. Children rapidly learn that the ice cube essentially stays still while they jerk the tray under it.

2. A game that children have fun with is "bombardier." The object of the game is to

drop a "bomb," a lump of clay, dough, bean bag, or other soft object, into a bowl while moving. Hold the "bomb" at shoulder height or higher and walk rapidly toward and past a bowl on the floor. Without changing the speed of your pace or the position of your hand, release the bomb so that it falls into the bowl as you walk by. You can make it increasingly challenging by walking faster and using a target with a smaller opening. Or, you can mark a target on the floor and have a contest as to who comes closest to the bull's-eye.

Have students carefully observe at what point they must release the bomb in order to have it land in the bowl. They discover that they must release it before it is actually over the bowl. The faster they walk, the sooner they must release it. Further, have students observe carefully the falling object in relation to the body of the "bombardier." They will note that the object stays aligned with the bombardier's body as it falls (assuming she/he keeps moving at the same pace). In other words, the "bomb" not only continues to move forward when it is released; it continues to move forward at the same speed despite its also falling.

3. A common "trick" is yanking a sheet of paper from under a glass of water without spilling the water. (Practice this with things that won't spill until you get the technique down. The trick is to give a really sharp yank in an exactly horizontal direction so that the glass is not tipped as the paper is yanked.)

Interpretive Discussion

After one or more of the activities, bring children to focus on the elements that they observed. Through Q and A discussion bring them to recognize:

First, objects that are not moving tend to remain put. Thus, the ball or ice cube tended to remain in the same place (relative to the room) while they jerked the tray around under it. (The ball or ice cube may move around some after the jerk because friction between them and the tray set them in motion to some extent.) The same can be said for yanking the paper from under the glass. The glass remained stationary as the paper was yanked from under it.

(In each of these cases, if the tray or paper is moved very slowly, the object will move with it. This is because friction between the object and the surface is enough to hold the object to the surface. It does require a fast sharp jerk.)

Second, objects in motion tend to keep moving in the same direction and at the same speed. (You will have to explain how gravity complicates this picture. The key is having students note that the "bomb" continued forward (the same direction) and kept pace with the bombardier. The falling is a separate motion added on. Have students visualize that, without gravity, the bomb wouldn't fall, but it would continue moving in the same direction and at the same speed as when it was released.

Have students cite additional examples from their own experience that illustrate these two ideas. You may have to stress that in all cases we are considering inanimate things that do not have their own motive power or steering capability. Regarding things staying put, their examples may include anything that has remained where it was put. Indeed, this is so much a part of our common experience that if something is not where it was last seen, it is cause for thinking that someone/something took or moved it. An inanimate thing moving about by itself occurs only in the realm of imagination and fantasy.

Regarding things tending to keep moving after the initial "push," examples may include spinning tops, spinning wheels, coasting on a bike (on level ground), swings, pendulums, anything thrown, etc. Of course, nothing here on earth remains in motion for long; without additional pushes, everything gradually slows down and stops—swings, spinning wheels, rolling balls, etc. This seems to be contrary. Explain that a factor invariably working against all moving objects on Earth is greater or lesser amounts of friction. Friction, explain, is simply the resistance of one surface to move over another. This includes resistance to moving through air or water. The extreme case is when the object crashes into the ground or a barrier of some sort. Without friction, it would keep going indefinitely. (We will address friction in more detail in Lesson C-6.)

Have children consider the rotation of the Earth and its orbiting around the sun. In moving through the void of space, there is no friction or other force impeding motion. Thus, these motions and those of other heavenly bodies do remain essentially constant for millions, even billions, of years.

As students grasp the idea that motionless things remain in the same place and moving things keep moving (until/unless they are affected by something else) they will be ready for the final step. Simply instruct them that this phenomenon is called INERTIA. Have them recall the three attributes of matter they learned in Lesson C-4: a) matter has a particulate nature; b) it occupies space; and c) it has mass (weight). Now, we are ready to add a forth attribute, its tendency to remain still if still and to keep moving if moving. We refer to this attribute as INERTIA. Saying this the other way about, INERTIA refers to the phenomenon that a still body remains motionless and a moving body keeps moving. To get something to change its motion (start moving, speed up, slow down, or stop), we must overcome its INERTIA.

Have students use the word, INERTIA, in describing common actions. For example: To overcome the INERTIA of a soccer ball, I must kick it. To overcome the INERTIA of a wagon, I must push it, and so on. Don't leave out the other side of the coin. The ball keeps going after I throw it because of its INERTIA. I can coast on my bike after I get it going because of INERTIA. I am pitched forward when the car is braked suddenly, because of the INERTIA of my body, and so on.

In the course of this discussion, students may bring up examples that appear to be exceptions. Be prepared for the following:

1. Things spontaneously start rolling or sliding downhill. Point out that gravity is constantly exerting a downward pull. If whatever is holding them up gives way, they will proceed to go downhill. Likewise, children may cite a flowing river.

Point out that the river is actually flowing downhill although the slope may be slight.

2. Air moves (wind) without anything apparent pushing it. Point out that there actually is something pushing it; solar heating produces differences in pressure from one location to another (Lesson A-6). Solar energy also perpetuates the water cycle.

Part 2. Relating Inertia to Energy

The final part of this lesson is to relate inertia to energy. Pose the question, "How does inertia relate to energy?" Allowing think time, guide students in drawing on their previous knowledge that movement is a basic form of energy along with heat, light, and electricity (Lesson C-1). Further, energy cannot be created from nothing; it must come from somewhere (Lesson C-3). From here, guide students to reason that a still object will not be set into motion until there is an energy input through a force: gravity, magnetic field, an actual push or pull from another object.) In short, the side of inertia that says a motionless object remains motionless is a reflection of the requirement for an energy input.

Turning to the other side of inertia, a moving object now has the kinetic energy of motion. It will not slow down or stop unless or until that energy is removed. The removal of movement energy may occur gradually through greater or lesser amounts of friction including wind or water resistance; it may occur abruptly as the moving object hits a barrier; or it may result from another force countering its motion such as gravity acting on a ball thrown up. If the movement energy is not taken away, the object will remain in motion indefinitely. As we have noted, this is the case for movements of the Earth and other heavenly bodies.

(I have seen many lesson plans in current use that confuse friction and inertia. Namely, a weight is dragged using a spring scale marked in units called newtons. The force required for the dragging is read on the scale and called a measurement of inertia (???). Nonsense! What is being measured is friction, the resistance to sliding/dragging one surface over another. We will investigate friction further in Lesson C-6. The fact that the measurement is made in units called newtons does not change the situation.

To measure the force required to overcome the inertia of an object, one would need to subtract out any contribution of friction. Then one would need to measure the force required to make the object accelerate (speed up) at a given rate. This obviously involves complexities far beyond our level here. It is sufficient and significant at this level to have students recognize that to set an object in motion requires an input of energy, a kick, push, or pull of some sort, quite apart from any friction involved.)

Questions/Discussion/Activities to Review, Reinforce, Expand, and Assess Learning:

Make a book illustrating inertia in action: how it keeps moving things moving or stationary things still; how an energy input is required make things move.

Set up an activity center where students can continue to test and gain experiential familiarity with the phenomenon of inertia.

Discuss how inertia is a reflection of the requirements for energy inputs and outputs.

Invite students to note experiences in their daily lives that demonstrate inertia at work. Invite them to report on these instances in the preclass show-and-tell/discussion period. A common instance is running and jumping on a skateboard. What keeps you going on the skateboard?

In small groups, pose and discuss questions, such as:

We kick or knock mud from our boots as a matter of course. How and why does this work? (Your boot stops suddenly and the mud keeps going because of inertia.)

How does a dog's shaking propel water from its fur?

Why do you feel yourself pitched forward as the car is braked suddenly? Why should you always wear seat belts?

Why do you feel yourself pushed back in the seat as the car is speeded up rapidly?

How does a hammer drive a nail, whereas you cannot just push the nail in?

There are many imaginary stories about stuffed animals and dolls coming to life and moving around. What does your _____(any toy or item) really do when you leave it? Why?

When you can't find something that you have been playing with, why is it most prudent to think of where you might have left it? Why should it be where you left it?

What is required to make inanimate things move? What is the word for what you must overcome?

When you trip, why do you always fall forward?

When you throw a ball, it keeps moving after it leaves your hand. Why?

Why do things not move by just thinking them into motion or by waving a wand at them?

Why does a bigger car require a bigger engine and more fuel?

To Parents and Others Providing Support:

In the routines of everyday life there will be many instances where you and your children will see/experience inertia at work. Common instances include:

Kicking mud from boots

Knocking "goop" from a spoon by hitting it the side of the pot

Hitting something with a hammer to make it go

Shaking out a rag

Being pitched forward as the car is braked suddenly

Telling children to buckle their seat belts

Take advantage of such teachable moments to have your child explain these things in terms of inertia, coaching as necessary.

Every sport has numerous instances, including movements of players themselves, where inertia may be seen to operate. Take whatever sport your children like to participate in or watch, pick out such instances, and have your children explain them in terms of inertia.

Connections to Other Topics and Follow-Up to Higher Levels:

Technological design: Considerations of inertia are crucial in every sort of machine.
Understanding the principle of inertia is the foundation for calculating momentum, which is critical in all mechanical engineering.
Application of inertia to sports, gymnastics, and dance
Understanding basic energy concepts, such as the conservation of energy and energy flow in technology and ecosystems.
Innumerable practical applications in critical thinking

Re: National Science Education (NSE) Standards

This lesson is a steppingstone toward developing students' understanding and abilities aligned with NSE, K-4:

Unifying Concepts and Processes
  • Systems, order, and organization
  • Evidence, models, and explanation
  • Constancy, change, and measurement

Content Standard A, Science as Inquiry
  • Abilities necessary to do scientific inquiry

Content Standard B, Physical Science
  • Properties of objects and materials
  • Position and motion of objects

Content Standard D, Earth and Space Science
  • Properties of earth materials

Content Standard E, Science and Technology
  • Abilities of technological design

Books for Correlated Reading:

Bradley, Kimberly Brubaker. *Forces Make Things Move* (Let's-Read-and-Find-Out Science, Stage 2). HarperTrophy, 2005.

Whitehouse, Patty. *Moving Machines* (Construction Forces). Rourke, 2007.

Lesson C-6

Friction

Overview:

   In this lesson, students will come to appreciate how FRICTION plays a role in nearly everything we do, how it works to our advantage in preventing slips and skids, and how it works to our disadvantage in wasting energy. They will further appreciate the significance of wheels, ball bearings, lubricants, and streamlined shapes in terms of reducing friction. As well as having innumerable practical applications, this provides an important block in the foundation for understanding the principle of energy flow, chemical and biological systems, and ecosystems.

Time Required:

   Part 1. The Concept of Friction (activity, 40-45 minutes, plus interpretive discussion, 20-30 minutes)

   Part 2. Wheels and Friction (activity, 25-35 minutes, plus interpretive discussion, 10-15 minutes)

   Part 3. Friction, Heat, and Energy (activities, 15-20 minutes, plus interpretive discussion, 15-20 minutes)

   Part 4. Wind and Water Resistance (interpretive discussion, 15-20 minutes)

Objectives: Through this exercise, students will be able to:

1. Understand and use the word FRICTION, as it applies in everyday situations.

2. Conduct tests that will demonstrate the relative amounts of friction between various surfaces.

3. Cite instances where we wish to reduce friction as much as possible.

4. Cite instances where we wish to maximize friction and use it to our advantage.

5. Demonstrate that friction produces heat; explain how this is the result of movement energy being converted to heat energy.

6. Tell how energy expended in overcoming friction is basically wasted.

7. Describe ways and things we use to reduce friction.

8. Show and explain how wheels serve to markedly reduce friction.

9. Describe how ball bearings reduce friction even more.

10. Understand how wind resistance and water resistance are special forms of friction.

11. Demonstrate what is meant by TURBULENCE.

12. Draw, describe, and recognize examples that illustrate a streamlined shape. Tell how that shape minimizes wind/water resistance.

13. Explain what would occur in the total absence of friction; cite things that illustrate this.

Required Background:

Lesson C-1, Concepts of Energy I: Making Things Go
Lesson C-3, Concepts of Energy II: Kinetic and Potential Energy
Lesson C-4, Concepts of Energy III: Distinguishing Between Matter and Energy
Lesson C-5, Inertia

Materials:

Part 1. Introducing The Concept of Friction
   A a ramp: any broad, flat, smooth surface mounted at an angle. The ramp should be broad enough so that items may be slid down two at a time, side-by-side, and steep enough so that most of the items will slide. A table with one end propped up will work fine.
   Various items to slide down the ramp, such as:
   Wood blocks
   Plastic blocks
   Books
   Chalkboard eraser
   Large rubber eraser
   Ice cubes
   3 x 5 cards and pencils

Part 2. Wheels And Friction
   In addition to the ramp and items to slide as above
   Marbles and/or small balls
   An actual wheel 6 to 12 inches in diameter, or a disc about 8 inches in diameter cut from A cardboard box
   A race of ball bearings (You can get a damaged or worn set, perfectly good for demonstration, from any bicycle repair shop.)

Parts 3 and 4. No special materials are required

Teachable Moments:

Set up a ramp and invite children to experiment with sliding various things down it. The ramp must be steep enough so that most of the items will slide down.

Methods and Procedures:

Part 1. The Concept of Friction

Set up the ramp with various items at hand and invite children to experiment with sliding various things down it. Which will slide the fastest? Which will slide the slowest? Which will not slide at all? In the process, children will want to test additional items from around the room and perhaps items they have with them, a tennis shoe for example. Allow them to do so at your discretion.

> Note: At this stage, don't use any items such as wheeled toys or marbles that will roll down the ramp. This part of the exercise is to focus on the relative friction of sliding alone. We will contrast sliding with rolling objects and wheels in Part 2.

At your option, you may incorporate a lesson in competitive ranking and recordkeeping as well. Have children put the name of each item on a separate 3 x 5 card. By sliding items two at a time, students can determine which is faster than the other. By always placing the card for the faster item on top of the slower item(s), they can gradually sort their cards from fastest to slowest. They may then make a record ranking the items.

After children have experimented with sliding things down the ramp for a time, settle them down for interpretive discussion. Explain that the resistance of one surface to slide over another is called FRICTION. An object that slides easily (fast down the ramp) shows little FRICTION. An object that sides slowly or not at all shows a lot of FRICTION.

Have children give the results of their experimentation using the word FRICTION. For example: The _____ went faster than the _____ because there was less friction. The _____ did not slide at all because there was a lot of FRICTION. In this discussion, explain and emphasize that friction does not refer to the item itself. It refers to the surfaces sliding. For example, a rubber eraser itself does have a lot of friction. The friction is in rubber sliding over the surface (wood, plastic) of the ramp. Extend this discussion into practical applications in everyday life:

> Where do we wish to have a lot of friction? Question students to consider soles of shoes on floors and tires on pavement.
> Why are you apt to slip and fall on ice? Students have probably discovered that an ice cube was the fastest of all their sliding objects, showing little friction between ice and other surfaces.

Where do we wish to have little friction? Children should note slides, bikes, and skates, but bring them to consider engines and other sorts of machinery.

Part 2. Wheels and Friction

With the ramp, as in Part 1, and various items listed, ask children to compare "sliding items" with "rolling items." They will quickly discover that small balls, marbles, and other items that roll outperform those that slide. They go down the ramp faster. (The possible exception is that an ice cube on a very smooth surface will go down nearly as fast.)

Pose the question, "Why should it be that rolling items go down the ramp faster than sliding items?" From their experience in Part 1, students should reason that it must have to do with friction. Sliding objects have more friction; rolling objects have less.

But, why should rolling things have much less friction? Let's analyze a wheel. Have children take an actual wheel or disc cut from cardboard, mark a point on the edge, and slowly roll it forward on a tabletop observing how the point on the wheel contacts the table.

Is there any sliding? Allow them repeat the exercise as often as necessary to reach a conclusion. If done carefully, they will observe and conclude that the answer is NO! As the wheel is rolled, the marked point comes down, contacts the tabletop, remains stationary for an instant, and is then picked up again as the wheel rolls on. There is no sliding! With no sliding, friction is essentially zero. Thus, is it any wonder that rolling items go down the ramp faster than sliding items?

Extend discussion by asking, "How would we move things if we did not have wheels?" Dragging should be one response. What about the friction in that case? Lifting it up and carrying it might be another response. What about the work of doing that?

In conclusion, students will note that wheels eliminate the work of picking up and carrying and almost entirely eliminate the friction of dragging. Hence, wheels greatly reduce the work of moving things. Is it any wonder why the invention of the wheel was such a major step for civilization? Is it any wonder why we put wheels under anything we want to move? This includes moving ourselves on bicycles, skates, and all sorts of vehicles. Of course, wheels require a hard, relatively smooth surface to roll on, hence roadways, and one must still expend energy to go up hills, but then one may have a free ride coasting down.

Wheels do not eliminate all friction, however. Point out how there is still a sliding surface between the turning wheel and a stationary axle. But we can use "wheels" here as well.

Ball bearings amount to placing steel balls between the axle and the wheel itself, again greatly reducing sliding surfaces and hence, friction. Show students a race of ball bearings (bearings mounted in a ring). They may also be able to see ball bearings by peeking through the narrow space between the wheel and axle of skates and/or other pieces of equipment.

Friction, Squeaks, and Lubrication

There remain many cases in engines, machines, and mechanical joints where sliding surfaces are unavoidable and/or bearings are impractical. Even with ball bearings, there is still some sliding between the bearings and the "ring" that holds them in place. An activity can be to have children look for and find examples where sliding surfaces exist. Door hinges are one common example.

How do we reduce friction on such sliding surfaces? OIL, GREASE, or other LUBRICANT is the answer. It is tempting to suggest that students test this by putting a lubricant on their ramps and seeing how it enables things to slide more readily, but this gets extremely messy, so I won't suggest it. You might go so far, however, as to have students contrast rubbing two smooth wooden blocks together while pressing them together with some force. Then spread some butter (grease) on the sliding surfaces and try it again.

Point out that the friction between sliding surfaces without lubrication commonly causes squeaks that may become "screams" in severe cases. Oil or grease is the solution, and it is not just to lessen the annoying squeak. The noise also means that the parts are literally tearing at each other. An engine is rapidly destroyed in the absence of sufficient oil. You may be fortunate enough to have a squeaky hinge or some other "squeaky" item in your classroom. Have students observe how a bit of oil both eliminates the squeak and makes it work (slide) more easily.

With reference to automobile and other sorts of engines, some children may have confusion regarding the oil that is used for lubrication and the fuel that is burned for energy. Be sure they are clear on this. An engine has many sliding parts, most prominently the pistons sliding up and down in the cylinders. Therefore, there must be constant lubrication. Cars have an oil warning light that will give notice that oil is not circulating properly. If that light goes on, you should stop immediately and turn off the engine. As noted above, if your engine stops BECAUSE of running out of oil, i.e., insufficient lubrication, you not only stop; you will also have a repair bill of several thousand dollars to replace the engine. Oil for lubrication of the transmission and other parts is likewise critically important.

Most importantly, stress that lubricating oil does nothing to supply energy; it only makes the parts slide more easily. Gasoline, on the other hand, is the potential energy to make the car go. That potential energy is released as the gasoline is burned in the engine. Interestingly, the raw material for both lubricating oils and gasoline is crude oil.

Part 3. Friction, Heat, and Energy

Have students press down firmly with their open palm on the pant or skirt over the top of their leg and rub rapidly back and forth. Ask them what do they feel in addition to the rubbing sensation. They will respond that their hand gets warm, even hot, sometimes so hot that it will make them stop rubbing. Likewise, have them rub a wood block back and forth rapidly on a wood or cloth surface, then feel the rubbed surface of the block. It will feel warm. (As a side note, you may mention that ancient peoples learned to start fires by rubbing sticks together.)

The key question to pose and discuss is this: Why does friction generate heat? Review, as necessary, that movement is a form of energy, and energy is readily converted from one form to another (Lesson C-1). What do you see happening here? With Q and A discussion, bring students to recognize that by resisting movement FRICTION CAUSES MOVEMENT ENERGY TO BE CONVERTED INTO HEAT ENERGY.

Why, then, do we want to reduce friction as much as possible in any engine, machine, or other mechanical device? Guide children in reasoning that FRICTION CAUSES ENERGY TO BE WASTED. In peddling their bikes, for example, friction causes their energy of peddling to go into producing heat rather than into moving them. Similarly, in every other sort of machine, friction causes input energy (fuel, electricity, or human pushing) to go into heat rather than the desired outcome. Again, be sure students are clear that lubricating oil is NOT supplying any energy; it is only reducing the amount of energy required to slide one surface over another.

Have students consider this in terms of technological advancements. We have already noted the invention of the wheel itself, but the first wheels were simply wooden discs with a hole in the middle turning on a bare axle. What improvements have been added? They should think of lubricants, bearings, and tires. Point out that such technological improvements are ongoing. There are many careers in working on these improvements.

Friction and Inertia

Importantly, bring students to contrast friction with inertia (Lesson C-5). What did we learn about inertia? Inertia is the attribute of matter that still things remain still and moving things tend to keep moving. Now what about making something move? Guide students to recognize that there are two things that must be overcome: friction and inertia. Even without friction, there must be input of movement energy—energy to overcome its inertia. With friction, additional energy is required to overcome that.

Students can experience this in swinging a heavy door. True, there is some friction in the hinges, but it is minimal (assuming the hinges are lubricated). The push they need to give to make the door swing is what is required to overcome its inertia. Now, force a wad of paper under the door. Swinging the door requires a much harder push. Why? The extra push is required to overcome the friction of the paper against the floor. Thus, emphasize again that inertia and friction are different things. Both must be overcome in getting something to move.

Now, here is another aspect of the distinction between friction and inertia that will intrigue students. Ask, "What happens to the energy input needed to overcome friction? Guide students to note how their previous experience showed it was simply changed into heat. What about the energy that overcomes inertia? It becomes movement energy and tends to keep the object moving. Again, if no friction were present, the object would keep moving indefinitely, as seen in the movements of the Earth and other heavenly bodies. On Earth, however, some friction is always present; therefore, movement energy is gradually lost as heat and the moving body slows down and stops as this loss occurs.

(Note again the common confusion between friction and inertia discussed in Lesson C-5.)

Part 4. Wind and Water Resistance

The topic of wind and water resistance may come up at any point in the discussion of friction. Whenever it comes up, explain that wind and water resistance (resistance to moving through air or water) are special cases of friction. There are two aspects to this friction. First, there is a certain amount of friction in the air/water sliding over the surface. To minimize this friction, surfaces need to be smooth. Any bumps or "warts" on the surface will catch air/water and create more friction. (The dimples on a golf ball bring up a special case that may be saved for later.)

Second, moving through air or water creates what we call TURBULENCE. Point out that turbulence is the word for the waves and swirls you create as you swish your hand through a bathtub of water. Very simply, turbulence results from air/water being pushed aside and returning to place as the object moves. Now, point out that it takes energy to create that turbulence. It follows that an object will move through air/water with the greatest ease (least energy expenditure) if it has a body shape that minimizes turbulence. This is the STREAMLINED SHAPE, a pointed or rounded front leading to the widest point about one third of the way back and then a long tapering back like the shape of a typical fish. Indeed, you might well use looking at a fish as a teachable moment to discuss its shape in terms of minimizing resistance (friction) in moving through water.

Similarly, consider the design of cars, airplanes, and boat hulls in terms of minimizing wind/water resistance, i.e., streamlining to minimize turbulence. But note that it is often necessary to compromise an ideal streamlined shape for the practical aspects of carrying passengers and cargo. Of course, the faster you wish to go, the more important streamlining becomes.

Again, heat is the byproduct of wind/water resistance, but this usually goes unnoticed, except in extreme cases. One extreme case that students may be familiar with is space vehicles reentering the atmosphere. Their speed is so great that the friction from wind resistance would literally burn them up if it were not for a covering of special materials to act as a "heat shield." (Of course, the space orbiter, Columbia, did burn up on reentry, killing all on board, because of damage to its heat shield. Heat shields on space vehicles remains an ongoing concern.) A similar example is "shooting stars." Meteorites, on entering the Earth's atmosphere, encounter such friction from the air that they literally burn up, creating a streak of light in the sky.

Bernard J. Nebel, Ph.D.

Questions/Discussion/Activities to Review, Reinforce, Expand and Assess Learning:

Write a book illustrating various aspects of friction.

Set up an activity center where students can continue to experiment and test the relative friction among various sliding surfaces. Another avenue of testing relative friction, and which incorporates measuring angles with a protractor, is to determine the degree to which a ramp must be tilted to get various items to slide down.

Invite children to find examples that illustrate friction in their daily lives and report on them in the preclass show-and-tell/discussion session at the beginning of the day.

Measure friction of various sorts of shoes or different treads by dragging them with a spring scale. (Load each shoe such that weights are equal.)

By dragging variously shaped objects through a tub of water with a piece of string and estimating turbulence, students can gain appreciation for a streamlined shape.

In small groups, pose and discuss questions, such as:

Cite instances where we wish to maximize friction or minimize friction.
What happens when you apply the brakes on a bike or car? How do brakes work to slow you down?
What is the cause of a squeak or squeal from a machine or door hinge? How can it be fixed?
How can you reduce the friction between sliding surfaces? Why do we use oil in engines, wheel bearings, and other places where we wish to reduce friction?
Why does friction produce heat?
Why was the invention of the wheel such a major step forward in civilization?
What is the use of wind tunnels in developing new aircraft designs?
If friction were eliminated, would an object start flying around by itself? What other factor is involved?
Why do swings, spinning tops, etc. slow down and stop? What happens to their energy as they slow down?

To Parents And Others Providing Support:

Talk about friction as instances come up in the course of everyday life. Examples might include:

In playing on one slide they may zip down. You might say: Wow! You go fast; there is very little friction. On another side they may move slowly or not at all. You can say: Too bad, there is too much friction.
On taking extra care not to slip and fall on a slippery surface, you might say: Be careful, there is not much friction.
On putting oil on an axle or in an engine, you might say: The oil is to reduce friction.
In stopping quickly, you might say: Wow, it is a good thing there was a lot of friction.

Conversely, invite your children to explain whether or not they slide/slip in terms of friction.

Show children the rim brakes on a bike, how squeezing on the brake forces the brake pads against the rim producing friction and, thus, slowing the turning of the wheel. Similarly, in many cars and especially motorcycles you can look through the spokes of the wheel and see the disc brakes. Point out how putting on the brakes causes the calipers to squeeze down on the disc causing friction to stop the wheel.

Additionally, you may point out that braking lightly puts little pressure of the pads against the rim; hence there is little friction and stopping will be very gradual. Stepping on the brakes hard increases the pressure between the surfaces; friction is increased and stopping is more rapid.

Discuss the utility of streamlined shapes and features as you observe them in car design, boat hulls, etc. For example: Is there an advantage to making headlights of cars flush with the body?

In observing fish, point out and discuss their streamlined shape. Why is this advantageous to the fish?

Connections to Other Topics and Follow-Up to Higher Levels:

Understanding friction will have innumerable practical applications in everyday life:

Energy efficiency
Flow of energy
Lubricants/lubrication
Design skills to minimize friction and turbulence
Design skills regarding treads and materials, to maximize friction.

Re: National Science Education (NSE) Standards

This lesson is a steppingstone toward developing students' understanding and abilities aligned with NSE, K-4:

Unifying Concepts and Processes
  • Systems, order, and organization
  • Evidence, models, and explanation
  • Constancy, change, and measurement

Content Standard A, Science as Inquiry
  • Abilities necessary to do scientific inquiry
Content Standard B, Physical Science
  • Properties of objects and materials
  • Position and motion of objects
  • Light, heat, electricity, and magnetism

Bernard J. Nebel, Ph.D.

Content Standard D, Earth and Space Science
  • Properties of earth materials

Content Standard E, Science and Technology
  • Abilities of technological design
  • Understanding about science and technology

Books for Correlated Reading:

Bradley, Kimberly Brubaker. *Forces Make Things Move* (Let's-Read-and-Find-Out Science, Stage 2). HarperTrophy, 2005.

Cole, Joanna. *The Magic School Bus Plays Ball: A Book About Forces*. Scholastic, 1998.

Niz, Ellen Sturm. *Friction* (Our Physical World). Capstone, 2006.

Royston, Angela. *Forces and Motion* (My World of Science). Heinemann, 2002.

Twist, Clint. *Force & Motion* (Check It Out!). Bearport, 2005.

Trumbauer, Lisa. *What Is Friction?* (Rookie Read-About Science). Children's Press, 2004.

Whitehouse, Patty. *Good Friction, Bad Friction* (Construction Forces). Rourke, 2007.

Lesson C-7

Push Pushes Back

Overview:

Part of everyday experience is that when we push on something, we must lean forward into it and brace our feet because we find that our pushing forward pushes us back as well. In this lesson, students will come to appreciate this experience as revealing a basic Law of Motion and how it underlies the functioning of many things from balloons zipping about a room as they exhaust air to jet engines and rockets. This is a critical element in the foundation for understanding how things work, as well as for more advanced studies in space exploration, engineering, and physics.

Time Required:

Activities, 25 –35 minutes, plus interpretive discussion, 15-30 minutes

Objectives: Through this exercise, students will be able to:

1. Demonstrate that whenever you push on something, it effectively pushes back at you.

2. Use this concept (1 above) to explain how/why a balloon exhausting air is propelled.

3. Apply this concept further to explain the functioning of jet engines, rockets, and jet boats.

4. Apply the concept further to explaining the functioning of propeller-driven craft.

Required Background:

Lesson C-1, Concepts of Energy I: Making Things Go
Lesson C-5, Inertia

Materials:

Rubber party balloons

Bernard J. Nebel, Ph.D.

Teachable Moments:

Introduce this lesson with the game/activity described below.

Methods and Procedures:

Have one of your students stand close to a door that is free to swing. He/she should be facing the door with feet together and toes touching or almost touching the door. Have him/her raise his/her hands and give the door a sharp push. Others, as well as the "pusher," should observe what happens.

The door will swing, but also the pusher will have to step back to avoid falling backwards. Have children cite and, so far as practical, demonstrate similar activities. For example, in pushing to move a heavy piece of furniture, can you stand vertically and just push it? If you try to push something heavy while standing on a slippery surface, what happens? What happens if you stand vertically at ease and give a ball a hard throw from that position?

Conduct Q and A discussion to guide children to derive the common element of all these experiences. The conclusion should be to the effect that, whenever you push on something, it effectively pushes back at you. Hence, you have to lean forward and brace your feet to prevent yourself from falling back.

How hard are you pushed back? Students can demonstrate with the initial swinging door activity. If they give the door a very gentle slow push, they are not pushed back to an appreciable extent. When they give the door a fast, hard push, they must quickly step back to catch themselves. Have children note that their experience is similar for pushing or throwing other sorts of things. The harder they have to push on something, the harder the effective push back. Thus, guide children to expand the conclusion: Push pushes back, and it does so with equal force.

If you have had a course in physics, you may recognize that we are demonstrating one of the basic laws of motion: Every action has an equal and opposite reaction. But those words make the simple experience difficult to understand. We find that children gain the understanding much more easily by putting it in the simple words: PUSH PUSHES BACK WITH EQUAL FORCE.

(If you describe this as one of the basic "laws" of motion, be sure to add that things don't behave in this way because humans passed a law. We call it a "law" because all our experience and tests show that movement behaves according to this principle.)

This phenomenon is so much a part of everyday life that we learn to compensate for it in nearly everything we do. We don't attempt to pitch a ball, shove a piece of furniture, or otherwise push on something without leaning forward into it and bracing our feet. Thus, it frequently goes unnoticed.

As children gain an appreciation for the basic concept that push pushes back, you can move on to how it is utilized in jet engines, rockets, and various other things. The simplest demonstration is to have children blow up balloons, release them, and watch them go hissing about the room.

After they have had the fun doing this, use Q and A discussion to bring them to analyze what is happening in terms of the "push pushes back" principle. Guide them in reasoning that the balloon is pushing air out in one direction. But, push pushes back! Thus, the balloon is being pushed in the other direction.[24]

Now consider rockets and jet engines. What is the jet or rocket engine doing? Guide students in noting that it is burning fuel and blasting the byproducts of combustion out the back. We said "blasting," but see this really as a violent pushing of stuff out the back. Again, push pushes back. The push of stuff out the back pushes the vehicle forward with equal force. The distinction between a jet engine and a rocket is: the jet uses oxygen in the air to burn fuel. The rocket uses fuel that does not require oxygen to combust, or liquid oxygen is carried on board. Thus, a rocket will function in space where there is no air, whereas a jet won't.

Have students analyze other examples. Why does a gun kick back as it is fired? What is happening as you paddle/row a boat? Propellers may be analyzed in the same way. In each case, whatever is being pushed in one direction also pushes in the other direction.

There are many toys and model rockets of various degrees of sophistication on the market. A lesson plan in common use is to have students make a model rocket using a two-liter soda bottle and an air pump. The bottle is filled half full of water; the air above is pressurized and off it goes as air pressure pushes water out the opening. Purchase or have children construct such things at your discretion. In every case, however, the key point will be in having children analyze how/why it works. It will always be according to the push-pushes-back principle, but their analysis may go further. It follows that push forward is increased according to push back. The push back is proportional to the amount (mass) of material and the velocity at which it is pushed. Increasing the pressure in the soda bottle, for example, gives a harder push on the water being expelled, propelling the "rocket" higher.

A question that commonly comes up is this: Does the stuff coming out the back of a jet need something, namely the air, to push against in order to propel the airplane forward? The answer is, No! for the following reason. Have students recall our friend, inertia (Lesson C-5). It takes a push to overcome inertia and start something moving. Thus, a strong push is necessary to send all that stuff out the back at great velocity. The push to

---

[24] You will find that physics textbooks generally analyze the effect of a balloon being propelled by exhausting air as a pressure difference. The pressure at the "head end" is greater than at "tail end" where the air is escaping. The difference in pressure drives the balloon forward. Again, please recognize that this is simply a more complex way of looking at the same thing. Children gain the idea more easily by visualizing that air is pushed out of the balloon in one direction; this pushes the balloon in the other direction.

overcome the inertia of the material being sent back pushes the vehicle forward with equal force. Whether or not the material being sent back pushes on anything in turn is immaterial. Have children note that rockets operate in space where there is no air.

Questions/Discussion/Activities to Review, Reinforce, Expand, and Assess Learning:

Make a book illustrating what happens when one thing pushes on another. (It should show how any push actually pushes equally in opposite directions.)

A great apparatus, if you or some of your support people can build it, is a simple 2' x 3' platform of 3/4 inch plywood with roller bearing casters attached at the four corners. On a smooth hard floor, children sit on the platform (one at a time) and propel themselves backwards by throwing cushions or pillows forward. The cushions are retrieved and handed back by teammates. The potential for contests is self-evident. Be sure that children are able to explain the principle of what they are doing. Additionally, students may discover their propulsion increases with the weight (mass) of the cushion, the speed (force) of their throw, and the lightness of the person throwing. How/why these factors influence the result as they do may be discussed, but will be the focus of another lesson.

In small groups, pose and discuss questions, such as:

What happens as you try to push something heavy while standing on a slippery surface? Why?

Why do you have to lean forward in pitching a ball? What would happen if you didn't? Why?

An inflated balloon hisses about the room as the air is released. Explain what is going on.

What is the similarity between the propulsion of a balloon, a jet engine, a rocket, and a jet boat?

If you are in a rowboat loaded with balls but no oars, how might you propel yourself forward (without touching the water)?

Wiley Coyote loads a huge cannon ball into a lightweight cannon and fires it at Roadrunner. What happens? Why? (As the gunpowder explodes, it pushes in both directions. The lighter cannon is what will move the most.)

To Parents and Others Providing Support:

Seeing jet airplanes, pictures/videos of rocket launchings, and playing with rubber party balloons all provide teachable moments. Take advantage of such moments to have your children explain what is happening in terms push pushes back.

If your children are so motivated, facilitate their building model rocket devices, within the limits of practicality and safety, of course. Use their work as a teachable moment in reviewing the principle of their operation.

In news concerning space exploration, there is often mention of using rockets or jets to speed up, slow down, or alter the path of a vehicle, as well as their use in the initial launching. Call your children's attention to these notations and ask them to explain them in terms of what they have learned.

Connections to Other Topics and Follow-Up to Higher Levels:

Relation to momentum (mass x acceleration = mass x acceleration)
Acceleration
More technical aspects of jet engines, rocket engines, rocketry, space exploration, and so on

Re: National Science Education (NSE) Standards

This lesson is a steppingstone toward developing students' understanding and abilities aligned with NSE, K-4:

Unifying Concepts and Processes
• Systems, order, and organization
• Evidence, models, and explanation
• Constancy, change, and measurement

Content Standard A, Science as Inquiry
• Abilities necessary to do scientific inquiry

Content Standard B, Physical Science
• Properties of objects and materials
• Position and motion of objects

Content Standard E, Science and Technology
• Abilities of technological design
• Understanding about science and technology

Books for Correlated Reading:

Bradley, Kimberly Brubaker. *Forces Make Things Move* (Let's-Read-and-Find-Out Science, Stage 2). HarperTrophy, 2005.

Cole, Joanna. *The Magic School Bus Plays Ball: A Book About Forces.* Scholastic, 1998.

Hewitt, Sally. *Forces Around Us* (It's Science!). Children's Press, 1998.

Mason, Adrienne. *Move It!: Motion, Forces, and You* (Primary Physical Science). Kids Can Press, 2005.

Murphy, Patricia J. *Back and Forth* (Rookie Read-About Science). Children's Press, 2002.

_____. *Push and Pull* (Rookie Read-About Science). Children's Press, 2002.

Nelson, Robin. *Push and Pull* (First Step Nonfiction). Lerner, 2004.
_____. *The Way Things Move* (First Step Nonfiction). Lerner, 2004.

Stille, Darlene. *Motion: Push and Pull, Fast and Slow* (Amazing Science). Picture Window Books, 2004.

Twist, Clint. *Force & Motion* (Check It Out!). Bearport, 2005.

Whitehouse, Patty. *Pushes and Pulls* (Construction Forces). Rourke, 2007.

Zoefeld, Kathleen Weidner. *How Mountains Are Made* (Let's-Read-and-Find-Out Science, Stage 2). HarperTrophy, 1995.

Thread D

Earth and Space Science

Pursue Threads A, B, and C in Tandem

For a flowchart of lessons and an overview of the concepts presented see pages 8-13

Lesson D-1

Gravity I: The Earth's Gravity; Horizontal and Vertical

Overview:

In this lesson, children will learn to interpret their experience of things falling down in terms of gravity, a force that pulls everything toward the center of the Earth. They will learn to measure weight, designate horizontal and vertical, and find that these are all related to gravity, as are the orbits of the Earth, the moon, and other heavenly bodies. They will learn that birds, planes, and rising balloons are not exceptions to gravity but are enabled by other factors countering the force of gravity.

Time Required:

Part 1. The Concept of Gravity (interpretive discussion, 20-30 minutes, plus practice weighing things as desired)

Part 2. Horizontal and Vertical (interpretive discussion, 10-15 minutes, plus activities as desired)

Part 3. Gravity and the Orbits of Heavenly Bodies and Satellites (interpretive discussion/demonstration, 15-20 minutes)

Objectives: Through this lesson students should be able to:

1. Recognize and use the following words in their proper context: GRAVITY, HORIZONTAL, VERTICAL.

2. State that things tend to fall down because of a force we call gravity.

3. Describe how and why people anywhere on Earth experience gravity in the same way.

4. Measure and express weight as gravity's pull on the given object.

5. Show what is meant by vertical and horizontal and how these "angles" relate to gravity.

6. Describe how all heavenly bodies (sun, moon, other planets, etc.) have their own gravity depending on their size/mass.

7. Describe how the orbits of heavenly bodies and satellites result from a balance between the pull of gravity and the motion of the body.

Required Background:

No special background is required.
Lesson A-3, Air Is A Substance, should be integrated.

Materials:

Cooking oil
Pipette
Clear-sided glass or jar of water
Globe mounted on a stand
Bathroom scale
Plumb bob (any weight on the end of a string will do)
Carpenter's level
Video (or pictures) of astronauts bounding or performing other amazing feats on the moon (may be found on the Internet)
Soft item on the end of a string that can be swung around without danger

Teachable Moments:

Whenever something (inconsequential) is dropped or falls down.

Methods and Procedures:

Part 1. The Concept of Gravity

Incidentally drop an inconsequential item and say, "Oops, GRAVITY pulled it down." Do this several times and then pose the question, "What do I mean by gravity?" Ask questions such as: Is gravity pulling on the table? You can lift a corner a bit and let it drop showing that the answer must be yes. Why is it not falling down? Because, it is held up by its legs. By doing this with a number of items and having children cite their own examples, bring them to the conclusion that gravity is constantly pulling down on everything. Hence, everything that is not supported falls down. Further, we refer to this constant pulling as a force. Thus, we can say that the force of gravity constantly pulls everything down.

Children are likely to raise questions regarding birds, airplanes, and helium-filled balloons. Point out that by flapping their wings, birds and other flying animals push down on the air with sufficient force to keep themselves up. An airplane's wings and engines do the same. Have children note what happens when wings or engines stop functioning.

For helium-filled balloons, demonstrate how drops of oil float to the top of water. Use a pipette to inject oil below the surface of water in a clear-sided glass or jar, and have children observe how the oil drops float toward the surface. Point out that the same principle holds for a helium-filled balloon. Helium is a lighter gas than air. Therefore, the helium-filled balloon floats up through air just as oil floats up through the water. Gravity is still pulling on both; it is just that one is lighter than the other. Exactly the same idea holds for hot air balloons. The lighter object/material floats toward the top.

But which direction is up and which is down? With a globe in hand, review how it is a model of the Earth just as a toy car is a model of a real car. Looking at the globe, have students consider: Why don't people who live "down under" in Australia fall off the Earth? Let them know that, in fact, people in Australia and everywhere else on Earth experience gravity exactly as we do here. So what can we say about gravity? With Q and A discussion, bring children to reasonconclude that gravity is a force that pulls everything toward the center of the Earth. In traveling around the world, there is no danger of falling away from the Earth. (You may include that before the time of Columbus, the common belief was that if you ventured too far out in the ocean, you might indeed fall off the Earth, or, at the very least, you would go down hill around the curve of the Earth so far that you couldn't get back. As well as discovering the New World, Columbus proved to people that gravity remained constant regardless of their location on the Earth.)

(You may know that there are small variations in gravity, especially with increasing elevation. And, gravity decreases markedly with increasing distance from the Earth, but this can wait for later lessons.)

Move on to explain that the weight of an object is really a measure of gravity's pull on it. Have children use a bathroom scale to weigh various objects and themselves and practice saying, "Gravity's pull on _____ is _____ pounds (or kilograms)." The simultaneous lesson in teaching children to read the scale should not be overlooked.

Part 2. Horizontal and Vertical

Introduce children to the words horizontal and vertical as they apply to everyday situations. Simply explain that VERTICAL means straight up and down and HORIZONTAL means level. Demonstrate and have children identify and list objects, lines, and surfaces in and about the room that are horizontal and vertical.

As students master the meaning of horizontal and vertical, pose the question, "How do these "lines" relate to gravity?" Demonstrate how we actually measure the vertical with a plumb bob. (Any weight on the end of a string will do.) The string with the weight hanging on the end is the vertical. Thus, being vertical means that you or any other object is lined up exactly with the force of gravity.

Why do we make walls and mount poles in the vertical position? Let students ponder. They will probably come up with answers to the effect that things, such as walls, would look silly if they were not straight, and this is true. However, guide them to the "structural" answer. Have them consider how their block towers readily fall down if they are not vertical. Guide them to observe that when things are not vertical, the pull of gravity pulls them down in the direction of the lean; when they are vertical, the pull of gravity is directly onto the footing under them and they do not fall—unless they are pushed, of course.

Horizontal, on the other hand, can be readily tested by placing a ball on the surface. If it remains stationary, the surface is horizontal—assuming there are no bumps or other factors impeding the ball's rolling. If the ball rolls, it means that the surface is tilted in the direction of the ball's roll. In terms of gravity, have kids observe that gravity only pulls things straight down. If the surface is exactly horizontal, there is no downhill direction, and the ball remains stationary. Why do we take pains to make floors and tabletops horizontal? What would happen if they were not horizontal?

Have students consider the surface of still water. If the water is higher at one end of a tub than at the other, what happens? Which direction will water go? In short, draw children to reason that water flows downhill until the water surface is level. Thus, the surface of still water is always a measure of the exact horizontal. Have children observe that the bubble in the glass of a builder's level utilizes the principle of the water always seeking the horizontal.

Explain that we learn to simply "eyeball" horizontal and vertical with a fair degree of accuracy, but to be precise, builders routinely use plumb bobs, the rolling-ball technique, and levels.

As children gain proficiency with numbers, you may go on to instruct them in how we refer to the horizontal as zero degrees, the vertical as 90 degrees, and various "slants" as angles with degrees from 0 to 90. We also refer to a 90 degree angle as a RIGHT angle. However, don't try to give too much too fast. Be sure kids have mastered the concept of vertical and horizontal and gained proficiency with numbers before getting into this.

Part 3. Gravity and the Orbits of Heavenly Bodies and Satellites

Watch a video or view pictures of astronauts leaping about on the moon. Ask and discuss how this is possible? Discussion should bring out that gravity on the moon is less because the moon is smaller. With less gravity, astronauts and everything else weighs less. With less weight but the same strength, they can jump higher and farther than is possible on Earth. Imagine what you could do on the moon!

Carry the discussion on to say that every heavenly body (the sun, stars, planets, the moon, etc.) has its own gravity according to its size (mass). The larger (more massive) the body, the greater its gravitational pull or force. Bodies smaller than the Earth have less gravity; bodies larger than the Earth have greater gravity. Imagine living on a planet much larger than the Earth and with a gravity five times greater. Could you even stand up? (You may know that, in fact, gravity is a property of all masses. But with relatively small masses, the gravity is so slight that it goes unnoticed. It is only detectable with very sensitive instruments. Again, this can wait for more advanced lessons.)

Introduce children to the concept (most will already be familiar) that the Earth and other planets orbit the sun and the moon orbits the Earth. Pose the question, "What holds planets and moons in orbits as opposed to their going off into space?" Guide them in reasoning that it must be the mutual pull of gravity. Then why doesn't gravity cause the Earth to crash into the sun?

You may demonstrate that when you swing something around on the end of a string and then let go, it immediately goes off away from you. Thus, guide students to reasonexplain that without the mutual pull of gravity between the Earth and the sun, the Earth would go off into space by itself. On the other hand, without the forward motion of the Earth in its orbit, the pull of gravity would cause the Earth to fall into the sun. Thus, the orbit of the Earth around the sun is a perfect balance between Earth's tendency to go off into space and its tendency to fall into the sun. Exactly the same can be said for the orbits of the moon and human-made satellites about the Earth and the orbits of other planets. In every case, the orbit is a balance between falling due to gravity and moving away as would occur without gravity.

You may choose to name and talk about other planets in the solar system here, but we present this as another lesson.

Questions/Discussion/Activities To Review, Reinforce, Expand and Assess Learning:

Make books illustrating: things falling due gravity; the fact that gravity pulls everything toward the center of the earth; horizontal and vertical.

Have students use the words "horizontal" and "vertical" in identifying respective lines/surfaces/objects in their surroundings.

If you travel to distant locations on Earth, how will you experience gravity? Why?

What did Columbus prove, in addition to discovery of the Americas.

Measure the weight of items and describe their weight in terms of gravity.

In small groups, pose and discuss questions such as:

Many of Columbus' sailors were fearful of what? Why?
Do birds and airplanes "turn off" gravity? How, then, do they fly?
Are helium-filled balloons not affected by gravity? How, then, do they rise?
How do horizontal and vertical relate to gravity?
Why/how is an astronaut able to jump higher and farther on the moon than on Earth?
How would you feel on a planet much larger/more massive than the Earth? Why?
What keeps the moon and human-made satellites in an orbit around the Earth and the Earth and other planets in an orbit around the sun? What two things/forces are involved? What would happen if gravity were absent? … if the earth were not moving forward in its orbit?

Bernard J. Nebel, Ph.D.

To Parents and Others Providing Support:

You may easily conduct and review any and all portions of this lesson in your home environment.

Connections To Other Topics And Follow Up To Higher Levels:

The Earth's atmosphere
How things fly
Distinction between weight and mass
Gravity as a property of all masses
Variations in gravity with elevation and difference in mass
Inclined planes
Things fall at the same speed regardless of weight
Center of gravity
Density
Gravity and pendulums
Gravity, inertia, and momentum
Trajectories of baseballs and other such objects
Measuring angles

Re: National Science Education (NSE) Standards

This lesson is a steppingstone toward developing students' understanding and abilities aligned with NSE, K-4:

Unifying Concepts and Processes
 • Systems, order, and organization
 • Evidence, models, and explanation
 • Constancy, change, and measurement

Content Standard B, Physical Science
 • Properties of objects and materials
 • Position and motion of objects

Content Standard D, Earth and Space Science
 • Properties of earth materials

Books for Correlated Reading:

Bailey, Jacqui. *Up, Down, and All Around: A Story of Gravity* (Science Works). Picture Window Books, 2006.

Bradley, Kimberly Brubaker. *Forces Make Things Move* (Let's-Read-and-Find-Out Science, Stage 2). HarperTrophy, 2005.

Branley, Franklyn M. *Gravity is a Mystery* (Let's-Read-and-Find-Out Science). HarperCollins, 2007.

Cobb, Vicki. *I Fall Down*. HarperCollins, 2004.

Nelson, Robin. *Gravity* (First Step Nonfiction). Lerner, 2004.

Niz, Ellen Sturm. *Gravity* (Our Physical World). Capstone, 2006

Trumbauer, Lisa. *What Is Gravity?* (Rookie Read-About Science). Children's Press, 2004.

Lesson D-2

Day and Night and the
Earth's Rotation

Overview:

In this lesson, students will learn that the passage of day and night is the result of the Earth rotating on its axis. This is the cornerstone for understanding directions of north, east, south, and west; telling time; and many other aspects of earth and space science.

Time Required:

Activities plus interpretive discussion, 40-50 minutes

Objectives: Through this lesson, students will be able to:

1. Use a globe to explain how the Earth rotates on its axis.

2. Model and explain how our experience of the passage of day and night is due to the Earth's rotation.

3. Describe sunrises and sunsets in terms of the Earth's rotation.

4. Point out how sunrises and sunsets occur at opposite horizons and tell why.

Required Background:

Lesson D-1, Gravity I: The Earth's Gravity; Horizontal and Vertical
Lesson D-3, Reading and Drawing Maps (should be given concurrently or follow soon after)

Materials:

A globe mounted on a stand
A small, easily moved table or stand to set the globe on
A darkened room
A lamp with a naked bulb on a stand
Toothpaste
Some rice grains
Pad of sticky notes

Teachable Moments:

Draw children to consider: Where is the sun now? Where was it earlier in the day? Where will it be later in the day? Pose the question, "How is it that we see the sun rising, moving across the sky, and setting every day?"

Methods and Procedures:

Prepare by having a lamp with a naked bulb on a stand near the center of your room and a globe on another small table ten or so feet away. Have the room so that it can be darkened easily. Make sure that when the room is darkened and the lamp is turned on, it is conspicuous that one side of the globe is well illuminated and the side away from the lamp is in shadow.

Have children to consider their experience of where they see the sun in relation to the passing of a day. Where is the sun as they are getting up and coming to school? Where is it around lunchtime? ... in the evening?

Students' experience should yield the general conclusion that the sun rises at one horizon, gets progressively higher reaching a high point around lunchtime, and then moves gradually lower in the sky until it sets on the opposite horizon. If there are students who have never noted this, give them the assignment to begin doing so. However, admonish them to never stare directly at the sun. You can include making such observations at various breaks during the school day.

Move on by posing the challenge: How can we explain this (apparent) movement of sun? You will probably have children who declare that the sun is going around the Earth. Grant them that this was the belief through ancient times until more careful observations by Copernicus (1473-1543) and Galileo (1564-1642) led to a different conclusion. Other students will already know the "different conclusion" that the Earth rotates on its axis. Invite all to examine the situation more closely.

Turn on the lamp, darken the room, and have students group around the table with the globe where they can see more closely, but not shade the globe. Explain that the light represents the sun, and the globe is the Earth millions of miles away in space. With a dab of toothpaste (toothpaste because it works well and is easily removed with a damp cloth) stick a rice grain on end somewhere near the equator. The rice grain represents a person standing on the Earth. Children can name her/him; we will call him "Joe."

Slowly turn the globe counterclockwise—the direction is significant because this is the direction the Earth does turn, and it becomes critical in determining east and west from sunrises and sunsets, which will be the topic of Lesson D-3A. As you slowly turn the globe, have students observe and describe how Joe would perceive the sun. If necessary, have each child put his/her eye right next to the globe so that they get a better perspective of how Joe sees the sun in relation to the horizon of the Earth.

As the Earth's turning brings Joe from the darker side toward the lighted side, it should become evident that he will first see the sun, i.e., sunrise, on the horizon. (It is the eastern horizon, but save that for another lesson.) As the Earth turns further, Joe will perceive the sun as getting higher and higher in the sky, until he will be viewing it more or less overhead. As the Earth's turning takes Joe toward the dark side, he will see the sun as getting progressively lower in the sky. As the Earth's turning takes Joe into the darker side, he will see the sun as going down to and then below the horizon, i.e., sunset.

Summarize and emphasize the conclusion. The sun is actually stationary (with respect to the solar system); it is the turning of the Earth that gives us the impression of the sun rising, crossing the sky, and setting on the opposite horizon. Don't expect students to absorb this immediately. Children have probably had fun playing with a globe, spinning it on its axis so they are likely to have a false notion of how fast the Earth turns. Emphasize that the Earth actually rotates very slowly; it takes a full day and night to make one complete turn. With this emphasis, repeat the activity having individual students pretending to be where the rice grain is, and describe how they would perceive the position and apparent movement of the sun as the Earth turns. Have them relate this to their daily activities and actual experience of the sun's location at those times. For example, as their rice grain moves into the lighted side, the child might say things such as: I see the sun rising; I am waking up, getting dressed, having breakfast, and going to school, and so on over the course of a day.

Finally, use a sticky note with an arrow to mark the direction on the globe in which Joe sees the sunrise, and a second arrow to show the direction in which he sees the sun set. Call children's attention to the fact that the arrows point in opposite directions on the Earth. Have them relate this to the fact that they see sunrises and sunsets on opposite horizons. (This is why it is very important to turn the globe, and emphasize that the Earth does turn, only in the counterclockwise direction.)

If you ask children to point to where they see the sun rise, you will probably see fingers pointing in every direction. This is because it is natural to "feel" one's home location as a reference and point according to that orientation. It will be significant to emphasize that they need to pay attention to where the sun actually does rise as seen from where they presently are. On the surface of the Earth, this will always be the same direction, that direction pointing counterclockwise around the earth. Similarly, the direction of a sunset will always point clockwise around the Earth. This will be true regardless of where they may be located (except at or near the poles, but this is another matter to be saved for later).

You can see this lesson readily moving into a discussion of compass directions, and we will address that in Lesson D-3A. You may also interject that the Earth moves (orbits) gradually around the sun as well as rotating on its axis, but that will also be the focus Lesson D-5.

Questions/Discussion/Activities to Review, Reinforce, Expand, and Assess Learning:

Make a book illustrating how day and night result from the Earth's turning on its axis.

Set up an activity center with a globe and lamp to facilitate children modeling the day-night cycle.

Have students note and report on how they experience the (apparent) movement of the sun and how this experience really results from the turning of the Earth.

In small groups, pose and discuss questions such as:

Using the globe and lamp, have students describe what "Joe" is probably doing and how he perceives the sun as you slowly rotate the globe.

If Joe is in Florida and Sally is in California—Position rice grains accordingly—who will see the sun rise first in the morning? Who will see it set last? Thus, how is it that there is a three-hour time difference between Florida and California? (Each time zone on the Earth is set so that 12 noon occurs when the sun is approximately at its highest point.)

Would we ever see the sun going down in the morning? Have children ponder and discuss how we set our time and activities according to the turning of the Earth.

How is it that sunrise and sunset are seen on opposite horizons?

To Parents and Others Providing Support:

Repeat and review the globe and lamp activity as described.

Take advantage of and make opportunities to view sunsets and sunrises, and discuss them in terms of the Earth's turning. For example, on viewing a sunset, ask, "Is the sun really going down, or are we turning away from the sun?"

Call your children's attention to the apparent movement of sun during the day. Ask and discuss, "Is the sun moving, or is it the Earth's turning that gives the impression of the sun's movement?"

Call your children's attention to how the sun rises and sets on opposite horizons, and discuss why.

Connections to Other Topics and Follow-Up to Higher Levels:

All further aspects of the solar system and astronomy
Telling approximate clock time from the position of the sun and/or shadows
Telling N, E, S, and W from the location of the sun at given times of day
Seasons and the Earth's orbit around the sun
The sun, temperature, and climates

Re: National Science Education (NSE) Standards

This lesson is a steppingstone toward developing students' understanding and abilities aligned with NSE, K-4:

Bernard J. Nebel, Ph.D.

Unifying Concepts and Processes
- Systems, order, and organization
- Evidence, models, and explanation
- Constancy, change, and measurement

Content Standard A, Science as Inquiry
- Abilities necessary to do scientific inquiry
- Understanding about scientific inquiry

Content Standard D, Earth and Space Science
- Objects in the sky
- Changes in earth and sky

Books for Correlated Reading:

Bailey, Jacqui. *Sun Up, Sun Down: The Story of Day and Night.* Picture Window Books, 2004.

Branley, Franklyn M. *What Makes Day and Night* (Let's-Read-and-Find-Out Science, Stage 2). HarperTrophy, 1986.

Fowler, Allan. *The Sun Is Always Shining Somewhere* (Rookie Read-About Science). Children's Press, 1992.

Gibbons, Gail. *Sun Up, Sun Down.* Voyager Books, 1987.

Hall, Margaret. *Day and Night* (Patterns in Nature). Capstone, 2007.
_____. *The Seasons of the Year* (Patterns in Nature). Capstone, 2007.

Olson, Gillia M. *The Phases of the Moon* (Patterns in Nature). Capstone, 2007.

Rau, Dana Meachen. *Hot and Bright: A Book About the Sun* (Amazing Science). Picture Window Books, 2006.
_____. *Night Light: A Book About the Moon* (Amazing Science). Picture Window Books, 2006.

Saunders-Smith, Gail. *Sunshine.* Capstone, 1998.

Tomacek, Steve. *Moon* (Jump Into Science). National Geographic Children's Books, 2005.
_____. *Sun* (Jump Into Science). National Geographic Children's Books, 2006.

Lesson D-3

Reading And Drawing Maps

Overview:

This lesson will set children on the pathway of developing map reading and map drawing skills. The ability to read and draw maps is a crucial component of studies in geography, ecology, history, and numerous other social and natural sciences. Furthermore, it serves to develop a child's special/directional awareness and perception. Its importance as a general life skill hardly needs mentioning.

Time Required:

Part 1. Drawing Maps (initial activity, 40-50 minutes)
Part 2. Reading Maps (initial activity, 40-50 minutes)

The ability to read and draw maps is a skill that will be developed through practice on repeated occasions. Therefore, plan to revisit this exercise repeatedly. Revisits can be made into art projects and/or recreational activities.

Required Background:

Lesson D-2, Day and Night and the Earth's Rotation
This lesson dovetails with Lesson D-3A, North, East, South, and West. The two lessons should be conducted concurrently and integrated.

Objectives: Through this exercise, students will be able to:

1. Draw maps of local areas that they are familiar with.

2. Identify the features (roads, rivers, parks, etc.) shown on a map.

3. Associate the features shown on a local map with the actual streets, rivers, parks, etc. that they are familiar with.

4. Orient a local map to correspond with the actual landscape.

4. Use a local map in following a route from one location to another.

Bernard J. Nebel, Ph.D.

Materials:

Pencils or markers, and paper
Local maps showing streets and other landmarks in your local area. (such maps may
    be downloaded from the Internet)
A map of your city
An atlas and/or an assortment of other maps including a world map

Teachable Moments:

Invite students to begin a special sort of drawing/art project.

Methods and Procedures:

Part 1. Drawing Maps

Instruct students that they are going to make a map of their classroom. Demonstrate
the general principle by drawing on the chalkboard and explaining that a map is a sort of
picture of how the room appears from looking down on it from overhead. Thus, it will
show the room as a rectangle with breaks showing the location of doors and windows.
Within the room rectangle will be smaller rectangles indicating the location of desks,
tables, counters, etc.

Quickly erase your drawing from the board and ask students to commence making
their own map of the room. (Erasing your own drawing is significant because you want
children to exercise their own capacity to observe the room and translate what they see
into a map, not just copy your drawing.)

A key point at the start will be to have them make the room-rectangle large so that it
occupies most of the paper. Otherwise, various features will be microscopic. Further,
have them label the top side of the rectangle as the front of the room. Coach students
individually, as necessary, in putting in doors, windows, and so on in their proper relative
locations. For example, if the entry door is at the front-right side of the room it should
appear on the upper-right side of their room rectangle.

Help children perceive their gross errors: The door is there; your map seems to say
that it is over there; what about that? Allow them to make changes or start over as they
wish. However, be sensitive to the ability level of each child.

Keep in mind that your real goal is to develop children's ability to take what they see
from the horizontal perspective and translate it into a "picture" of how it might look from
the perspective of looking down from above. Therefore, anything that shows the
approximate layout of the room should be taken as satisfactory at the K to grade 1 level.
(Third grade lessons might well include making actual measurements and using graph
paper to draw things to scale.)

A splendid activity that you may add is to have students make maps of alternative room arrangements—chairs in a circle, small groups, etc. Then allow students to select a map, discuss the layout that it shows, and arrange the room accordingly.

As children master the ability to make a make a map of your classroom, you may move on in subsequent lessons to have them make a map of the school yard showing the relative locations of the school building, play areas, and so on. Then their maps may include broader areas including adjacent buildings and streets.

As students make maps of areas beyond the classroom, they will face the problem of orientation. Which part of the schoolyard should you put at the top of the map? Here is where lesson D-3A, North, East, South, and West, should be integrated. For the initial lessons will be best to set this complexity aside and make the orientation as convenient.

Let it be known throughout this lesson that making maps is an ongoing activity. It is a career that they might choose to go into. It is also an essential part of numerous other professions.

Part 2. Reading Maps

As children gain the concept that a map is a bird's-eye view of a certain area, move them on to reading maps. Start with local maps that show streets in your vicinity. Help them orient the map so that directions on the map correspond to those in the real-world (Lesson D-3A). Point out the keys and show them how the key enables them to decipher various symbols used on the map. Have them work in groups and help each other pinpoint on the map the locations of their homes, school, parks, religious buildings, and other features that they are familiar with.

Here is a great learning activity. Spread out on the floor a map that includes your present location. Have students orient it so that directions on the map correspond to the real world. For example, turn the map so that the streets on the map run in same directions as the real streets outside. With the map oriented, ask, "What direction should we go to get to _____ (another place indicated on the map)? Point!"

With some coaching, students will catch onto the "trick." When the map is properly oriented and they know their present location, the map shows the direction to anyplace else. For example, if they draw a line on the map from their present location to where they desire to go, that line shows the direction they should point. (Of course, choosing available roads/paths to actually get there is another step.) Create opportunities for students to put this into practice in the real world on field trips and recreational outings.

Emphasize that the key point in this activity is knowing your present location and identifying that point on the map. It should be evident that if you don't know where you are, it will be impossible to know what direction you should go to get anywhere else including home. The admonishment is this: Get a map and note your present location before you start out, and keep track of your "wanderings" on the map so that you always know where you are. Only then will the map, oriented properly, show you which direction to go to get to the next location. If you wait until you are lost and don't know

Bernard J. Nebel, Ph.D.

where you are, a map and even a compass will do you little good.

The overlapping of this lesson with D-3A is conspicuous. Conduct the two lessons together to gain mutual reinforcement and synergism.

Questions/Discussion/Activities to Reinforce and Expand and Assess Learning:

Have some students make a map of the schoolyard, hide a "treasure," and indicate its location on the map. Have classmates use the map to find the treasure. Discuss: Was the map sufficiently accurate to be oriented and read to enable finding the treasure?

Set up an activity center with an assortment of maps that students can peruse at their leisure.

Have students make maps of an imaginary "mystery island," putting in whatever features they wish and identifying those features in a key. Include directions for finding a hidden treasure.

Have children study local maps and mark various routes they might take in getting from here to ____.

Integrating Lesson D-3A, add compass directions to the exercises in drawing, reading, and following maps.

In small groups, pose and discuss questions such as:

How should we orient a map in order to follow it properly?
Point at various symbols on a map and ask: What does this mean? (Allow students to consult the key as necessary.)
How should we indicate special features such as a fence, tree, woodland, or stream on a map? Show students how we may use any symbols as long as we identify them in a key.
Show students samples of maps with errors, and ask them to identify the error and how to correct it. (The maps should not belong to identifiable persons.)

Add challenges involving compass directions as students learn about them (Lesson D-3A).

In conjunction with developing numerical skills students may:

Make actual measurements of floor plans and use graph paper to make drawings to scale.
Add up distances shown on a typical road map to determine total distances between points.
Use the scale on a typical road map to calculate distances.

To Parents and Others Providing Support:

Facilitate and encourage children in making maps of their room, the floor plan of your home, your yard, and the neighborhood so far as they are familiar with it.

With your children, study the map that is (usually) in the entryway of a shopping mall. Show them it indicates where they are in relation to the mall and where various stores are located. As they learn the technique, have them guide you.

With your child, study a local map that includes your neighborhood. Coach them as necessary in identifying the location of:

Your home
A friend's home
A local park
School ground
Religious building
Store where you shop for groceries
Other locations they are familiar with

Help them orient the map and use it to determine the actual direction they would need to go to get to a given place.

Guide children in observing street names and route numbers on signposts and relating these to those shown on the map.

On repeated occasions, pull out a map, and with your children, use it to determine how you may get from home to a given location (even if you already know the way). Gradually, you can ask them to use the map and be the navigator for you.

Obtain and make generous use of maps on trips, hikes, etc. At stops, examine the map; point out where you started, where you are, and where you are headed. Have children assist in reading road signs and keeping you on course.

In going on car trips, invite your children to use maps and help decide the route to take. Of course, travel time will be a factor. You may decide to use interstates for most of the trip, but with one or more scenic diversions along the way.

Ask for a map of the park you may be visiting. Point out how it shows various features. Let your children have a voice in deciding points of interest you may visit and trails you may take. Coach them as necessary in orienting the map to correspond to the landscape (N, E, S, and W), reading it, and leading the way in getting to the desired location.

Bernard J. Nebel, Ph.D.

Connections to Other Topics and Follow-Up To Higher Levels:

Measuring and drawing maps to scale on graph paper
Using the scale to determine actual distances
Determining locations by latitude and longitude
Reading topographic maps
History: Study how maps and mapping were an integral part of world exploration.
Geography, Ecology, Archeology and many other subjects draw on maps and mapping to
    say nothing of its being a practical life skill.

Re: National Science Education (NSE) Standards

This lesson is a steppingstone toward developing students' understanding and abilities
aligned with NSE, K-4:

Unifying Concepts and Processes
    • Systems, order, and organization
    • Evidence, models, and explanation

Content Standard A, Science as Inquiry
    • Abilities necessary to do scientific inquiry

Content Standard B, Physical Science
    • Position and motion of objects

Books for Correlated Reading:

Aberg, Rebecca. *Latitude and Longitude* (Rookie Read-About Geography). Children's
    Press, 2003.
    _____. *Map Keys* (Rookie Read-About Geography). Children's Press, 2003.

Bredeson, Carmen. *Looking at Maps and Globes* (Rookie Read-About Geography).
    Children's Press, 2002.

Fitzgerald, Joanne. *This is Me and Where I Am*. Fitzhenry & Whiteside, 2004.

Hartman, Gail. *As the Crow Flies: A First Book of Maps*. Aladdin, 1993.

Leedy, Loreen. *Mapping Penny's World*. Owlet, 2003.

Sweeney, Joan. *Me on the Map*. Dragonfly Books, 1998.

Wade, Mary Dodson. *Map Scales* (Rookie Read-About Geography). Children's Press,
    2003.
    _____. *Types of Maps* (Rookie Read-About Geography). Children's Press,
    2003.

Lesson D-3A

North, East, South, and West

Overview:

Knowing and being able to determine the directions north, east, south, and west is critical to reading and drawing maps, relating maps to the real world, and just finding one's way about without getting lost. This lesson will lay the foundation of this important life skill.

Time Required:

Part 1. Designating Directions on the Globe (activity plus interpretive discussion, 20-30 minutes)

Part 2. Connecting Maps to the Globe (activity plus interpretive discussion, 30-40 minutes)

Objectives: Through this exercise, students will be able to:

1. Identify on a globe the North Pole and the South Pole.

2. Identify on a globe directions north, east, south, and west.

3. Recognize and show how maps are "pictures" looking down on greater or lesser portions of the globe.

4. Point out how N, E, S, and W on maps correspond with N, E, S, and W on the globe.

5. Identify north, east, south, and west at their home and in other locations using the sun as an indicator.

6. Orient maps according N, E, S, and W and follow them to given locations.

Required Background:

Lesson D-2, Day and Night and the Earth's Rotation

Lesson D-3, Reading And Drawing Maps, should be given concurrently and integrated

Bernard J. Nebel, Ph.D.

Materials:

Globe mounted on a stand
Pad of sticky notes
World map
One-page map of the U. S.
Map of your state
Local map of your area
Atlas or other assorted maps
Directional compass

Teachable Moments:

In the course of using a globe to review Lesson D-2, Day and Night, extend instruction into this lesson.

Methods and Procedures:

Part 1. Designating Directions On The Globe

In the course of using the globe to review Lesson D-2, emphasize that the Earth does not rotate any which way; it continually turns about the same points, as modeled by the globe turning on its stand. Hasten to add, however, that there is no axle through the world; it is just the way the Earth turns. Likewise, emphasize again that the rotation is always in the counterclockwise direction as viewed from above.

Go on to explain that the top point of rotation of the globe and of the real world is called the NORTH POLE; similarly the bottom point of rotation is called the SOUTH POLE. Tracing your finger on the globe, point out how directions are designated. Upwards toward the North Pole is the direction NORTH; traveling down toward the South Pole is SOUTH. EAST is toward the right as you are facing the globe, the same as the direction of rotation. WEST is to the left, counter to the direction of rotation. Have students respond to the directions as you move your finger on the globe until they have them down.

Make a compass cross (a cross with arrows at the ends of the arms labeled N, E, S, and W respectively) on sticky paper and place it on the globe as a reminder.

Now ask, "In what direction will you always see the sun (apparently) rise?" Allow think time and revisit Joe on the globe (Lesson D-2) as necessary. Students will gradually catch on that because the Earth is always turning toward the east, Joe and everyone else on Earth will see the sun rise in the east. Similarly, guide children in reasoning that the sun will always be seen as setting in the west.

Next, guide students in relating these directions to the real world. Where do you see the sun rise? What direction must that be? East! And where do you see the sun set? What direction must that be? West!

Point out that if they know just one direction, they can determine the other directions. Have them ponder how they would do this. Guide them in figuring that it only takes knowledge of the compass cross. If they can determine the direction of one arm of the cross, they will have the other directions as well.

A fun activity is to teach children to use their bodies as the compass cross. If they stand straight, raise the right hand to the side, and turn so that it is pointing to the location of the rising sun (east) their nose will be pointing north, their left hand will point west, and south will be behind them. Challenge students to perform variations; if they face the setting sun and raise their arms to the sides, which directions will they be pointing? And so on.

Facilitate students in marking out a large compass cross on the floor of your classroom and/or in the schoolyard using the sun to determine directions.

There is likely to be some error in using the rising or setting sun as an indicator of direction. The sun only rises exactly in the east and sets exactly in the west on the equinoxes (March 21/22 and September 21/22). From March to September it rises and sets somewhat to the north of true east/west. From September to March it rises and sets somewhat to the south of true east/west. The reasons for this have to do with the tilt of the Earth and the way it progresses in its orbit around the sun, but this is for later lessons. Also, as the sun rises, it proceeds at an angle toward the south as well as up.

You may be tempted to say, "Forget it. Just use a compass and be done with it." However, in real life one is often without a compass. Most importantly, however, using a compass does nothing to convey an understanding of the relationship between the Earth's turning, the sun's apparent movement, and the way we designate directions. Further, a compass does nothing toward helping children develop a natural sense of direction; knowing directions from the sun is paramount in this regard. Therefore, teaching children to approximate directions from the sun has many advantages that should not be overlooked.

Another method of determining directions from the sun is that, at noontime, the sun is more or less to the south of directly overhead. Thus, noontime shadows point north. (This does assume, of course, that you are in the northern hemisphere, and for summer months north of the Tropic of Cancer.) If you do this exactly at solar noon, the high point of the sun in the sky, shadows will point exactly true north. Otherwise this will also be an approximation.

Of course, you may add in using a compass, and compare your estimations of directions from sun those from the compass. But, recognize here that a compass is not an exact indicator of true north either. This is because the magnetic north pole does not correspond exactly with the geographic northpole, which is the true north pole. Therefore, the compass indication of north needs to be adjusted accordingly. Just how much adjustment will vary with your location. A map showing the required adjustments across the United States can be found at

Bernard J. Nebel, Ph.D.

http://education.usgs.gov/common/lessons/how_to_use_a_compass_with_a_usgs_topogr
aphic_map.html

Part 2. Connecting Maps to the Globe

The next step is to guide students in making the connection between the globe, maps, and the real world. Facilitate children in comparing and contrasting a globe with a world map. Have them find and compare respective continents, oceans, etc. on both the world map and the globe (although they are likely to be of different scale/size).

The potential for making this a geography lesson is self-evident. However, the main point here is to have children understand that a world map is really a flat "picture" of a round globe. Indeed, some students may note certain differences, especially in northern and southern regions. Greenland, for example, appears much larger on a world map than on a globe. Point out that it is impossible to make a flat map of a curved surface without stretching the edges and/or shrinking the center. Nevertheless, maps have conspicuous advantages over globes in terms of use.

Similarly, have students make the connection between a map of the United States and the United States on the globe and the same comparison between a map of your state and its location on the globe. What is a map, really? Children should come to understand that every map is really a "picture" looking down on a greater or lesser portion of the globe or portion of the Earth. By making a large picture of a small portion of the globe, much greater detail can be included.

The "much greater detail" is even more conspicuous on a local map. Have children examine a local map that includes their particular area. Help them identify on the map streets, parks, and other features that they are familiar with as described in Lesson D-3. Then, help them pinpoint as accurately possible their local area on the globe. Some children may want to use a magnifierhand lens to examine the globe to see if there area is actually shown. Allow them to do so, but explain that the huge difference in size scales does not allow globe makers to include such detail. Still, every map is a "picture" looking down at a certain portion of the globe/Earth.

Emphasize, as well, that the correspondence between the globe, maps, and the real world goes both ways. Every point on the globe depicts a location in the real world; every map of the real world, including maps they might draw of their neighborhood (Lesson D-3), depicts a certain location on the globe.

The last portion of this lesson is to have children observe that directions on a map correspond to those on the globe and in the real world. Have them contrast, again, a map of the United States and the United States as it is seen on the globe. Does the map show the U.S. as upside down or reversed right to left from its position on the globe? No! Students should observe that it is the same as on the globe. What about directions on the map then? Guide children in noting that maps are always (with few exceptions) drawn such that directions on the map correspond to those on the globe, north is toward the top, east is to the right, south is down, and west is to the left.

Explain and show students how this, in turn, enables us to turn maps so that they correspond to directions in the real world. That is, if we turn the map so that the right-hand side is toward the location of the rising sun (east) or the left side is toward the setting sun (west), then other directions fall into place accordingly. Most importantly, when the map is oriented accordingly, directions on the map correspond to directions in the real world. (Refer to the activity described in Lesson D-3.)

Students should exercise this lesson in all further activities involving drawing of maps. That is, they should first determine the compass directions in the area they wish to map. (Revisit the activity of using your body as a compass cross described in Part 1.) They should then draw their map so features they depict in the north are put at the top of their map. Additionally, they should add a compass cross to their map showing which direction is north, i.e., toward the top if it has been drawn correctly.

(Of course, you will find maps with compass crosses indicating north to the right, left, or even down. In other words, it appears that the compass cross was added as an afterthought. Directions should not be an afterthought. It will greatly strengthen the learning experience if students are required to determine directions first and draw their maps accordingly.)

Questions/Discussion/Activities to Review, Reinforce, Expand, and Assess Learning:

Make books/maps illustrating directions (N, E, S, and W) according to the sun and/or shadows.

Move your finger in various directions on the globe and on maps, and ask: What direction is this? When students master N, E, S, and W, you may add northeast, southeast, and so on. Have them point in corresponding directions in the real world.

Using a map and standing on a compass-cross on the floor, ask children to point toward Canada, the North Pole, the Atlantic Ocean, Europe, and other locations they are familiar with from geographic studies.

Have children follow a "treasure map" that gives instructions in compass directions.

In small groups, pose and discuss questions such as:

Can you go around the world repeatedly by going east? How so?
Can you go around the world repeatedly by going north? Why not?
Point east and ask, "How do I know that direction is east?" Similarly, have students reason out the other directions from the sun.
Explain why the sun is always seen to rise in the east.
When you see the setting sun, how can you identify all four directions?
Point in the direction you would go to get to the North Pole.
What is the relationship between a map and a globe? ... a map and the real world?
How can you determine directions from the sun?
Find your location on a local map, orient it, and use the map to point the direction to

other familiar locations.

To Parents and Others Providing Support:

Seeing a sunset, ask children to point to all four compass directions.

Coach your kids in marking out a compass cross at home using the sun for determining directions. Check it with a compass.

Have children follow a "treasure map" that gives directions as N, E, S, and W.

Get maps of parks and other locations you are visiting and have your children be your guide in showing where you are, orienting the map, and pointing in the direction you should go to get to another location, coaching them as necessary.

Give and have children follow directions according to N, E, S, and W. For example, to get to the store from here, go north to the corner, west to the traffic light, then north to the second traffic light, and so on.

Connections to Other Topics and Follow-Up to Higher Levels:

Correlating directions and time of day by shadows; making a sundial
Designating directions by degrees
Identifying locations by longitude and latitude
All studies of geography and much of history
Identification of land forms (deserts, mountains, rivers, etc.) in various parts of the world
(Lesson D-4)

Re: National Science Education (NSE) Standards:

This lesson is a steppingstone toward developing students' understanding and abilities aligned with NSE, K-4,

Unifying Concepts and Processes
  • Systems, order, and organization
  • Evidence, models, and explanation

Content Standard A, Science as Inquiry
  • Abilities necessary to do scientific inquiry

Content Standard B, Physical Science
  • Position and motion of objects

Content Standard D, Earth and Space Science
  • Changes in the earth and sky

Books for Correlated Reading:

Cobb, Vicki. *Sense of Direction: Up, Down, and All Around.* Parents Magazine Press, 1995.

De Capua, Sarah. *We Need Directions* (Rookie Read-About Geography). Children's Press, 2002.

Fowler, Allan. *North, South, East, and West* (Rookie Read-About Science). Children's Press, 1993.

Trumbauer, Lisa. *You Can Use a Compass* (Rookie Read-About Geography). Children's Press, 2004.

Lesson D-4

Land Forms and
Major Biomes Of The Earth

Overview:

This lesson will utilize and expand upon students' understanding of the relationships among the globe, maps, and the real world (Lessons D-3 and D-3A) It will put them on the path of learning major land forms of the earth (mountains, plains, ice caps, rivers, etc.) and biomes (deserts, tundra, forests, grasslands) occupying different regions. This will provide the foundation for understanding the different climates that are experienced in different parts of the world. In turn, students will begin to appreciate how different climates support different flora and fauna and the respective biomes or ecosystems that develop.

Time Required:

This will be an ongoing activity of having students make connections between landforms (mountains, plains, valleys, rivers, etc), biomes (forests, deserts, tundra, etc.) types of flora and fauna, climates, and their locations on maps.

Initial set-up and introductory explanation (40-50 minutes)
Posting pictures plus interpretive discussion (5-10 minutes two or three times a week as occasions arise).) Utilize the preclass show-and-tell/discussion periods described in Chapter 1 (page 19).

Objectives: Through this exercise, students will be able to:

1. Point out on a globe or world map the locations of major landforms such as mountains, rivers, and plains.

2. Point out on a globe or world map the general locations of ice caps, tundra, forests, grasslands, deserts, etc.

3. Describe the general types of plants and animals that inhabit the different regions noted in 2 above.

4. For particularly distinctive animals such as penguins, elephants, and monkeys, tell where they may be found in the wild.

5. Distinguish between weather and climate.

6. Give the predominant features of the climates of different regions (2 above).

7. Describe how the flora and fauna of a region are adapted to the given climate.

Required Background:

Lesson D-3, Read and Draw Maps
Lesson D-3A, North, East, South, West
Lesson B-4A, Identification of Living Things (should go on concurrently)

Materials:

World map 30 x 48 inches or larger
Wall space to mount the map with a foot or more of clear space around the map for mounting pictures
Colored yarn or string
Tacks or tape for mounting pictures
Globe on a stand. The globe should be one that is colored to show general topography and vegetation types such as tundra, forests, grasslands, deserts.

Teachable Moments:

The activity of bringing in pictures and mounting them on the map, as described, will create teachable moments.

Methods and Procedures:

With students' help, mount a world map on a wall where there will also be a foot or more of clear space surrounding the map for tacking up pictures.

With globe in hand and/or pointing at the world map, set the mood by "storytelling" to children that they are about to go on explorations to find and study various animals living in the wild. Where should they go to find lions living in the wild? ... elephants? ... monkeys? ... camels? ... moose? ... panda bears? ... polar bears? ... kangaroos? ... penguins? ... whales? etc.? (Let children suggest other animals that they might be interested in finding in the wild.) Don't expect students (or yourself) to know all the answers and don't try to give answers at this stage. This is only to perk students' interest in starting on the study as follows.

Instruct students that their assignment for the succeeding days will be to find pictures of various animals IN THEIR NATURAL HABITATS. Importantly, they don't want "zoo pictures" that just show the animal by itself; they want pictures that show the animal in the "landscape" of where it naturally lives—penguins on ice caps, monkeys in jungles, camels in deserts, lions in grasslands (savannah), kangaroos in Australia's outback, and so on. Landscape pictures of mountains, valleys, plains, rivers, etc. will be fine too. As well as the picture, they will need to identify from text associated with the picture, the general, if not the specific, location of the scene.

Solicit the support of parents and other caregivers in helping their children browse magazines, travel brochures, and other sources for suitable pictures and make copies as necessary to bring to class. You and caregivers may also help students find and download pictures from the Internet.

The general procedure will be to have children give a brief show-and-tell regarding a picture they bring in, post it beside the world map, and connect the picture to its location on the map with a piece of yarn or string. It is conspicuous that this will be an ongoing activity taking a few minutes, two or three times a week. Integrate it into your preclass show-and-tell/discussion period along with identification of species and other ongoing lessons.

It is obvious that when and what pictures are brought in cannot be scripted. It will occur as it does. However, as pictures accumulate, general patterns will emerge. Periodically, draw your students' attention to the patterns. Ice caps and associated animals occur at the poles; jungles (tropical rain forests) and associated animals occur in equatorial regions; deserts with their distinctive flora and fauna occur in subequatorial regions. Mid-latitudes are noted for temperate climates, a winter season that is markedly cooler/colder than the summer season. As one moves northward in the temperate regions, winters become increasingly long and severe and summers are progressively shorter and cooler.

Point out and emphasize how the emerging picture has two major elements. First is the climate; second is the distinctive flora and fauna.

(You will need to insert here some instruction regarding the difference between weather and climate. WEATHER is the particular conditions, mainly temperature and precipitation, which occur on any given day. CLIMATE is the regime of temperature and rainfall that occur, on average, throughout a year. Climate is usually "pictured" as a bar graph showing average rainfall that occurs each month, and a line graph showing average temperature for each month.)

Use Q and A discussion to bring students to recognize how climate and the predominant flora and fauna are connected. For example, equatorial regions of the Earth have warm temperatures and high rainfall all year around; this supports what we call tropical rain forests and associated animals. Climates with very little rainfall are renowned as deserts where only plants and animals adapted to minimal water can survive. Mid-latitudes have climates with winters that include periods of freezing temperatures and warm to hot summers; plant and animals are adapted to these extremes accordingly. The poles have climates of perpetual cold, but some animals are adapted to these conditions, notably polar bears in the Arctic and penguins in the Antarctic.

You can readily see that the amount of information regarding different climates of the Earth and how different species of plants and animals are adapted to cope with those climates is vast. Going into great detail here is impractical. Bear in mind that at this stage (grades 1-2) our objective is only to introduce students to the general concept that different regions of the Earth do have different climates and those different climates do support distinctive flora and fauna and preclude others. This picture/concept will emerge from connecting pictures of flora and fauna with respective regions of the Earth and associated discussion. Details will come and fit into place naturally as the study progresses. Note that this study may progress through all the grades. Indeed, the study of how particular plants and animals are physiologically and/or behaviorally adapted to climate is the focus of ongoing investigations for numerous scientists.

The same activity of connecting pictures to locations on a map may be used to help students learn to name, identify, and locate various landforms, such as mountains, ridges, river valleys, plains, volcanoes, and so on.

Finally, the climate of a region is also affected by factors beyond north/south latitudes. In particular, temperatures decrease with increasing elevations, and rainfall is markedly influenced by what is called the "rain shadow" effect. These, as well as the causes for the different climates, will be addressed in later lessons.

Questions/Discussion/Activities to Review, Reinforce, Expand, and Assess Learning:

Make a book illustrating the location of particular flora/fauna on the Earth.

Let the map and connecting pictures become ongoing in an activity center.

Read and discuss stories of explorers going to, discovering, describing, and photographing the flora and fauna of different regions.

Conduct the same sort of activity with the focus on the United States and/or your own state in particular.

Help students use the Internet to retrieve pictures of different regions/places where their interests take them.

In small groups, pose and discuss questions such as:

What patterns emerge from the map and pictures? Are the locations of tropical jungles, ice caps, tundra, great deserts, etc. random about the globe, or do they fall into a global pattern? What is that pattern?

What does this picture (select any given landscape picture) indicate about the climate (temperature and rainfall) of this region?

Where should you go to find _____ living in the wild?

What is the difference between weather and climate? What is the climate of your region?

How do climates vary according to north/south location on the globe?

Pointing to given locations on the globe, ask, "What sort of climate would you expect to experience here? What sort of flora and fauna would you expect to find?"

Bernard J. Nebel, Ph.D.

To Parents and Others Providing Support:

Aid your children in searching for, finding, and taking pictures to school as described above.

Spend time in a library with your kids and with a globe. Browse travel magazines and identify the locations of scenic pictures on the globe. Discuss the climate of the location as deduced from the picture, and the type of flora and fauna so far as it may be shown.

You may decide to mount a world map and conduct the same exercise as described above with your kids at home.

Connections to Other Topics and Follow -Up To Higher Levels:

Connections to studies of geography, history, and culture
Causes for different climates occurring in different regions of the earth
How geographic isolation, as well as climate, influences flora and fauna of a region
Adaptations of specific plants and/or animals to the climate and other conditions of their environment

Re: National Science Education (NSE) Standards

This lesson is a steppingstone toward developing students' understanding and abilities aligned with NSE, K-4:

Unifying Concepts and Processes
- Systems, order, and organization
- Evidence, models, and explanation
- Form and function

Content Standard A, Science as Inquiry
- Abilities necessary to do scientific inquiry

Content Standard C, Life Science
- Characteristics of organisms
- Organisms and environments

Content Standard D, Earth and Space Science
- Changes in earth and sky

Content Standard E, Science and Technology
- Abilities to distinguish between natural objects and objects made by humans

Content Standard G, History and Nature of Science
- Science as a human endeavor

Books for Correlated Reading:

Brenner, Barbara. *One Small Place by the Sea.* HarperCollins, 2004.

_____. *One Small Place in a Tree.* HarperCollins, 2004.

Hablitzel, Marie and Kim Stitzer. (Numerous titles). Barker Creek.

Kalman, Bobbie. *What is a Biome?* (The Science of Living Things). Crabtree, 1998.

Silver, Donald M. *African Savanna* and numerous additional titles. One Small Square series. McGraw-Hill/Contemporary Books, 1997.

Stockland, Patricia. *Sand, Leaf, or Coral Reef: A Book About Animal Habitats.* Picture Window Books, 2005.

Lesson D-5

Time and the Earth's Turning

Overview:

In Lesson D-2, students learned that the daily cycle of night and day results from the Earth's turning on its axis. In this lesson, they will learn that the Earth's turning is further used to measure the time of day. They will learn to approximate the time of day from the position of the sun and/or shadows and they will make a crude sundial. Beyond providing the basis for telling time, this provides the foundation for interpreting many other observations concerning the solar system and the heavens above.

Time Required:

Part 1. Relating Time to the Earth's Turning (15 minute activity at the beginning and end of an outdoor recreational period followed by interpretive discussion, 30-40 minutes)

Part 2. Making and Using a Sundial to Tell Time (Making the sundial, 40-50 minutes; calibrating it, a few minutes each hour during the course of a day)

Objectives: Through this exercise, students will be able to:

1. Describe and show how shadows shift with the passing of time.

2. Use a globe to model and describe how shifting of shadows are caused by the Earth turning on its axis.

3. Tell how one full day (day-night cycle) equates to one turn of the Earth on its axis.

4. Explain how we divide the time of the Earth's turning into 24 parts, hours, and tell time accordingly.

5. Tell how clocks are set according to the sun.

6. Explain why and how we divide the country and the world into time zones.

7. Tell what does and doesn't happen with daylight savings time.

8. Make a sundial, calibrate it, and use it to tell the time of day.

9. Explain the discrepancies between solar noon, clock noon, and daylight savings time.

10. Cite and explain the reason for the discrepancy between the length of one day (24 hours) and the time it takes the Earth to make one exact turn on its axis (11 hours, 56 minutes).

Required Background:

Lesson D-2, Day and Night and the Earth's Rotation
Lesson D-3, Reading and Drawing Maps
Lesson D-3A, North, East, South, and West
Some familiarity with telling time is assumed

Materials:

Part 1. Relating Time to the Earth's Turning
    Sunny location and chalk to trace shadows
    Globe and lamp as in Lesson D-2

Part 2. Making and Using a Sundial to Tell Time
    Piece of poster paper about 12 inches square
    Soda straw
    Glue
    Window sill or other location that receives direct sunlight for the greater part of
        the day
    Tacks or tape
    Pencil and straight edge

Teachable Moments:

This lesson can be inserted anywhere in the course of children learning to tell time. Invite children to trace shadows at the beginning and at the end of an outdoor recess/recreation period and note the shift.

Methods and Procedures:

Part 1. Relating Time to the Earth's Turning

At the beginning of an outdoor recreational period on a clear day, have children trace each other's shadow with chalk on pavement. Be sure the person casting the shadow stands straight and still. Trace the exact position of his/her feet, as well as the rest their shadow. Specifically, place a dot at the top and center of the head of the shadow. At the end of the period (at least an hour later) have the "shadow casters" stand in exactly the same place and position while others retrace their shadow, again paying special attention to the location of the shadow's top center.

Pursue with Q and A discussion: What did you observe about the two shadows you have just traced? (The shadow moved.) What caused the movement of the shadows? (The sun moved.) Did the sun actually move or is the Earth's turning? Review Lesson D-2 as necessary to refresh memories that apparent movements of the sun are caused by the rotation of the Earth. Thus, bring students to reason that the movement of the shadow over the period is the result of the Earth's gradual and continual turning on its axis.

How many times does the Earth turn in one day? ... seven days? Such questions will reinforce the understanding that each day is equivalent to one complete turn of the Earth. When we count days until a birthday or holiday: What are we really counting? (Turns of the Earth!) But we wish to count more than just days. We want know when we should get up, go to school, have lunch, and so on.

Explain that, therefore, we divide each day-night cycle (commonly referred to as one day) into 24 parts or hours (two 12 hour halves) and split each hour into minutes and seconds. (How much time you spend on teaching students to tell time and count minutes and seconds will be up to you, and it will depend on your students' background.) Here, it is most significant to have children recognize that the Earth is continually in motion turning at constant speed, neither speeding up nor slowing down (at least not over time periods of human experience). Dividing a day (one turn of the Earth) into 24 hours was a totally arbitrary decision made by people sometime back in history. Importantly, the Earth does not turn according to our clocks. We construct our clocks to correspond to the turning of the Earth. (This is a point that will bear repeated emphasis.) Furthermore, official time is still set and kept according to the Earth's turning.

How do we set our clocks according to the Earth's turning? Explain that we use the point at which the sun reaches its ZENITH, its highest point in the sky, halfway between sunrise and sunset. This is called SOLAR NOON, and we set our watches at 12 noon to correspond—but more is involved. Review how Sally in California sees the sun rise and set three hours later than Joe in Florida, because of her location further west. Therefore, having 12 noon on our watches correspond exactly to solar noon would mean that everyone east and west would be on different times, perhaps only minutes or seconds earlier or later.

Before the building of the transcontinental railroad, which was completed in 1869, each town and city did keep its own time, setting their clocks so that twelve o'clock did correspond to solar noon. With the railroad enabling relatively rapid east-west travel, this meant constantly adjusting watches and confusion with schedules was terrible. Therefore, it was decided to break the country, and the rest of the world as well, into TIME ZONES. Everyone within a given time zone keeps the same time. At roughly the center of each time zone, 12 noon does correspond to solar noon, but to the east and west there is a difference of up to a half hour. Still, having everyone within a time zone on the same time, and making an even one-hour change from one time zone to the next, saves much confusion.

Daylight savings time is another matter. Emphasize that the actual time of the daylight hours remains exactly the same, since that depends on the Earth's turning. We only shift our clocks so that the daylight hours fit more conveniently with our hours of greatest activity.

Nevertheless, official timekeeping is still set according to the sun reaching its zenith. Emphasize again that the time from the zenith of one day to the zenith of the next is broken into 24 equal segments, or hours. Hours are broken into minutes and seconds and we tell time accordingly.

We have said that each day corresponds to one turn of the Earth on its axis and a day is from the zenith of one day to the zenith of the next. Actually, there is a slight discrepancy between these two figures, because of the Earth's progression in its orbit around the sun. This may be demonstrated as follows. Use a globe with "rice-grain Joe" and a "sun" in the center of the room as in Lesson D-2. Adjust the globe so that Joe would be seeing the sun at its zenith, i.e., Joe's point on the globe is toward the sun. Now move about one quarter of the way in the "orbit" around the "sun," and rotate the globe one exact turn on its axis (360 degrees). Have students observe that this results in Joe facing to the side with respect to the sun. To bring Joe back to the zenith (facing the sun) we must turn the globe one-quarter turn more.

In other words, because of the Earth's progression in its orbit around the sun, the zenith of one day to the zenith of the next demands a rotation of the Earth on its axis of slightly more than 360 degrees; it must turn nearly 361 degrees. Thus, the zenith of one day to the zenith of the next provides our measurement of one day, 24 hours. But, one exact rotation of the Earth, 360 degrees, is accomplished in 23 hours and 56 minutes. Depending on the abilities of your students, you may choose to leave this aspect out of the lesson for the time being. It will become more evident as students observe the apparent movement of the stars in later years.

Part 2. Making and Using a Sundial to Tell Time

Can we tell time without a clock? Use Q and A discussion to lead students in reasoning how this might be done. Then invite them to actually do it, i.e., make a sundial.

With glue, mount a straw vertically near the center of a piece of poster board. Tack or tape the poster board on a level surface where it will get the sun for all or most of the day. (If it is possible for you and your students to mount a pole in a sunny location in the schoolyard and visit it periodically, this will gain even more interest.) Proceed to have students calibrate their sundial by tracing the shadow of the straw/pole each hour at 9:00, 10:00, etc. for all the hours you can. Especially, mark the end of each shadow and label it with the time.

Ask students to describe what they observe. In addition to the change in direction of shadows, they should note that they get progressively shorter during morning and lengthen again in the afternoon. The point where the shadow is the shortest is when the sun is highest in the sky and corresponds to the zenith, or solar noon. It will correspond to roughly 12 noon clock time but probably not exactly because of your location in your time zone as discussed in Part 1. Also, if you are on daylight savings time, it will occur around 1 p.m. because of having shifted our clocks.

Invite students on successive days to tell the time from their sundials.

When the shadow is between the hour lines, they can approximate the fractions of an hour. They will have fun seeing how close they come to the correct clock time. Emphasize again that the sundial is not behaving according to our clocks. We have designed our clocks to correspond to the apparent movement of the sun, i.e., the turning of the Earth. Likewise, you may point out that the materials and technology for making accurate clocks and watches only became available in the last 500 years or so. Before that sundials were the exclusive means of telling time and, even now, the official time that is used to set all our clocks is still determined by the sun's zenith.

Facilitate children using their sundials as long as there is interest. However, if they use it for more than a couple weeks, they will begin to see an error between their sundial and clock times in the hours before and after noon. To work consistently throughout the year, the pole of the sundial must be parallel to the axis of the Earth's rotation. Otherwise the shadow "wobbles" as the Earth progresses in its orbit about the sun. The situation may be corrected by tipping the straw/pole toward the line of solar noon, the shortest shadow, such that the angle between the ground and the pole is equal to your latitude. (Use a protractor to make this adjustment. This effectively makes the pole parallel to the Earth's axis of rotation. Thus, its shadow will move uniformly with the turning of the Earth throughout the year.)

Students should also note that, as days lengthen in the spring, they can extend their sundials to show shadows and times both earlier and later than before. Also, they should note that the shadow at noon gets progressively shorter; the sun at noon is higher in the sky. One point remains constant, however. The direction of the shadow at solar noon remains constant. Emphasize that this direction is TRUE NORTH. Following this direction long and far enough would take them to the North Pole. From true north, other directions can be determined as described in Lesson D-3A. (Of course, this assumes you live in the northern hemisphere north of the Tropic of Cancer. South of this, shadows can still be used to determine time and direction, but certain adjustments are necessary. In the southern hemisphere, for example, noonday shadows actually point south.)

Questions/Discussion/Activities to Review, Reinforce, Expand and Assess Learning:

Make a book illustrating how the time of day is connected to the Earth's rotation.

How and why do shadows change in direction and length during the day?

How are movements of the Earth connected to our telling time? … counting days? … counting years?

Read the time from a sundial. Compare it to clock time. What are the reasons for discrepancies?

Does the Earth behave according to our clocks, or are our clocks constructed to move at the same rate the Earth turns? Explain.

In small groups, pose and discuss questions such as:

Imagine being stranded on an island with no watch or compass. How might you keep track of time and determine directions?

You are responsible for calling someone at about noon, but your watch is broken. How can you determine noontime?

How can you determine true north, and other directions, without a compass?

If you spent all day walking toward the sun, where would you end up?

How could you keep on a continual westward trek using the sun as your guide?

How is movement related to time? Are there other situations where you measure time according to movements or measure movement according to time?

Is there a discrepancy between solar noon on two successive days and an exact 360 degree rotation of the Earth? Explain the discrepancy and why it occurs.

To Parents and Others Providing Support:

Facilitate children drawing their shadows at different times during the day.

Discuss how shadows relate to the turning of the Earth, the time of day and directions.

Facilitate children making a sundial and telling time with it.

On a day's outing, invite children to keep track of time and directions by observing shadows.

Connections to Other Topics and Follow-Up to Higher Levels:

History of keeping time and exploration
Determination of latitude and longitude
Math: innumerable problems involving the connection between the Earth's turning (degrees) and time (The Earth rotates 15 degrees per hour.)
The Earth's orbit around the sun
Changing seasons
Apparent movements of stars
The moon
Navigation

Re: National Science Education (NSE) Standards

This lesson is a steppingstone toward developing students' understanding and abilities aligned with NSE, K-4:

Unifying Concepts and Processes

- Systems, order, and organization
- Constancy, change, and measurement
- Evidence, models, and explanation

Content Standard A, Science as Inquiry
- Abilities necessary to do scientific inquiry
- Understanding about scientific inquiry

Content Standard B, Physical Science
- Position and motion of objects

Content Standard D, Earth and Space Science
- Changes in earth and sky

Content Standard G, History and Nature of Science
- Science as a human endeavor

Books for Correlated Reading:

Bailey, Jacqui and Matthew Lilly. *Sun Up, Sun Down: The Story of Day and Night* (Science Works). Picture Window Books, 2004.

Bulla, Clyde Robert. *What Makes a Shadow?* (Let's-Read-and-Find-Out Science, Stage 1). HarperCollins, 1994.

Dorros, Arthur. *Me and My Shadow.* Scholastic, 1990.

Karas, G. Brian. *On Earth.* G. P. Putnam's Sons, 2005.

Schuett, Stacey. *Somewhere in the World Right Now* (Reading Rainbow book). Dragonfly Books, 1997.

Sweeney, Joan. *Me Counting Time: From Seconds to Centuries.* Dragonfly Books, 2001.

Zolotow, Charlotte. *Over and Over.* HarperTrophy, 1995.

Lesson D-6

Seasonal Changes and The Earth's Orbit

Overview:

This lesson will bring students to an experiential understanding of how the passing months are really a depiction of the progression of the Earth in its orbit around the sun, one year equating to one circuit. Students will correlate their first-hand experience of changes of both physical parameters of the environment (temperature, day length, and rain/snowfall) and changes in flora and fauna with the Earth's orbit. Skills of observation, recording, and recordkeeping will be constantly called upon and developed.

Time Required:

This will be an ongoing study that may easily be included in the preclass show-and-tell-discussion period (see Chapter 1, page 19). Integrated with the ongoing study of identification of living things (Lesson B-4A) there will be synergistic reinforcement of both. Additionally it will serve in developing writing and recordkeeping skills.

Preparing for the study (Part 1) will take about 2 hours that may be divided as desired.

Objectives: Through this exercise, students will be able to:

1. Demonstrate with a globe how the Earth orbits the sun, as well as rotates on its axis, and name the months of the year in relation to the Earth's circuit.

2. Name the seasons and describe the general changes in temperature, rain/snowfall, and day length that occur in your region in the course of the Earth's orbit.

3. Describe and correlate with the seasons changes/events in the biological world, such as the sprouting of bulbs, trees coming into leaf, certain plants flowering, birds migrating/nesting, animals breeding, and fall coloration that are experienced in your region.

4. Give examples of how changes in the biological world (3 above) are adaptations to the physical world (2 above).

5. Exercise skills of making observations, recording, and keeping records.

Bernard J. Nebel, Ph.D.

Required Background:

Lesson D-2, Day, Night and the Earth's Rotation
Lesson D-3A, North, East, South, and West
Lesson D-4, Land Forms and Biomes, should continue concurrently
Lesson D-5, Time and the Earth's Turning
Pursuing this lesson concurrently with Lesson B-4A, Identification of Living Things, will create a natural synergy.
Reading, writing, and measuring skills will come into play but are not necessary at the outset. This activity will serve as an avenue for the development of these skills.

Materials:

Globe on a stand on a small table or high stool that can be moved conveniently about the room
Poster board cut into strips 4 x 18 inches
Markers
Tape/thumbtacks
Beach ball
String
Two three-ring note books (2-3 inch)
Supply of notebook paper (three-hole punched)

Teachable Moments:

At the end of a show-and-tell session involving some aspect of their natural environment, tell students that they are about to commence a study of how things such as temperature, rain/snow and flowers blooming are connected to the Earth's travel about the sun.

Methods and Procedures:

Part 1. Preparing for the Study

With globe in hand review Lesson D-2, namely how our experience of day and night is caused by the Earth turning on its axis. Explain/review how, in addition to the rotating on its axis, the Earth is also moving in a great circle or orbit around the sun. Demonstrate this by having a student stand in the in the center of the room holding a beach ball to represent the sun and walk around the perimeter of the room with the globe. Then, pose the question, "How long does it take the Earth to make one complete circle or orbit?"

Students should recall, but reemphasize as necessary, that the answer is exactly one year, exactly. Explain that this is no accident or coincidence. We base our year on the time that it takes the Earth to complete one orbit. You may drive this point home by having children consider how many times they have ridden the Earth around the sun? For any person, it will always be the same as their age. Also point out how we count the years from the (estimatedestimatd) time of Jesus Christ's birth to give the year part of dates, but eons of time existed before then, as well.

(If the question of leap years comes up, you may instruct students that the number of days it takes the Earth of make one complete orbit does not come out even. More exactly, it takes 365 and one quarter days to make a complete circuit of the sun. There is no practical way, however, to stick a quarter day into our calendar for a year. Therefore, we make the adjustment of adding one day (4 fourths) every four years. This adjustment makes our calendar continue to correspond to the Earths orbit, i.e., one year equaling the time it takes the Earth to make one orbit around the sun.)

To children, a year seems like a very long time and they may conclude that the Earth must be moving incredibly slowly. Explain that the Earth is so far from the sun (93 million miles) and, hence, it is so far around the "track" (584 million miles) that the Earth is actually traveling at about 70 thousand miles per hour to make the circuit in one year.

Instruct students that they are about to embark on a project that will model the movement of the Earth around the sun and connect it to the months and seasons of the year. Organize children as you see fit to have them participate in preparing the room.

1. In large, clear letters, write the name of each month on a strip of poster paper (one month on each of 12 strips).

2. At the position of the room determined to be north (Lesson D-3A) pose December. To the west, post March. To the south, post June, and to the east, post September. Post the other months evenly in between. Posting the months just barely below will keep them out of the way of other things and still enable them to be seen clearly. Note that the end result is a counterclockwise, circular "calendar" around the room.

3. Suspend a beach ball representing the sun in the center of the room just below the ceiling where it will be out of the way.

4. Place a globe, on high stool or stand near the side of the room under the current date, so far as it can be approximated.

5. Adjust the globe so that its tilt is toward north (December).

The ongoing lesson will involve periodically calling children's attention to the globe, its movement, and the calendar date, and having them shift the position of the globe a corresponding amount. Emphasize that the Earth does not move in such "jumps," but you have no practical way of modeling its actual, continuous, even motion.

In the process of moving the globe, it will be important to always keep its tilt toward north (December), and call children's attention to this orientation. This is to model what is true in nature; the Earth is tilted with respect to the plane of its orbit and the direction of this tilt is maintained throughout the year. (This is verified by the fact that the north star, as seen from Earth, remains in the same place throughout the year, i.e., over the North Pole.)

Maintaining this orientation, causes the result that the North Pole is tilted away from sun in the winter months and toward the sun in the summer months. In turn, it is responsible for the yearly change of seasons, but I recommend saving a detailed explanation of this for a later lesson. It will come much more easily and logically after students are fully familiar with the concept of the Earth maintaining its tilt in the course of its orbit and correlating the orbit with the change of seasons, which they experience.

The final step is to prepare a notebook of calendar pages for recordkeeping. Have students make a page of three-hole notebook paper for each day of the year, writing the month, day, and year in bold letters at the top and leaving the rest blank for records. File the pages in a notebook such that the starting page is your current date.

Part 2. Getting On With The Exercise

The plan, which will be ongoing, will be to keep the current week of calendar pages posted where it will be easily reachable and readable. Each day, students will add notes of their observations to the appropriate page. At the end of week, or first thing Monday, the pages for the past week should be filed in a second notebook and pages for the upcoming week posted. Having students fill in pages for weekends and holidays will be your option. Posting the new week's pages may also be the time for addressing the ongoing movement of the Earth and moving the globe accordingly.

All of this activity may be incorporated into the preclass show-and-tell/discussion session described in Chapter 1. It can well be a part of children bringing in samples and/or citing their observations concerning "happenings" in the natural world. It will create strong synergy among the biological, physical, and social aspects of the world. Paramount, it maintains the bridge between school and real-life experience.

The extent and detail of what students record will depend entirely on their ability level and the amount to time you choose to devote to this activity. Note that developing children's skills of measuring and recording data can be a key component of this activity. Therefore, have children take turns being the recorder while others discuss and agree on what should be written down. Coach and assist the recorder as necessary while admonishing others to practice patience, respect, and support.

The things that children should be encouraged to observe and record (again to the extent of their abilities) are the following:

1. Factors regarding the physical environment
   a. Temperature
   b. Rain/snowfall
   c. Day length (This may be partially experiential as days get shorter in the fall and lengthen again in the spring. Precise day length can be calculated from

times of sunrise and sunset obtainable from the weather page of a newspaper or from the Internet. This parameter is particularly important, because it is key in causing the change of seasons, as will be elucidated in a later lesson.)

2. Factors regarding the biological world—this will depend entirely on the flora and fauna of your region and what you and your children observe. Look for:

   a. When particular garden and wildflowers come into bloom, cease blooming, and die back.
   b. When trees burst into leaf and lose their leaves
   c. When particular fruits/berries ripen
   d. Behaviors of particular animals, including birds, insects, etc. Beyond sightings, things such as nesting, rearing young, migrating, and so on may be noted.

3. Allow children to record birthdays, holidays, and other events that are significant in their lives as well.

Periodically draw children's attention to and discuss how events in the biological world correlate with changes in the physical environment and how both correlate with the progress of the Earth in its orbit around the sun. In turn, discuss how the behaviors of plants and animals are adaptations to changes or impending changes in the physical environment. For example, many birds migrate south, and squirrels gather nuts in preparation for the impending winter.

Likewise, periodically mention how such observing, recording, and recordkeeping may extend all the way into professional careers. The number of species and the ways in which each is adapted to cope with the changing seasons provides an infinite body of material for exercising all aspects of inquiry. Allow yourself and your students to experience and express the awesome quality of it all as you delve ever deeper.

Questions/Discussion/Activities To Review, Reinforce, Expand, and Assess Learning:

The ongoing nature of this activity will provide ample reinforcement and opportunities for discussion and assessment. Further note that this activity serves as an avenue toward developing observing, measuring, writing, recordkeeping, and various social skills, as well as the central objective of becoming experientially familiar with the changes that occur with the seasons. For example, with respect to weather, students might start by simply recording, WARM and SUNNY; but this may mature to measuring and recording temperature, amount and type of cloud cover, rainfall, wind, and so on. We emphasize again that this is an ongoing activity that may well be carried on through all the elementary grades.

Make books illustrating how the Earth's orbit around the sun is connected to the months and seasons.

Make books illustrating how changes in the physical environment and/or changes in the biological world correlate with the change of seasons and the Earths orbit.

Note and discuss the following particularly noteworthy points in the Earth's orbit, as they occur:

Spring equinox, March 21/22, day and night are equal
Summer solstice, June 21/22, the longest day of the year (in the northern hemisphere)
Fall equinox, September 21/22, day and night are again equal
Winter solstice, December 21/22, the shortest day of the year (in the northern hemisphere)

Add observing and recording phases of the moon.

As students proceed with this activity, work with them in advancing their skills of observing, writing, and recordkeeping. Such activities may include:

Mounting a thermometer outside a window and teaching students how to read it.
Mounting a rain gauge and teaching students how to use it.
Observing and recording day-length (time from sunrise to sunset)
Recording and graphing changes in these parameters should follow in course.

In small groups, pose and discuss questions such as:

What time of the year (month(s)) are we most likely to see/experience ___ ? (Cite things such as: warming temperatures, tulips blooming, hot days, cold days, trees turning color, chickadees building nests, or any other biological or physical factor that children have experienced and recorded. Help them look at their notebook of records to find answers.)
How are changes in the biological world correlated with changes in the physical environment? How are both connected to the Earth's orbit?

To Parents and Others Providing Support:

In the course of recreational outings and even day-to-day routines such as going shopping, there will be innumerable opportunities to note particular "happenings" in the biological world: flowers blooming, trees leafing out, a bird building a nest, a squirrel gathering nuts, etc. Take advantage of such opportunities to discuss and consider with your kids how the given sighting is connected to the particular time of year, and the changes in the physical environment that are occurring. Facilitate your child in recording such things/experiences and taking them to school for sharing.

With a globe and pretend sun, have your children model the Earth's orbit and tell how the seasonal changes are related to its progress in the orbit, coaching as necessary.

To the extent of their interest, help your children set up their own thermometers and other instruments and coach them as necessary in making measurements, recording results, and keeping records. They may well take an interest in focusing on one particular aspect, and this interest may change from month to month. Facilitate them in pursuing their interest as they may wish.

Connections to Other Topics and Follow-Up to Higher Levels:

How seasonal changes stem from the Earth maintaining its "tilt" as it orbits the sun
Ecology, any and all aspects
Astronomy: how the night sky (positions of various constellations) changes with the passage of seasons. This change is caused by the fact that you are looking out into different parts of the universe as the Earth makes its way around the sun.
Plant and animal physiology: how plants and animals perceive the changing seasons and adapt their growth, reproduction, and habits accordingly.
History: the history of calendars and discovering the relationship between a year and the Earth's orbit.
Meteorology
Climatology

Re: National Science Education (NSE) Standards (p. 8-11)

This lesson is a steppingstone toward developing students' understanding and abilities aligned with NSE, K-4:

Unifying Concepts and Processes
  • Systems, order, and organization
  • Evidence, models, and explanation
  • Constancy, change, and measurement

Content Standard A, Science as Inquiry
  • Abilities necessary to do scientific inquiry
  • Understanding about scientific inquiry

Content Standard C, Life Science
  • Characteristics of organisms
  • Life cycles of organisms
  • Organisms and environments

Content Standard D, Earth and Space Science
  • Objects in the sky
  • Changes in earth and sky

Content Standard F. Science in Personal and Social Perspectives
  • Changes in environments

Content Standard G. History and Nature of Science
  • Science as a human endeavor

Bernard J. Nebel, Ph.D.

Books for Correlated Reading:

Bernard, Robin. *A Tree for all Seasons*. National Geographic, 2001.

Branley, Franklyn M. *Sunshine Makes the Seasons* (Let's-Read-and-Find-Out Science, Stage 2). HarperTrophy, 2005.

Fowler, Allan. *How Do You Know It's Fall?* (Rookie Read-About Science). Children's Press, 1992.
_____. *How Do You Know It's Spring?* (Rookie Read-About Science). Children's Press, 1991.
_____. *How Do You Know It's Summer?* (Rookie Read-About Science). Children's Press, 1992.
_____. *How Do You Know It's Winter?* (Rookie Read-About Science). Children's Press, 1991.

Gans, Roma. *When Birds Change Their Feathers* (Let's-Read-and-Find-Out Science*)*. Thomas Y. Crowell, 1980.
_____. *How Do Birds Find Their Way?* (Let's-Read-and-Find-Out Science, Stage 2). HarperCollins, 1996.

Gibbons, Gail. *The Reasons for Seasons,* Holiday House, 1996.
_____. *The Seasons of Arnold's Apple Tree*. Voyager Books, 1988.

Griffin, Sandra Ure. *Earth Circles*. Walker and Company, 1989.

Llewellyn, Claire. *Paint a Sun in the Sky: A First Look at the Seasons*. Picture Window Books, 2004.

Maestro, Betsy. *Why Do Leaves Change Color?* HarperTrophy, 1994.

Zolotow, Charlotte. *When the Wind Stops*. HarperTrophy, 1997.

Lesson D-7

Gravity II: Rate of Fall, Weightlessness in Space, and Distinction Between Mass and Weight

Overview:

In this lesson, students will discover that their intuition that heavy objects will fall faster than light ones is erroneous. They will discover that objects fall at the same rate (except where air/wind resistance comes into the picture). They will go on to relate this to weightlessness in space and make the distinction between weight and mass. This lays the foundation for understanding orbits of heavenly bodies, satellites, and many aspects of space flight.

Time Required:

Part 1. Things Fall at the Same Rate (activity, 30-40 minutes or more as desired, plus interpretive discussion, 20-30 minutes)

Part 2. Weightlessness in Freefall and in Space (activity, 30-40 minutes, plus interpretive discussion, 10-15 minutes)

Part 3. Distinction Between Weight and Mass (demonstration plus interpretive discussion, 20-30 minutes)

Objectives: Through this exercise, students will be able to:

1. Demonstrate that objects of different weights fall at same rate.

2. Express why a sheet of paper falls much more slowly than a solid ball.

3. Demonstrate how a sheet of paper will fall just as rapidly as a solid ball, if air resistance is reduced sufficiently.

4. Cite how/where we use air/wind resistance to our advantage.

5. Describe why light and heavy objects fall at the same rate (neglecting wind resistance).

6. Express and demonstrate how things are weightless when they are in a state of freefall.

7. Contrast weightlessness in freefall with weightlessness in space.

8. Distinguish between MASS and WEIGHT; tell how weight changes with gravity and how mass remains the same; name two units of mass.

9. Tell how mass is measured. Name and describe the instrument used.

Required Background:

Lesson D-1, Gravity I: The Earth's Gravity; Horizontal and Vertical
Lesson A-3, Air Is a Substance
Lesson A-4, Matter I: Its Particulate Nature
Lesson C-5, Inertia
Lesson C-6, Friction

Experience with a see-saw (teeter-totter) and weighing themselves on a bathroom scale is assumed.

Materials:

Assortment of differently sized objects that may be dropped without causing damage: lumps of clay or dough are ideal, but other items work well also: small (thumb-size) and large (fist-size) stones, small and large marbles, rubber balls of various sizes, various sized wooden blocks, etc.
Piece of notebook paper
Rubber party balloons
Empty tin can and water
Bathroom scale with a damped dial or digital scale that prevents it from bouncing back and forth as it comes to a reading
Video clip showing astronauts and their equipment in the weightless conditions in space
Soft item on the end of a string that can be swung around and let go without danger
Access to a balance

Teachable Moments:

Introduce this lesson as a game to discover which of an assortment of objects will fall the fastest.

Methods and Procedures:

Part 1. Things Fall at the Same Rate

With a number of various "droppable" items of different size and weight in hand, pose the question, "Which will fall the fastest when dropped?"

Allow students to make guesses (hypotheses) as they will, but then remind them that there is nothing like the experimental test to determine the answer. How should we do the test? Allow them to suggest ways to perform the test, but boil suggestions down to:

Dropping any two items side by side and observing which hits the floor first.

Have students take a prone position on the floor where they can gain a good view of the objects hitting the floor when you drop them. Of course, students may do this in small groups, having one of their members be the "dropper" and others observing. Things to watch for in the dropping are: Hold the two objects a few inches apart such that the lowest side/edge of each of the objects are even with one another. Release them simultaneously just letting them fall. Any flip or toss will introduce errors. For this part of the exercise, use items that are made of solid material (wood, metal, clay, rubber) and more or less spherical or cubic in shape. Hollow, spongy, or fluffy items such as ping-pong balls, cotton balls, sponges, or inflated balloons will be slowed down significantly by wind resistance as will be addressed later in this lesson.

A student who does not drop the two items exactly simultaneously and evenly should invite a discussion concerning scientific honesty. In science, we are trying to discover what is really true, what really occurs. Any way in which we bias the results gives us false answers and leads to confusion or worse. Pushing objects simultaneously off the edge of a shelf or a high table may help eliminate suspected bias in dropping.

By dropping items two at a time, students can contrast any combination of different weights and different materials: a glass marble, and a wooden block, a wooden block and a lump of clay, and so on. Students can make a table and record each pair of items they drop and which one falls fastest and hits the floor first. Or, do they hit the floor at the same time?

In comparing the falling rate of the solid, more or less spherical or cubic items, students will discover that they have great difficulty discerning which hit the floor first. There may be disagreement and perhaps argument. Retesting should always be the resolution. Gradually, they will come to the conclusion that both objects, regardless of material or weight, hit the floor together. In short, bring students to the conclusion that THINGS FALL AT THE SAME RATE! Neither the weight of the object nor the material from which it is made make any difference. They still fall at the same rate and hit the floor at the same time. (Again, we stress that this is true so long as air/wind resistance does not become a significant factor.)

A fun and instructive reinforcement to this conclusion is to present students with an open tin can of water and ask, "What will happen as the can of water falls?" (Do this outdoors and/or be prepared for the spill.) Again, let children make guesses (hypotheses) as they will, but then emphasize that the answer will be in the test. Drop the can of water with no throw or toss as students observe. It is always quite surprising to note that water does not spill out of the can until it hits the ground. Why? They should now be able to reason that both the water and the can fall at exactly the same rate; therefore, there was no spilling until the can hits the ground.

The Factor of Wind/Air Resistance

For the testing described above, we have stressed that it is important to use objects made of solid material that are more or less spherical or cubic in shape. Otherwise, air resistance becomes a significant factor and confuses the picture. Indeed, some students may have protested during the above experiments that things such as pieces of paper, inflated party balloons, and cotton balls really do fall more slowly. Here, you may demonstrate that this is true by dropping a piece of paper with a solid ball. It will be conspicuous that the ball hits the floor while the paper is still gliding down.

Pose the question, "Why should this be the case?" Proceed with Q and A discussion to bring children to reason out the answer. If necessary, jog their memories—that air is a substance (Lesson A-3) and friction plays a role (Lesson C-6)—to suggest that the answer may lie in air/wind resistance. Air catches the paper causing it to float one way and another while the solid object falls straight down through the air. Have students reflect: it should follow that if we reduce wind/air resistance sufficiently, the piece of paper will fall just as rapidly as the ball.

Let's test this idea. Make the paper into a streamlined airplane by folding the two top corners to the center, creasing, and folding the same "corners" again to the center. Fold the two sides together along the centerline, and finally fold out the two sides about an inch from the centerline to make wings. Now make the comparison. Hold the airplane by its tail so that its nose is even with the bottom side of the ball and drop the two together as students observe. Both children and adults are generally struck by the fact the paper airplane and the ball now hit the floor together. Indeed, when air resistance is sufficiently reduced, a sheet of paper falls just as rapidly as anything else.

Students may bring up additional examples of certain things falling more slowly. Guide them in analyzing each such case and observing that the answer lies in relative resistance to moving through the air. Of course, the solid objects tested above also encounter a certain amount of wind resistance. But with solid objects, the amount of wind resistance is relatively small compared to the force of gravity pulling them through the air. Hence, a significant difference in rate of fall was not observed.

Go on to point out and have students consider how air resistance is used as an advantage in many instances. What would happen to a parachutist if there were no air resistance against the parachute? Would the flight of birds, bats, insects and airplanes be possible without air resistance? It is wings pushing against the resistance of air that keeps the body aloft. Without this, they would never get off the ground. (Rockets use the principle of "push pushes back," Lesson C-7.)

Still, small light objects falling at the same rate as large heavy objects (neglecting air resistance) seems counter intuitive. Students may well ask why this should be the case? You can help them visualize the answer as follows.

Review how everything is made of particles (Lesson A-4). Gravity necessarily pulls on each and every particle with the same force. Hence, each should be expected to fall at the same rate. Will it make any difference whether those particles are clumped together into a large mass or a small mass? No!

Part 2. Weightlessness In Freefall And In Space

Pose the question, "How much will things weigh as they are falling?" Let's find out.

Place a heavy book of several pounds on a bathroom scale. Hold the loaded scale over a pillow on the floor and have it such that students can see and read the dial as you drop the loaded scale onto the pillow. (The pillow will avert damage to the scale.) It does require quick eyes, but you and your students will observe that the dial returns to zero as soon as the scale is released and remains at zero until, with the book, it hits the pillow. In other words, according to the scale, the book weighed nothing as it and the scale were falling. (You do need a scale with a damper on the dial for this. Otherwise you will just see the dial bouncing back and forth about zero as it falls.)

Use Q and A discussion to guide students in reasoning out why this should be. The key idea that should come out is this: If the book and the scale are both falling at the same rate neither will be pressing against the other. Hence, the scale registers zero. Said another way, THINGS ARE WEIGHTLESS WHEN THEY ARE IN A STATE OF FREEFALL. Weight is only experienced when gravity is pulling us and other objects down against the ground or other another surface that is NOT falling.

Our most common experience with weightlessness is in the arena of space flight. If possible, view a video clip that shows astronauts working weightlessly in space. Ask, "How is it that things are weightless in space?" Proceed with Q and A discussion.

Is it because gravity is absent? No! The Earth's gravitational field, although it decreases with distance from the Earth, extends out indefinitely. Indeed, the astronauts/satellite would proceed to simply fall back to Earth were it not for another factor. What is another part of the picture? In addition to taking the astronauts/satellite up, the launching rocket has also put them into an orbit such that they are traveling around the Earth. Does this have something to do with it?

Demonstrate, again, that when a weight is swung around on the end of a string and let go, it goes off in a straight line in the direction it was traveling the instant it was released. (SAFETY: Use a soft object as the weight in this demonstration.) Thus, for astronauts/satellites in space, two things are going on together. First, gravity is pulling them and they are consequently falling toward the Earth. Second inertia (Lesson 5A) is tending to take them in a straight line that would take them off away from the Earth. The orbit is such that these two forces are in exact balance.

The following activity may help you and children visualize the situation. Mark a large circle 10 or more feet in diameter and place an object in the center. The object in the center represents the Earth and the circle represents an orbit. Have a student be a satellite traveling around the orbit. Now, from any point in the circle, have the satellite/student walk *straight* forward a few steps, rather than following the path of the circle. Just like the swinging weight released from the string, this is the direction that inertia would take them.

To get back on the circle/orbit, he/she needs to sidestep. This "sidestep" represents falling toward Earth under the pull of gravity. Children enjoy putting words to this. "Inertia is sending me this way," as they go in a straight line, which causes them to leave the circle. Then, "Help, I am falling." as they sidestep back to the circle. Initially these two steps are obviously separate but guide students to make the intervals between them shorter and shorter until students see that the two are actually going on together. The conclusion is that an object in orbit is constantly in a state of freefall. Only, its rate of fall and the distance it falls is exactly equal to the rate and distance that inertia would take it away from the circle.

Being in a state of continuous freefall, the satellite, the astronauts, and all their equipment are in a state of weightlessness—but not without mass.

Part 3. Distinction Between Weight and Mass

Up to this point in our studies, we have not attempted to make a distinction between mass and weight, because for all practical purposes, children experience them as the same thing and attempting to make a distinction too often leads to bewilderment. Now, with the observation that objects are weightless when they are falling, the distinction becomes unavoidable.

Draw students to ponder the situation that the same object my weigh 10 pounds here on Earth, less than two pounds on the moon, and nothing at all when it is in state of freefall, as we noted for the book and scale falling and for objects in orbit. But the object, itself, is unchanged. It is only the measurement of weight that is haywire.

The measurement of weight changes with gravity, because of the mechanism we use. In a typical bathroom or kitchen scale, the weight on the platform pushes against a spring that is connected to the dial. If possible, take the cover off a scale to show students this mechanism. How much the spring is bent by a given weight will obviously depend on gravity. In short, a typical spring scale will give us erroneous readings if there is any variation in the pull of gravity.

What should we do about this? We need a measurement that expresses the "heftiness" of an object, but one that does not change with different gravitational pulls. Let students struggle with this thought for a time and then instruct them:

The solution that scientists devised was to invent a new term, MASS, and use it in place of weight. Mass is equivalent to weight here on earth where gravity is present, but, unlike weight, mass is defined as remaining constant regardless of more gravity, less gravity, or none at all. For example, we observed that the book on the scale became weightless as it fell. But, the mass of the book remained the same regardless.

In turn, scientists had to devise a different set of units. Have students reflect that it wouldn't do to measure mass in pounds because that would confuse it with weight, which changes with gravity. Further, scientists had to devise a method of measuring mass that would not depend on the strength of gravity. The decision they came to was to designate one liter of water (the volume held by a cube-shaped container 10 centimeters on each side) as having a mass of one KILOGRAM.[25] The mass of all other items would be determined by comparing them to the mass of water.

For example, a mass equal to the mass of one liter of water would, by definition, be a mass of one kilogram; a mass equal to half a liter of water would be 0.5 kilogram; a mass equal to two liters of water would be two kilograms, and so on.

But, how can we compare something with the mass of a liter of water?

Some children may already have the answer—a balance. In any case show children a balance (a simple toy-store balance will do) and draw on their familiarity with a see-saw (teeter-totter). Point out and demonstrate how equal weights on the two ends balance, like two children of equal weight on the ends of a see-saw. Thus, a set of weights is made which equal the mass of a liter of water and various decimal fractions thereof. Placing the object on one end of the balance and adding known masses to the other end until it balances gives the mass of the object.

Have children reflect on how this system of measuring is independent of gravity. If the two, equal-weight children and the see-saw were on the moon, where gravity is much less than on Earth, they would still balance. If they were on a planet where gravity was twice as great, they would still balance. The reason, students may surmise, is because gravity affects both ends of the balance equally regardless of its force.

One additional factor comes into the picture, and students may have already brought it up. It is their experience that two children of unequal weights can balance if the heavier child moves closer to the fulcrum (pivot point of the balance), or the lighter child moves further away. Many, indeed most, balances are constructed to utilize this principle. Rather than adding actual weights to make the balance, one slides weights along respective scales to achieve the balance. Emphasize that, despite this difference, one is still measuring the mass of an object by balancing it with a known mass equivalent to a certain volume of water.

Children may enjoy noting that the scales they are weighed on in a doctor's office is really a balance. The connections are hidden from view, but the platform you stand on is hooked to one end of the balance arm, the sliding weights are on the other.

Questions/Discussion/Activities To Review, Reinforce, Expand, and Assess Learning

Make a book illustrating things falling at the same rate, weightlessness in space, and/or other aspects of the lesson.

---

[25] Water does expand and contract with temperature. Therefore, the mass of a liter of water changes accordingly. Likewise, any impurities present in will affect its mass. Therefore, the technical definition of a kilogram includes chemically pure water at 4 degrees Celsius, its most dense point. However, such technicalities can be set aside for later lessons.

Encourage students to show their friends how things fall at the same rate (except where wind/air resistance enters the picture).

Set up an activity center where students can redo and try additional variations on activities described including weighing things on a balance.

In small groups, pose and discuss questions such as:

Which of _____(name any two objects) will fall faster, or will they fall at the same rate? Explain why.

Why does a sheet of paper fall more slowly than a coin? How can the sheet of paper be made to fall at the same rate (nearly) as the coin.

Are birds and airplanes changing gravity? If not, how do they fly?

When things are falling, do they exert any weight on one another?

Why does freefall produce a state of weightlessness?

How do satellites stay up?

Why are satellites always put into orbits? What would happen to them if they were not moving forward in an orbit?

Astronauts in orbit are weightless. How so? Is their no gravity?

If you were to weigh yourself on a SCALE here on Earth and then on the moon, would get a different reading? What about on a BALANCE? Explain.

What is the distinction between weight and mass?

How can we get a consistent measure of mass when gravity is different?

How does a balance work? Why is it independent of gravity?

To Parents and Others Providing Support:

Conduct further tests to show that things fall at the same rate, except where wind/air resistance becomes a factor.

When any mention of a satellite, the space telescope, or the space lab comes up, ask and discuss how it stays up. (Note that they are always in orbit. Why? What is the role of orbiting in keeping them up?)

Any time weightlessness in space is noted, ask and discuss why/how astronauts and their equipment are weightless?

When your children are being weighed in a doctors office, have them observe that a balance with sliding weights is used. Would this balance give the same reading on the moon where gravity is less? Discuss why/how the answer is yes.

Discuss the distinction between weight and mass.

Connections to Other Topics and Follow-Up to Higher Levels:

Acceleration due to gravity
How things fly
Principle of Balances: mass x distance = mass x distance
The metric system

Re: National Science Education (NSE) Standards

This lesson is a steppingstone toward developing students' understanding and abilities aligned with NSE, K-4:

Unifying Concepts and Processes
- Systems, order, and organization
- Evidence, models, and explanation
- Constancy, change, and measurement

Content Standard A, Science as Inquiry
- Abilities necessary to do scientific inquiry

Content Standard B, Physical Science
- Properties of objects and materials
- Position and motion of objects

Content Standard D, Earth and Space Science
- Properties of earth materials

Content Standard E, Science and Technology
- Abilities of technological design
- Understanding about science and technology

Books for Correlated Reading:

Branley, Franklyn. *Floating in Space* (Let's-Read-and-Find-Out Science, Stage 2). HarperTrophy, 1998.

Curry, Don L. *What Is Mass?* (Rookie Read-About Science). Children's Press, 2005.

Lesson D-8

Rocks and Fossils

Overview:

Starting with hands-on examination of fossils, students will view and gain an understanding of the processes that lead to the entrapment of organisms in rock (erosion, transportation, and sedimentation). They will learn to identify different kinds of sedimentary rocks, distinguish them from igneous rocks, and understand why only sedimentary rocks contain fossils. They will learn the principle of how the age of rocks is determined and how this enables the construction of a timeline regarding both geological changes in the Earth's surface and the sequence of development of biological species. This will provide a foundation underlying all further studies in geology and evolutionary biology.

Time Required:

> Part 1. Erosion, Sediments, and Fossils (activities, 40-50 minutes, plus interpretive discussion, 20-30 minutes)
> Part 2. Limestone (activities, 20-30 minutes, plus interpretive discussion, 20-30 minutes)
> Part 3. Contrasting Sedimentary and Igneous Rock (activity, 20-30 minutes, plus interpretive discussion and review, 20-30 minutes)
> Part 4. The Stories Rocks Tell (viewing pictures/videos and activities, 40-50 minutes, plus interpretive discussions, 30-40 minutes)

Objectives: Through this exercise, students will be able to:

1. In a suitable rock sample, point out fossils; explain what they are.

2. Tell how fossils come to be imbedded in rock; in the course of this description, explain EROSION, TRANSPORT OF SEDIMENTS, SEDIMENTATION, and SEDIMENTARY ROCKS.

3. Recognize and distinguish among different types of sedimentary rocks, especially sandstone, shale, and limestone; describe the origin of each.

4. Explain the factors that cause sand to settle in different locations from finer particles (clay).

5. Tell how limestone differs from other sedimentary rocks.

6. Separate sedimentary rocks from igneous rocks; point out distinguishing features; tell how they differ in origin; tell why one would never find fossils in igneous rocks.

7. Give one or more examples of "stories" that rocks tell about the past history of the Earth and creatures that inhabited the Earth in the past.

8. Tell how we know that dinosaurs inhabited the Earth 250 to 65 million years ago; describe, in principle, the method of telling how old rocks and fossils are.

Required Background:

Lesson D-4, Land Forms and Biomes
Lesson A-10, Rocks, Minerals, Crystals, Dirt and Soil

Materials:

Part 1. Erosion, Sediments, and Fossils
Samples of rock containing fossils (may be purchased)
Additional samples of fossils and/or photographs of fossils as may be available
Dirt (2-3 quarts that contains a variety of differently sized particles— pebbles, sand, silt, and clay)
Sprinkling can of water
Dish draining pad (light color if possible)
Wide-mouth gallon jar
Samples of sandstone and shale
Pocket magnifiers
Lump of clay as is used in making pottery
Glass or clear plastic jar (about 1 quart) of water with a tight lid

Part 2. Limestone
Sample of fossil-rich limestone that students can handle and observe
Assorted seashells, including pieces of coral that students can handle and observe
Sample of shale
Dropper bottle of acid if available; vinegar will work

Part 3. Contrasting Sedimentary and Igneous Rock
Samples of sedimentary and igneous rocks
Pocket Magnifier

Part 4. The Stories That Rocks Tell
Photographs and/or videos of the Grand Canyon
Hourglass timer (one per group if possible)

Bernard J. Nebel, Ph.D.

Teachable Moments:

Show and let students examine whatever fossils may be available, photographs of fossils, photographs of reconstructed dinosaur skeletons etc. Most children will be fascinated. Pose questions concerning how and where such fossils were found and what they may tell us about the past history of the Earth.

Methods and Procedures:

Part 1. Erosion, Sediments, and Fossils

Show and let students examine whatever fossils, and photographs of fossils may be available. Specimens should include fossils still imbedded in stone. (If you have areas near you where fossils may be found and collected, a field trip is definitely in order.) After they have examined the fossils for a time, pose the question, "What are they?"

Students will recognize that they have features that identify them as once-living things. But, how did they get there, imbedded in the rock, which was found _____ (give location if possible)? What do they tell us about the history of the Earth? Let's investigate this.

(It will be best to conduct the following activity as a demonstration with students crowded around. If children do it, the messes will be horrendous.)

Put a small bucket full (2-3 quarts) of moist dirt on a dish drainer pad, and pack it down tightly with your hands. Adjust the drainer so that water will drain into a large glass or clear plastic jar. Ask, "What happens when rain falls on bare earth?"

Have students observe and describe what they see as you sprinkle water onto the pile of dirt. Call their attention specifically to what is described in the following. (You may have them record notes.)

As rivulets of water run down the pile of dirt, they carry dirt particles away with them. Explain that we call this effect EROSION. The running water ERODES the dirt. Then, the dirt particles that are carried along with the water are collectively referred to as SEDIMENT. Putting this into a single statement, flowing water erodes the earth and carries the particles along as sediment. (We are going to use "particles" to refer to the specks of soil, sand, silt, and clay, but emphasize the these dirt particles are different from the fundamental chemical particles that we learned about in lessons of Thread A. Even the tiniest visible dirt particle contains thousands of fundamental chemical particles.)

Have students observe the movement of the sediment very closely as it washes across the drain pan. Ask, "Are all the particles carried along evenly, or is there a separation according to the size of the particles?" They will observe that it is the latter. As the water flows over the drain pad, the particles of dirt are separated according to size (and density). Very fine particles (clay) are carried along readily; sand-sized grains are tumbled along more gradually; pebbles and stones remain on the pile. (You will have to adjust the rate of your sprinkling carefully to observe this separating effect clearly. If you sprinkle water on the pile too fast, everything may be carried away so fast that observing the effect will be impossible. Of course, it also assumes the dirt you have chosen to use is a loam, which contains a variety of differently sized particles. Test some of your dirt ahead of time to be sure that this is the case.)

Use Q and A discussion to help students relate this to their real-world experience. During and after heavy rains, they have probably seen streams/rivers flowing full with very muddy looking water. What is going on? They should reason that the rain has caused erosion. The muddy looking water is the sediment being carried down the river.

Ask where the stream/river goes, and what happens to its load of sediment when it gets there. From Lesson D-4, children should be familiar with the fact that streams/rivers empty into lakes and eventually into the ocean. What happens there? This is where the collecting jar under the drainer comes in. Have children observe how the sediment (note it is still called sediment) settles to the bottom. With Q and A, guide students in reasoning that the same thing occurs in the real world. When streams/rivers enter the larger body of water, the water becomes relatively still and sediments settle out.

Now we come back to fossils. What happens if fish, shellfish, seaweeds, and other organisms are buried under the incoming sediment? Students will readily recognize that they may become permanently buried, only to be found eons later as fossils.

Why is it that we usually find only shells or bony remains as fossils, not the whole fish or dinosaur for example? Guide students in reasoning that fleshy material rots away soon after being buried. Only bony parts remain to become fossils.

But more is involved. How do the fossils end up in rock rather than in just loose sediment? Explain: as more and more sediment is piled on top, lower layers are pressed down harder and harder. This plus chemical factors cement the particles of sediment, including fossils, together to make rock. The type of rock is called SEDIMENTARY rock, because it is formed from sediments.

Is there just one kind of sedimentary rock? Show and have children examine pieces of sandstone and shale using a magnifier.hand lens. Explaining that that these are two kinds of sedimentary rock. (Save limestone for later, Part 2.) What is the difference? They will report that sandstone appears as its name implies; it looks like sand grains cemented together. Explain that that is exactly what it is.

In shale, however, they can't see any distinct grains; it has a uniform, smooth, texture throughout, similar to a mineral, but it has neither the luster, nor does it tend to break at particular angles like a mineral. Finally it is softer than most minerals. It is quite easily shattered and pulverized with a hammer as you may demonstrate. (Safety: Have children wear eye protection if you do this.) Explain that shale is derived from settled clay particles, which are the smallest of the dirt particles.

(Children, and adults too, are apt to think of clay as the moldable, sticky "mud" used for making pottery. Explain that clay is actually microscopic mineral particles. When moist, the particles may slip and slide over one another on films of water. Thus, the clay particles become a moldable mass. But, the individual nature of the clay particles is easily demonstrated. When you mix a small lump of clay into a glass of water, the particles separate and make the water look muddy. Then, gradually over the course of several hours, the clay particles will settle to the bottom, leaving the water looking more or less clear again.)

Students will readily recognize that critters may be buried under settling particles regardless of the size of the particles. Therefore, it follows that fossils may be found in any kind of sedimentary rock, but not in igneous rocks.

An obvious but curious fact is that we find shale and sandstone separately. What does this imply? Guide students in reasoning that it means that sand and clay settled out in different locations or at different times. Does this make sense? Guide students in considering how they just observed that different sizes of sediment particles were separated as they were carried along by water. Explain that the same thing occurs as the stream/river enters the larger body of water. As the water slows and stills, larger particles, namely sand, settle quickly near the mouth of the river; smaller particles (clay), which take longer to settle, are carried and settle further from the river's mouth.

It is worth demonstrating this effect. Put a handful of your dirt in a jar three fourths full of water, cover tightly, and shake vigorously until all the dirt particles are separated. Then set the jar where students can crowd around and observe the settling closely. They will note that sand and larger particles settle very quickly. Finer sand settles more slowly. The very fine particles of clay that make the water appear muddy seem to not settle at all; the water maintains its muddy appearance. But set the jar aside and have students observe it the next day. They will find that most of the clay has settled, leaving the water more or less clear. Observing the settled sediment at the bottom of the jar, students will note a layered appearance, coarse sand at the bottom, then successively finer particles, and finally a layer of clay at the top.

Have children visualize that if the water was flowing very slowly, this effect would result in the particles settling at different locations as noted above.

Some students may ask why the dirt particles should settle at different rates. Use Q and A to guide them to reason this out. Which will weigh more, a sand grain or a microscopic clay particle? (The sand grain.) Then, which will be pulled by gravity down through the water with the most force? (The sand grain, again.) So which gets to the bottom first? You may have students note that it involves the same principle as they observed in Lesson D-7. Without wind or air resistance, gravity does cause things to fall at the same rate. But resistance to moving through the air causes things such as pieces of paper and cotton balls to fall more slowly. Here it is resistance to moving through water.

(There is, as may be observed in the settled dirt, actually a spectrum of every sized particle ranging from coarse sand to fine sand, then to what are called silt particles, finally to the finest clay particles. Scientists have created a scale that classifies the particles according to their exact dimensions. However, this added complexity causes confusion and adds nothing to the conceptual picture of sedimentary rock being formed from settling sediment and burying organisms in the process. Therefore, we have chosen to omit getting into the grades of particles and resulting sedimentary rock types except to look at sandstone and shale. If siltstone is common in your region, however, you should certainly bring it into the discussion. The same goes for conglomerates, rock masses formed from the cementing together of various assortments of stones, gravel, pebbles sand and even clay. Limestone, however, is a special kind of sedimentary rock that does need to be addressed.)

Part 2. Limestone

Have students note that the sedimentary rocks considered thus far (sandstone and shale) were formed by the cementing together of various mineral particles derived form eroding dirt. You may remind them that the mineral particles of dirt were originally derived from the weathering of rocks (Lesson A-10). Explain that there is another form of sedimentary rock, often exceedingly rich in fossils, which has a different origin. That is limestone.

Show and let students examine pieces of fossil-rich limestone. Then let them examine an assortment of seashells (clam, oyster, snail, and others, including pieces of coral). Ask them to ponder the relationship between the fossil-rich limestone and the assortment of shells. They will note that many of the fossil shells are quite similar to the current shells.

Explain that all this shell material and also limestone is made from the same chemical, calcium carbonate. (They may recall that their own bones are another calcium chemical, calcium phosphate.) Calcium carbonate, like salt, is a mineral that dissolves and is thus present in solution in seawater. The critters inhabiting the shells (there are plants as well as animals) are able to take the calcium carbonate from water and make it into their shells. When they die and rot away, their shells remain. Over millennia, this shell material accumulates layer upon layer. There may be intact shells, but innumerable bits of broken shell material fill in the spaces between. There may also be direct chemical crystallization of calcium carbonate in the manner they observed when growing their salt crystals (Lesson A-9). The final result is thick layers or beds of limestone rich in fossils as they are observing.

How can you distinguish limestone from shale or a fine sandstone? Aside from differences in color and texture that students may note, explain that there is one sure test. Put a drop of acid on both a piece of limestone and a piece of shale. Students will note that on limestone, acid causes a fizzing; on shale it doesn't. Explain that the acid attacks the limestone and releases the carbonate as carbon dioxide; it is the bubbles of carbon dioxide that cause the fizz. You may remind students that it is the same sort of reaction they observed when vinegar (acid) was added to baking soda to produce carbon dioxide back in Lesson A-7. Shale and sandstones do not contain carbonate minerals, and, hence, there is no fizzing.

SAFTEY: (You may use vinegar in this demonstration but don't expect a great fizzing; bubbles will form relatively slowly. A stronger acid will cause greater fizz, but then be sure to keep it out of reach of children. Strong acid is very hazardous. Be sure to wash used portions away with plenty of water.)

Part 3. Contrasting Sedimentary and Igneous Rock

Have students reexamine igneous rocks, which they studied in Lesson A-10, and contrast them with the sedimentary rocks that they are observing here. (You may ask them to make a table comparing features of the two. Seek the help of a local naturalist or geologist to help you and your students discern the differences.)

Tables should include the fact that igneous rock is comprised of irregular, crystalline granules of two or more distinct minerals, whereas sedimentary rock is comprised of evenly-sized grains throughout. The grains in sedimentary rock may be so tiny (microscopic) that they are not discernable as individual grains. Rather, the rock has an even, smooth texture as in shale.

Most importantly, have students ponder how the differences in appearance reflect the difference in origin. Review how igneous rock results from the cooling and crystallization of magma. Sedimentary rock comes from rock that has been weathered into dirt particles, transported by flowing water (or in some cases wind), deposited, and cemented together. Or, in the case of limestone, deposited through the action of organisms as described above.

Significantly, one will only find fossils in sedimentary rocks. Have student ponder: Why? Would they ever find a fossil in igneous rock? Why not?

Part 4. The Stories that Rocks Tell

What can rocks and fossils tell us about the history of the Earth and life on Earth? Show pictures or a video of the Grand Canyon and emphasize how we see that the Colorado River has eroded its way down through and exposed successive layers of sandstone, shale, and limestone, each with respective fossils. These layers were laid down over the course of 300 million years. What does this tell us about this region of the United States?

Children and adults, too, may find it hardly believable, but the evidence is irrefutable. These rocks, especially limestone, only result from sedimentation of material under water, as we have noted above. Therefore, the conclusion must be that this region was once sea bottom. To allow that sedimentation to occur, it must have been much lower in elevation (below sea level), then must have been gradually raised to its current elevation, which is more than a mile above sea level.

(Examples of rocks and topographic features that portray changes in the earth's surface, including the movement of continents (plate tectonics) are innumerable and can quickly boggle children's minds. Therefore, bear in mind that the only objective at this level (second grade) is to introduce students to the concept that the Earth is a dynamic system, always changing. But changes are so slow that, barring great earthquakes and huge volcanic eruptions, they are imperceptible in the course of a human lifetime. Nevertheless, over the course of millions of years, indeed hundreds of millions of years, the changes are tremendous. Continents change positions, new mountain ranges are lifted up, old ones are eroded away, etc. But there will be plenty of time to explore the details further in later lessons. Here, we should only try to expose children to the ideas and let them percolate as they will.)

There is one further point of significance however. We just mentioned hundreds of millions of years. How do we have any idea how old rocks are? How do we tell their age? You should raise such questions even if students don't. This is critical to all our understanding of the Earth's history. You can explain it through the following analogy.

Show and let children play with a simple hourglass, timing it with a watch. Then pose the question, "Even if you didn't see the hourglass started, could you calculate when it was started?" Use Q and A discussion to guide children in reasoning: By observation, they can determine how fast sand is flowing through the neck. They can also measure/estimate how much sand is in the upper half and how much is in the lower half of the glass. These three figures will give the answer. For example, if it is a three-minute timer and they observe that half the sand is still in the top, then it must have been started one and half minutes before. If only one third of the sand is left in the top (two thirds in the bottom), then it must have been started two minutes before.

The same idea holds for the dating of rocks. Certain chemical elements contained in rocks are subject to gradual change. The starting chemical breaks down to form another chemical, but very, very slowly over the course of millions of years. They are the "hourglass." Scientists have determined their rate of breakdown, slow as it is. They measure the amount of starting chemical, the amount of ending chemical, and that gives them the starting time, the age of the rock. The age of a fossil is given by the age of the rock in which it is imbedded. Or, in some cases, actual bone for example, the age of the fossil itself can be determined in like manner. The process of such determinations is referred to as radioactive dating, but further description must await more advanced knowledge of chemistry.

By such dating, the evidence points to more than just the fact that changes in the earth have taken place. We can date them and put them into a geological-historical time line of when they occurred.

The most significant and important aspect of this study is that it provides us with a chronological record of the history of life on Earth. We don't just see that different life forms existed on Earth. We can tell the time period in which they existed and put them in a sequence. This shows us clearly, for example, that fish existed long before any land vertebrates. Dinosaurs existed during a period from about 250 to 65 million years ago. Mammals started about the same time as dinosaurs became extinct. (Except there is mounting evidence that dinosaurs are still with us. It is only that now we call them birds.) The earliest evidence of hominids (human-like animals) date from about 4.5 million years ago, about 60 million years after (traditional) dinosaurs disappeared from the scene. And so on.

Bernard J. Nebel, Ph.D.

Let students know that this line of study that can be pursued indefinitely, including all the way through life careers. There is much here still to discover and learn.

Questions/Discussion/Activities To Review, Reinforce, Expand, and Assess Learning:

Make a book illustrating/describing the steps, starting with erosion, that lead to fossils being imbedded in rock.

Invite children to bring in any fossils they might have in their possession and give show-and-tells regarding them.

Take advantage of opportunities to visit exhibits of fossils in museums and nature centers. If it is possible to take a field trip to an area where fossils can be found and collected, so much the better.

Show videos, display charts, and so on that describe the geological history of the earth. Especially discuss how that information was obtained by studying rocks and fossils.

Set up an activity center where students can continue to examine various types of rocks and fossils.

Invite a local naturalist or geologist to talk to your class about the geological history of your region and how that history is ascertained from rocks, topographic features, and fossils found in your region.

In small groups, pose and discuss questions such as:

What are fossils?
How do such creatures come to be embedded in rock? List and describe all the steps, starting with erosion.
What are sedimentary rocks? How were they formed?
Separate an assortment of rocks into categories: igneous and sedimentary. Sort sedimentary rocks into sandstone, shale and limestone.
How do sandstone and shale differ from limestone in origin?
If both come from the settling of dirt particles, how is it that sandstone is found separate from shale?
A friend wants you to help him search for fossils in a formation of rock that has a distinctly speckled appearance, the specks showing a distinct crystalline structure. What would you tell him/her? Why?
How do we know that dinosaurs existed 250 to 65 million years ago?
How do we know that humans did not exist at the time of dinosaurs?
What is the technique for determining the age of rocks and/or fossils?

(There are different processes of fossilization leading to different types of fossils that may come up in the course of this lesson. A fossil may be the actual bone or shell. In other cases, the original biological matter has disintegrated and been replaced with other material. Thus, one sees the impression of the bone, shell, or plant stem, but not the original material. In other cases there may be an impression, a footprint for example, of the ancient animal in soft earth. This dried and another layer was deposited over the top with no fusion between the layers. Later, the layers separate again exposing the impression as a fossil. Pursue these avenues according to your own inclination and your students' interest.)

To Parents and Others Providing Support:

Repeat the "erosion activity" described (outdoors) having your children particularly note how flowing water separates particles according to size and how particles of different size settle at different rates. Discuss why this occurs.

When you see a muddy stream/river, discuss: What is causing the water's muddy appearance? Where did the "mud" come from? Where is it going to end up?

Take advantage of opportunities to visit displays of fossils. Discuss how they came to be embedded in the rock, what they represent, and so on.

On excursions, take time to examine rocks. Are they sedimentary? If so, can we expect to find fossils? Search for fossils as is permissible. (Removing fossils from parks is generally strictly prohibited. If you do find one, point it out to a ranger. They may wish to put it in park collection for display.)

Avail yourselves of opportunities to hear presentations regarding local geology and what it tells about past geological history of your region. Do the same on trips to learn about the geological history of regions you are visiting.

Avail yourselves of opportunities to view and study charts, films, etc. of the earth's history, including different animals and plants that existed in different epics. Discuss how we know such things. What are the lines of evidence? (Fossils, rock types, and the ability to date them will always be key.)

Connections to Other Topics and Follow-Up To Higher Levels:

All aspects of geology
Plate tectonics
Paleobiology (the study and dating of fossils)
Evolutionary biology

Re: National Science Education (NSE) Standards

This lesson is a steppingstone toward developing students' understanding and abilities aligned with NSE, K-4:

Unifying Concepts and processes
  • Systems, order, and organization
  • Constancy, change, and measurement

Bernard J. Nebel, Ph.D.

- Evidence, models, and explanation
- Evolution and equilibrium

Content Standard A, Science as Inquiry
- Abilities necessary to do scientific inquiry
- Understanding about scientific inquiry

Content Standard B, Physical Science
- Properties of objects and materials
- Position and motion of objects

Content Standard C, Life Science
- The characteristics of organisms
- Organisms and environments

Content Standard D, Earth and Space Science
- Properties of earth materials
- Changes in earth

Content Standard E, Science and Technology
- Understanding about science and technology
- Abilities to distinguish between natural objects and objects made by humans

Content Standard F, Science in Personal and Social Perspectives
- Changes in environments

Content Standard G, History and Nature of Science
- Science as a human endeavor

Books for Correlated Reading:

Aliki. *Fossils Tell of Long Ago* (Let's-Read-and-Find-Out Science, Stage 2). HarperTrophy, 1990.

Bailey, Jacqui. *Monster Bones: A Story of a Dinosaur Fossil* (Science Works). Picture Window Books, 2004.
_____. *The Rock Factory: A Story About the Rock Cycle* (Science Works). Picture Window Books, 2006.

Dussling, Jennifer. *Looking at Rocks* (My First Field Guides). Grosset & Dunlap, 2001.

Ewart, Claire. *Fossil*. Walker & Company, 2004.

Gans, Roma. *Let's Go Rock Collecting* (Let's-Read-and-Find-Out Science, Stage 1). HarperCollins, 1997.

Morgan, Ben. *Rock and Fossil Hunter* (Smithsonian Nature Activities). DK, 2005.

Rosinsky, Natalie M. *Rocks: Hard, Soft, Smooth, and Rough* (Amazing Science). Picture Window Books, 2003.

Walker, Sally M. *Fossils* (Early Bird Earth Science). Lerner, 2006.
_____. *Rocks* (Early Bird Earth Science). Lerner, 2006.

Lesson D-9

Resources: Developing an Overview

Overview:

The idea of resources being required for anything we make has come up in a number of earlier lessons. Here, students will develop the concept of resources into a larger mental framework. They will learn to organize resources into five basic categories and diagram how resources from one or more categories are required for anything we make or produce. In turn, they will gain appreciation that human resources are indispensable, and that there is a synergism between human resources and natural resources. This will provide a foundation for numerous follow-up studies in areas of technology, history, and social issues.

Time Required:

Initial activity and interactive discussion, 45-60minutes, but plan on revisiting this lesson on future occasions to expand and add additional parameters.

Objectives: Through this exercise, students will be able to:

1. list 5 categories of resources;

2. diagrammatically show and discuss what and how various resources are required in the production of a given product;

3. show and discuss the relation between human resources and other resources;

4. discuss how human resources are a critical part of obtaining all other resources;

5. discuss the relationship between human resources, technology, and obtaining other resources.

Required Background:

Lesson B-2, Living, Natural Nonliving, and Human Made Things
Lesson A-5, Distinguishing Materials
Lesson A-10, Rocks, Minerals, Crystals and Soil

376

Lesson B-12, Plants, Soil, and Water
Lesson C-1, Energy I, Making Things Go

The general concept that should be reviewed and stressed is that nothing is made from nothing. One must always have starting material(s). The basic starting materials, termed RESOURCES, are things found naturally in/on the Earth.

Materials:

Typing paper (8 1/2 x 11)
Pencils
Poster making materials if you wish to have students make posters

Teachable Moments:

Review the general concept of resources as may be necessary. Then, hold up a picture of an item or an actual item and ask children to ponder what resources were required to make or produce it.

Methods and Procedures:

Start with a discussion of what is required to make/produce this _____. (Hold up a picture or an actual item.) If students give an answer such as wood for a piece of furniture, push them to consider where wood comes from. (Trees, a biological resource.) If they say metal for a pair of scissors, take them to: Where do we get metal? (Iron or other metallic ores, a mineral resource, Lesson A-10)

Conduct this discussion in a manner that guides students to conclude that all resource can be assigned to one of five categories: Soil, Water, Mineral Resources, Biological Resources (any flora and/or fauna); and Energy Resources. Emphasize that all resources go back to things that humans have found occurring naturally on/in the earth. (Yes, all fruits, vegetables, grains, and domestic animals, were derived through selective breeding of species originally found in nature.) From here, move into the following activity.

On a sheet of 8 1/2 x 11 typing paper, have each student make a circle about 4 inches in diameter. Equally spaced around the perimeter of the circle, they should write the names of resource categories: water, soil, mineral, biological, and energy. In the center, make a smaller circle about one inch in diameter.

The gist of the activity is to put a particular product/item in the center. Then, if a resource category is used in making that item, draw arrow from the resource (outer circle) to the inner circle. If the resource is not used there will be no arrow from it to the inner circle. Along the arrow, they may write particulars of the resource category.

For example, suppose "FOOD" is placed in the inner circle. What resources are used in producing food? Without much hesitation students will recognize that all resource categories are involved—soil and water are most evident, but what about the tools and machinery used to do the farming? That takes iron from iron ore, the mineral category. What about the plants and animals themselves? This brings in the biological category. Finally, what about fuel to run the machinery, and light for photosynthesis? There is the energy category.

A "HOUSE" placed in the inner circle would entail a line from the biological category—trees for lumber; a line from the mineral category noting iron, copper, and aluminum for nails, plumbing, wiring, and fixtures as well as perhaps clay for bricks; a line from the energy category noting power needs in sawing lumber as well as for power tools. Water and soil would also come into the picture as students note their need in growing trees.

"IRON" in the center circle would entail a line from the mineral category noting iron ore itself. Then there would be a line from the energy category noting fuel needs for mining equipment, as well as for smelting the iron itself, and a line from the water category—a great deal of water is used in the production of iron.

It is evident that this activity can be pursued to any extent and conducted on multiple occasions as desired. Also, there may be numerous variations. Students may choose an item they wish to analyze in this manner or you can challenge them with an unknown drawn from a hat. They can work individually or as a team and make a poster. You may have students give individual or group presentations regarding the resource requirements of any given item, and so on.

Human Resources

To take this whole exercise to a higher level, have students draw a larger circle around the "resource circle" and label this, HUMAN RESOURCES. With discussion, it readily becomes apparent that obtaining any and all resources depends on human inputs: figuring out/discovering how to do it; designing and making the necessary tools, machinery, and processes, and finally the actual labor for doing the work involved. Resources never become finished products by themselves. There is always human ingenuity and labor involved, that is, HUMAN RESOURCES.

In turn, students should observe that human resources do not exist independently; all the other resources go into supporting humans. Students may show this on their charts with double headed arrows between the resource circle and the human resource circle. Thus, there is a constant interaction and synergism between human resources and other resources.

Some students may see a "chicken-or-egg" problem here. Humans depend on resources; yet, making things from resources depends on humans. Which came first? Emphasize the historical perspective here.

History can be seen as innumerable steps, some large, some small, in discovering how to make new products/materials from existing, natural resources. In short, human resources and the utility of other resources develop hand in hand. Each advance in technology involves learning to use a resource to make a new product. The benefit of the product enables humans to go on and discover new ways to use resources for further benefit and so on.

At risk of being absurdly simplistic, consider that technology started as humans learned to chip stone tools and arrowheads from certain rocks (mineral resources) and fasten them to poles to make spears. The success of this innovation enabled humans to progress to another stage, learning to smelt metals from ores, and so on. In more recent times (just the past 40 years or so) humans learned to make tiny electrical circuits from chips of silica (still a mineral resource) and this ushered in the age of computers. Computers in turn, enable us to discover and use resources in ever more innovative ways. You may incorporate this concept throughout your teaching of history.

It is evident that when everything is considered, the resource circle can become quite a web of nearly everything depending on everything else. Be careful not to overwhelm children with everything at once. Keep it to what they can master. The exercise can be revisited any number of times to add additional features.

Questions/Discussion/Activities To Review, Reinforce, Expand, and Assess Learning:

Make copies of the "resource wheel" with the resources labeled but the rest blank. Assign something to the center and have students draw arrows and add specifics as described above.

Have students give show-and-tells pointing out how a given item/product depends on one or more of the resource categories.

To Parents and Others Providing Support:

This can be quiet time activity. Consider items around the house/apartment such as a piece of furniture and fill out a resource wheel about it as described above.

Connections to other topics and follow up to higher levels:

Renewable and non-renewable resources
Resource shortages/limits
Resource conservation
Technologies of converting resources to products
Recycling

Re: National Science Education (NSE) Standards

This lesson is a steppingstone toward developing students' understanding and abilities aligned with NSE, K-4:

Unifying Concepts and Processes:
  • Systems, order, an organization

Bernard J. Nebel, Ph.D.

Content Standard F, Science in Personal and Social Perspectives;
  • Types of Resources

Content Standard G: History and Nature of Science
  • Science as a human endeavor

Books for Correlated Reading:

See list under Lesson A-5.

Numerous titles in "Rookie Read-About Science" series. Just for Kids Books

Numerous titles in "What Living Things Need" series. Heinemann Library.

Gibbons, Gail. *Milk Makers* (Reading Rainbow Book). Aladdin, 1987.

Kerley, Barbara. *A Cool Drink of Water*. National Geographic Children's Books, 2006.

Krull, Kathleen. *Supermarket*. Holiday House, 2001.

Mitchell, Melanie. *Cloth*. Lerner, 2003.
_____. *Metal*. Lerner, 2003.
_____. *Plastic*. Lerner, 2003.
_____. *Wood*. Lerner, 2003.

Twist, Clint. *Materials* (Check It Out!). Bearport, 2006.

Appendix A

Science-Writing Synergy

There is a strong synergy between learning material and writing about it. Writing forces the author to focus attention, recall information, reflect on what he/she has learned, seen, or done, raise questions, and organize thoughts coherently. Even more, it brings the writer to recognize what she/he doesn't fully know or understand and where further efforts are in order. In short, writing develops *metacognition*, a key learning trait described in Chapter 2. Then, from the teacher's view, students' writing reveals errors and gaps in knowledge or thinking and enables further instruction to be focused most effectively and efficiently.

The major objective may be on teaching writing skills. Still, the synergy between writing and learning subject matter should not be overlooked. Any topic can be used as the format for teaching writing skills. Science topics provide a particularly good format in that beginning stages are invariably describing what is seen or experienced first hand. This lends itself especially well to everything from putting down single words to construction of descriptive sentences. As students advance in science lessons, they can advance writing skills as well. As students progress, the synergy between writing and scientific thinking becomes even more pronounced.

Recognizing the learning-writing synergy, nearly every lesson in this volume includes a suggestion in the section, "Questions/Discussion/Activities To Review, Expand, and Assess Learning," "Make books describing/illustrating ..." A "book" is made by folding a sheet of regular 8 1/2 x 11 paper top to bottom. The sheet is cut in half along the seam; the two pieces are placed on top of each other, folded side to side, and stapled along the fold. This makes a booklet with 8 pages if both front and back are used. More pages can be added if desired. The front "cover" should bear a title, the student's name, and the date.

Making these "books" is most appropriate for grades K-2. They can accommodate any level of writing skills from crude illustrations and a few letters to accurate, descriptive sentences. Allow children to have a choice in the pencils, markers, crayons, and colors they use. Whatever a child puts down provides a view of his/her writing ability and his/her comprehension of the lesson at hand. In turn, this shows where further instruction and/or personal assistance are required. Keeping the books on file provides a great visual display of the progress made by each student that you can show to parents, administrators, and others, as well as to the student him/herself. Beyond grade two, children will have gained sufficient skills and maturity that they can graduate to keeping a more traditional science notebook.

LaVergne, TN USA
07 April 2010
178391LV00001B/24/A